Lecture Notes in Computer Science 13131

More information about this subseries at https://link.springer.com/bookseries/7412

Esther Puyol Antón · Mihaela Pop ·
Carlos Martín-Isla · Maxime Sermesant ·
Avan Suinesiaputra · Oscar Camara ·
Karim Lekadir · Alistair Young (Eds.)

Statistical Atlases and Computational Models of the Heart

Multi-Disease, Multi-View, and Multi-Center Right Ventricular Segmentation in Cardiac MRI Challenge

12th International Workshop, STACOM 2021
Held in Conjunction with MICCAI 2021
Strasbourg, France, September 27, 2021
Revised Selected Papers

 Springer

Editors
Esther Puyol Antón
King's College London
London, UK

Carlos Martín-Isla
Universitat de Barcelona, BCN-AIM
Artificial Intelligence in Medicine Lab
Barcelona, Spain

Avan Suinesiaputra
King's College London
London, UK

Karim Lekadir ⓘD
University of Barcelona
Barcelona, Spain

Mihaela Pop
Sunnybrook Research Institute
Toronto, Canada

Maxime Sermesant
Inria - Epione Group
Sophia Antipolis, France

Oscar Camara
Pompeu Fabra University
Barcelona, Spain

Alistair Young
King's College London
London, UK

ISSN 0302-9743   ISSN 1611-3349 (electronic)
Lecture Notes in Computer Science
ISBN 978-3-030-93721-8   ISBN 978-3-030-93722-5 (eBook)
https://doi.org/10.1007/978-3-030-93722-5

LNCS Sublibrary: SL6 – Image Processing, Computer Vision, Pattern Recognition, and Graphics

This Springer imprint is published by the registered company Springer Nature Switzerland AG
The registered company address is: Gewerbestrasse 11, 6330 Cham, Switzerland

Preface

Integrative models of cardiac function are important for understanding disease, evaluating treatment, and planning intervention. In recent years, there has been considerable progress in cardiac image analysis techniques, cardiac atlases, and computational models, which can integrate data from large-scale databases of heart shape, function, and physiology. However, significant clinical translation of these tools is constrained by the lack of complete and rigorous technical and clinical validation, as well as benchmarking of the developed tools. To achieve this, common and available ground-truth data capturing generic knowledge on the healthy and pathological heart is required. Several efforts are now established to provide web-accessible structural and functional atlases of the normal and pathological heart for clinical, research, and educational purposes. We believe that these approaches will only be effectively developed through collaboration across the full research scope of the cardiac imaging and modelling communities.

The 12th edition of the Workshop on Statistical Atlases and Computational Modelling of the Heart, STACOM 2021 (http://stacom2021.cardiacatlas.org), was held in conjunction with the MICCAI 2021 international conference (Strasboug, France) as a virtual event. It followed the past eleven editions: STACOM 2010 (Beijing, China), STACOM 2011 (Toronto, Canada), STACOM 2012 (Nice, France), STACOM 2013 (Nagoya, Japan), STACOM 2014 (Boston, USA), STACOM 2015 (Munich, Germany), STACOM 2016 (Athens, Greece), STACOM 2017 (Quebec City, Canada), STACOM 2018 (Granada, Spain), STACOM 2019 (Shenzhen, China), and STACOM 2020 (Lima, Peru). STACOM 2021 provided a forum to discuss the latest developments in various areas of computational imaging, modelling of the heart, parametrization of models, application of artificial intelligence and machine learning to cardiac image analysis, electro-mechanical modelling of the heart, novel methods in preclinical/clinical imaging for tissue characterization and image reconstruction, as well as statistical cardiac atlases.

The topics of the 12th edition of the STACOM workshop included the following: cardiac imaging and image processing, machine learning applied to cardiac imaging and image analysis, atlas construction, artificial intelligence, statistical modelling of cardiac function across different patient populations, cardiac computational physiology, model customization, atlas-based functional analysis, ontological schemata for data and results, and integrated functional and structural analyses, as well as the pre-clinical and clinical applicability of these methods. Besides the regular contributing papers, additional efforts of the STACOM 2021 workshop were also focused on a challenge, namely the Multi-Disease, Multi-View & Multi-Center Right Ventricular Segmentation in Cardiac MRI Challenge (M&Ms-2). Along with the 25 regular papers accepted for presentation at STACOM 2021, 15 papers from this M&Ms-2 challenge (described in more detail below) were presented, which are included in this LNCS proceedings volume.

The M&Ms-2 challenge (https://www.ub.edu/mnms-2/) was proposed to promote the development of generalizable deep learning models for the right ventricle that can maintain good segmentation accuracy on different centers, pathologies, and cardiac

MRI views. A total of 15 participants submitted various solutions by implementing and combining methodologies such as data augmentation, generative models, and cross-view segmentation alternatives. The initial results show the promise of the proposed solutions but also the need for more research to enhance the generalizability of deep learning and multi-view analyses in cardiac MRI. The M&Ms-2 benchmarking dataset, totaling 360 cardiac MRI studies from three different clinical centers, has now been made open-access on the challenge website to enable further research and benchmarking on multi-disease, multi-view, and multi-center cardiac image analysis.

We hope that the results obtained by the challenge, along with the regular paper contributions will act to accelerate progress in the important areas of cardiac function and structure analysis. We would like to thank all organizers, reviewers, authors, and sponsors for their time, efforts, contributions, and support in making STACOM 2021 a successful event.

September 2021

Esther Puyol Antón
Mihaela Pop
Maxime Sermesant
Carlos Martín-Isla
Avan Suinesiaputra
Oscar Camara
Karim Lekadir
Alistair Young

Organization

Workshop Chairs

Mihaela Pop	Sunnybrook Research Institute, Canada
Esther Puyol Antón	King's College London, UK
Maxime Sermesant	Inria, Epione Group, Sophia Antipolis, France
Oscar Camara	UPF Barcelona, Spain
Avan Suinesiaputra	King's College London, UK
Alistair Young	King's College London, UK

STACOM Organizers

Esther Puyol Antón	King's College London, UK
Mihaela Pop	Sunnybrook Research Institute, Canada
Maxime Sermesant	Inria, Epione Group, Sophia Antipolis, France
Avan Suinesiaputra	King's College London, UK
Oscar Camara	UPF Barcelona, Spain
Alistair Young	King's College London, UK

M&Ms-2 Challenge Organizers

Carlos Martín-Isla	University of Barcelona, Spain
Karim Lekadir	University of Barcelona, Spain
Sergio Escalera	University of Barcelona and Computer Vision Centre, Spain
Carla Sendra	University of Barcelona, Spain
Victor Campello	University of Barcelona, Spain

Clinical Team

José F. Rodríguez Palomares	Vall d'Hebron University Hospital, Spain
Andrea Guala	Vall d'Hebron University Hospital, Spain

Publication Chairs

Mihaela Pop	Sunnybrook Research Institute, Canada
Maxime Sermesant	Inria, Epione Group, Sophia Antipolis, France

Webmaster

Avan Suinesiaputra   King's College London, UK

Additional Reviewers

Eric Lluch Alvarez
Jaume Banus Cobo
Teodora Chitiboi
Cristian Linte
Buntheng Ly
Tommaso Mansi
Felix Meister
Viorel Mihalef
Matthew Ng
Jesus Jairo Rodriguez Padilla
Yingyu Yang

Contents

Multi-disease, Multi-view and Multi-center Right Ventricular Segmentation in Cardiac MRI Challenge (M&Ms-2)

Regular Papers

Multi-atlas Segmentation of the Aorta from 4D Flow MRI: Comparison of Several Fusion Strategies

Diana M. Marin-Castrillon[1]([✉]), Arnaud Boucher[1], Siyu Lin[1], Chloe Bernard[2],
Marie-Catherine Morgant[1,2], Alexandre Cochet[1,3], Alain Lalande[1,3],
Olivier Bouchot[1,2], and Benoit Presles[1]

[1] ImViA Laboratory, University of Burgundy, Dijon, France
Diana-Marcela_Marin-Castrillon@etu.u-bourgogne.fr
[2] Department of Cardiology, University Hospital of Dijon, Dijon, France
[3] Department of Medical Imaging, University Hospital of Dijon, Dijon, France

Abstract. 4D flow Magnetic Resonance Imaging (4D flow MRI) provides information that improves the estimation of hemodynamic characteristics of the aorta and allows further flow analysis. However, reliable segmentation of the aorta, required as a preliminary step, is still an open problem due to the low image quality. Thus, an automatic segmentation tool could encourage the use of 4D flow MRI in clinical practice for diagnostic and prognostic decision-making. In this paper, we propose a fully automatic multi-atlas-based method to segment the aorta using the systolic phase of 4D flow MRI. The Dice similarity coefficient and Hausdorff distance were used to quantify the performance. In addition, a statistical significance test between the maximum diameters obtained with the manual and automatic segmentations was conducted to determine the reliability of the automatic segmentations. The results show that our method could be a first reliable step towards automatic segmentation of the aorta in all phases of 4D flow MRI.

Keywords: 4D flow MRI · Multi-atlas segmentation · Aortic diseases

1 Introduction

Phase Contrast Magnetic Resonance Imaging (PC-MRI) is a medical imaging technique developed to measure blood flow in the vascular system. 4D flow MRI modality provides both structural (magnitude images) and functional (phase contrast images) information [5]. Functional information allow analysis of the aorta flow patterns and the estimation of new and improved hemodynamic characteristics such as wall shear stress, relative pressure or turbulent kinetic energy [7].

4D flow MRI requires extensive pre-processing due to image quality and artifacts related with free breathing and arrhythmia. For this reason, this image

© Springer Nature Switzerland AG 2022
E. Puyol Antón et al. (Eds.): STACOM 2021, LNCS 13131, pp. 3–11, 2022.
https://doi.org/10.1007/978-3-030-93722-5_1

modality is currently used mainly in research. The segmentation of structures of interest is one of the most challenging preliminary steps in 4D flow MRI analysis since it is a time-consuming task made difficult by the low image quality and the high dimensionality of the images. To address this issue, semi-automatic and automatic segmentation methods have been proposed for aortic flow dynamics analysis.

Köhler et al. [4] proposed a semi-automatic graph-cut based segmentation of the aorta in 4D flow MRI, computing from the magnitude images a temporal maximum intensity projection (tMIP) on which the user indicates regions inside and outside the aorta to initialize the algorithm. In this approach, the number of inputs required to achieve the segmentation depends on the image quality and requires user experience to indicate the aorta. The proposed approaches for automatic segmentation of the aorta in 4D flow MRI are based mainly on the generation of a new image using the flow velocity information to improve vessel's identification. Roy van Pelt et al. [6] computed a temporal maximum speed volume (tMSV) with phase contrast images and implemented an active surface model by visually selecting a threshold to obtain the initial surface to segment thoracic arteries. Thus, the obtained segmentation is based only on phase images, without considering the morphology information of the aortic wall from magnitude images. Recently, a convolutional neural network was used for aortic segmentation on 3D phase-contrast MR angiogram image (PC-MRA) generated by combining the functional information from all phases of the cardiac cycle [1].

Bustamante et al. [2] presented a multi-atlas segmentation of the thoracic cardiovascular system by first generating a multi-atlas with eight 3D PC-MRA images. These images were generated by computing the PC-MRA image independently for each frame and aligning them with an intra-patient registration process. Nonetheless, the segmentation performance for the aorta was evaluated by calculating the stroke volume with a 2D plane located in the ascending aorta, which does not provide information on the quality of the 3D automatic segmentation to determine the possibility of using it to obtain hemodynamic characteristics.

In this context, we propose an automatic multi-atlas-based aortic segmentation approach that avoids the generation of PC-MRA images by using a volume from the magnitude image of 4D flow MRI in the systolic phase of the cardiac cycle, thus considering the artery wall morphology. This cardiac phase was selected because the image quality allows a more reliable segmentation that could be propagate to all 4D flow MRI time frames. Considering the importance of the pipeline definition for each of the multi-atlas-based segmentation stages, an exhaustive search of the registration parameters was carried out and different label fusion strategies were evaluated. For performance evaluation, we analyzed the results globally and locally, by dividing the aorta into three parts, ascending aorta including the aortic arch (AAO+AA), thoracic descending aorta (TDAO), and proximal abdominal aorta (PAAO). Additionally, the reliability of the automatic segmentation was evaluated with a statistical significance test on the maximum diameter difference between manual and automatic segmentation. To the best of our knowledge, this is the first automatic aorta segmentation method that uses directly the anatomical information of 4D flow MRI images and presents a detailed analysis about the performance of the obtained segmentation (Fig. 1).

2   Method and Experimental Settings

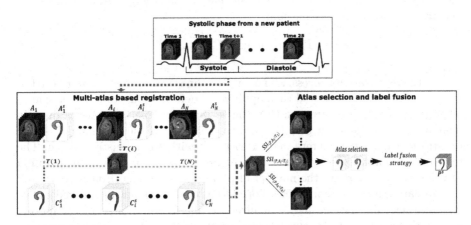

Fig. 1. Pipeline of the multi-atlas-based method to segment the aorta in the systolic phase from 4D flow MRI. The images A_i in the atlas are mapped to the target image P (Highlighted in red) with a transform T_i to produce candidate segmentations C_i^s that are merged after an atlas selection process to generate the target segmentation P^s. (Color figure online)

2.1   Data Acquisition and Dataset Building

For this study, 32 patients (22 men and 10 women) with aortic dilatation, average age of 62 years gave their written consent to undergo free-breathing 4D flow MRI acquisitions. The MRI acquisitions were done on a 3T magnet (Siemens Medical Solutions, Erlangen, Germany) with a phased thoracic coil. The sequence is ECG-gated, with an echo navigator to manage diaphragmatic movement. Retrospective gating provides 25 phases of the cardiac cycle with a spatial resolution of $2 \times 2 \times 2\,\mathrm{mm}^3$ and a temporal resolution between 24 to 53 ms according to the patient. The images were bias field corrected.

From the 4D flow MRI images was created a multi-atlas dataset of $N = 32$ patients consisting of an image (i.e. volume) of the systolic phase from the magnitude image A_i $(i = 1, 2, ..., N)$ and the corresponding manual segmentation A_i^s $(i = 1, 2, ..., N)$. The segmentations were performed using the ITK-SNAP software by an image analyst and reviewed by a clinical expert.

2.2   Registration

The first step of multi-atlas segmentation (MAS) is to register the magnitude image in systole from a new patient to be segmented P with each magnitude image A_i of the atlas to obtain a transformation T_i that spatially aligns $A_i \circ T_i$ with P. Then, the transformation T_i is applied to the corresponding manual segmentation A_i^s to generate a target candidate segmentation image $C_i^s = A_i^s \circ T_i$ $(i = 1, 2, ..., N)$. The quality of alignment between $A_i \circ T_i$ and P is assessed by computing a similarity measure S. The objective of the registration step is therefore to find the best transformation that spatially aligns $A_i \circ T_i$ and P by maximizing the following equation:

$$\hat{T}_i = \arg\max_{T_i} S(P, A_i \circ T_i) \tag{1}$$

For the registration process, affine and B-spline transformations were used to first map linearly P to A_i and then apply local deformations for a better spatial alignment. Taking into account that the segmentation performance is directly related to the registration performance, the selection of the affine and B-spline registration hyper-parameters were carried out with an exhaustive search. The tested affine parameters were the type of image pyramids (Gaussian pyramid or Gaussian scale space), the number of pyramid resolutions (from 1 to 4), and the similarity metrics (normalized correlation coefficient, mean squared difference, mutual information, and normalized mutual information). For the mutual information metric, 16, 32 and 64 number of bins were tested to compute the histograms. In total, 64 affine registrations were carried out per patient. The search for the best B-spline parameters was based on the optimization of the grid spacing value which is the most important parameter because it controls the flexibility of the local deformations. We varied this parameter from 10 mm to 46 mm with a step of 6 mm in each dimension independently. For the other registration parameters, we used a Gaussian pyramid with three resolutions and the normalized correlation coefficient metric. In total, were carried out per patient 343 B-spline registrations with the fixed parameters and all possible combinations of the spacing values in the three dimensions. The registrations were performed using the Elastix software version 5.0.

2.3   Label Fusion

The second step of MAS consists in merging the propagated segmented images C_i^s to generate the target segmentation. The fusion of candidate labels can be performed by global or local methods, using in some of them a similarity measure to assign a weight proportional to the quality of the registration between the target and the images in the atlas. In this work, the Structural Similarity Index (SSI) was used [8] as similarity measure and both types of label fusion algorithms were tested in order to evaluate the influence of these methods on the segmentation performance.

Majority Voting (MV) is the most straightforward label fusion method and uses a global strategy giving to all candidate segmentations the same relevance. For the target segmentation P^s, the most frequent label l ($l = 1, ...L$, with $L =$ number of labels) is assigned to each voxel x, such as

$$P^s(x) = \arg\max_{l \in 1,...,L} p_l(x) \tag{2}$$

where $p_l(x) = \frac{1}{N} \sum_{i=1}^{N} O_{i,l}(x)$ with $O_{i,l}(x) = 1$ if $C_i^s(x) = l$ and 0, otherwise.

Weighted Majority Voting (WMV) uses a global weight w_i proportional to the similarity measure between $A_i \circ T_i$ and P, to calculate the probability of a voxel x to be the label l as follows

$$p_l(x) = \frac{\sum_i^N w_i \cdot \delta[l, C_i^s(x)]}{\sum_i^N w_i}, \quad \forall l \in L \tag{3}$$

where $\delta[\cdot]$ is the Kronecker delta function. The segmentation P^s is generated with Eq. 2.

Patch Weighted Majority Voting (PWMV) is a local extension of WMV, where a different weight is assigned to each voxel x. The corresponding weight is proportional to a similarity measure calculated between a patch or kernel (a kernel size of $11 \times 11 \times 11$ was used in this work) centered on the voxel x of P and a patch of the same size and at the same position in $A_i \circ T_i$.

2.4   Atlas Selection

Instead of using all the C_i^s images for label fusion, atlas selection can be performed to avoid using irrelevant segmentations. According to the label fusion strategy used, the atlas selection for each voxel is made by considering the set SSI_T of global or local SSIs between P and all the deformed images $A_i \circ T_i$.

Let $max(SSI_T) = max(SSI(P, A_1 \circ T_1), ..., SSI(P, A_N \circ T_N))$ and $min(SSI_T) = min(SSI(P, A_1 \circ T_1), ..., SSI(P, A_N \circ T_N))$ be the maximum and minimum of SSI_T, respectively. The atlas selection is given by Eq. 4

$$\{C_i^s, \forall i \in 1, ..., N, SSI(P, A_i \circ T_i) > (max(SSI_T) - min(SSI_T)) * d + min(SSI_T)\} \tag{4}$$

where d is a number between 0 and 1 that indicates the percentage of atlases with the lowest SSI to be discarded.

2.5   Performance Evaluation

The performance of the segmentation was quantified with 3D Hausdorff distance (HD) to measure the maximum error of the contour and the Dice similarity coefficient (DSC) to know the spatial overlap between the manual and automatic segmentations.

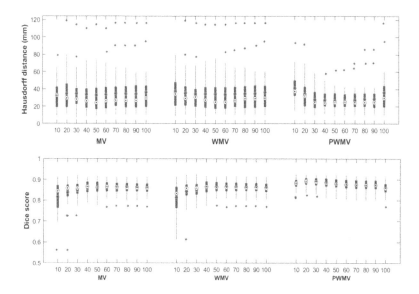

Fig. 2. Comparison of label fusion strategies with atlas selection. The boxplots display the median, maximum and minimum performance values for global DSC and global HD achieved with MV, WMV and PWMV and as functions of the percentage of atlases preserved. The red circle on each boxplot represents the average segmentation performance. (Color figure online)

In order to evaluate the reliability of the automatic segmentation, the statistical significance of the differences in the maximum diameter between manual and automatic segmentations obtained with the best pipeline was assessed with a t-test after checking the normality of the two groups with the Shapiro-Wilk test. Additionally, a scatter plot and a Bland-Altman plot were generated to analyze the correlation and agreement between the diameter obtained with manual and automatic segmentations. To calculate the diameter, the Vascular Modeling Toolkit[1] (VMTK) was used to extract the centerline from the aortic valve to the abdominal aorta and for continuous measurements of the aortic diameter using normal planes located at each segment of the centerline.

3   Results

Automatic segmentation of the aorta in the systolic phase of 4D flow MRI was generated for all 32 patients using the proposed method with a leave-one-patient-out strategy. Depending on the patient, the time required to obtain automatic segmentation was 20–25 min on a laptop workstation Dell Precision 7540 with Intel Core i7-9850H at 2.60 GHz processor. The best set of parameters for registration was selected considering the segmentation performance with the MV

---

[1] http://www.vmtk.org/.

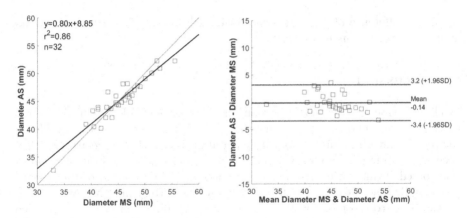

Fig. 3. Scatter plot (left) and Bland-Altman plot (right) of maximum diameters obtained with manual (MS) and automatic segmentations (AS)

fusion strategy. For the affine transform, the best parameters were Gaussian pyramid with three resolutions and the mutual information metric with a 64-bin histogram. For B-spline transforms, the best segmentation results were achieved with a grid spacing of 22 mm, 40 mm and 10 mm in x, y and z axes, respectively.

In Fig. 2 the global segmentation performance are presented for all the label fusion strategies with atlas selection. It can be observed that both in DSC and HD, the PWMV fusion strategy reduces the number of outliers and minimizes to a large extent the HD. The best PWMV performance is achieved with the use of about 50% to 60% percent of the most similar atlases with respect to the range of SSIs obtained for each patient. Concerning the local performance an average HD of 18.21 ± 7.68 mm, 7.43 ± 2.64 mm and 22.58 ± 13.03 mm and an average DSC of 0.88 ± 0.04, 0.90 ± 0.034 and 0.78 ± 0.12 were reached respectively for AAO+AA, TDAO and PAAO. We can highlight that compared to the average global HD and average global DSC (26.62 ± 11.19 and 0.88 ± 0.03), the method achieves robust performance in the segmentation of the aorta at the level of AAO+AA and TDAO, important parts for clinical evaluation of thoracic aneurysms. Furthermore, by visually inspecting the distance maps of all patients, it was found that in 75% of the patients, the highest error corresponding to the HD occurred in PAAO.

A p-value of $p = 0.8$ was obtained for t-test, approving the null hypothesis that there is no statistically significant difference between the maximum diameters measured with manual segmentation and the maximum diameters measured with automatic segmentation obtained with 50% of the most similar candidates segmentations and PWMV algorithm. Figure 3 shows a correlation graph of the diameters obtained with manual segmentation and the diameters obtained with automatic segmentation, in which a low dispersion of the data can be observed, obtaining a r^2 (i.e.: coefficient of determination) equal to 0.86. Additionally, a corresponding Bland-Altman plot is presented in which it can be seen that the mean difference of the values obtained with automatic segmentations and manual segmentations is 0.14 mm.

4   Discussion

In clinical practice, the aortic motion and the influence of flow stress on the aortic wall are important factors for the medical follow-up of a patient. This is the reason why obtaining an aorta segmentation that includes the artery wall is necessary for further analysis. Compared to previous studies, we evaluated multi-atlas-based segmentation method using anatomical information of the systolic phase of 4D flow MRI, which allows us to keep the artery morphology.

Contrary to most of the previous studies [2,4,6], we presented a detailed analysis about the segmentation performance both globally and locally. Therefore, it was possible to identify error-prone localization and challenging regions. In this sense, the results of aortic segmentation in some previous works were presented by calculating the blood volume with a 2D plane located manually or automatically in the AAO+AA. Thus, it was impossible to establish an overall segmentation performance, considering also that the calculation of the blood volume is highly sensitive to the orientation of the plane [3]. On the other hand, although the segmentation approach with deep learning [1] yielded an average Dice score of 0.951 and average Hausdorff distance of 2.80 mm, the images used for the segmentation were 3D PC-MRA. This kind of image facilitates the identification of the aorta, but does not preserve the information of the aorta position in each cardiac phase because it mixes information from all the 4D flow MRI frames, making a future analysis of the movement of the artery unfeasible.

It was observed in this study that label fusion methods with local weighting reached a better performance than globally weighted methods. Moreover, the use of a too small set of atlases or of the entire set of atlases in the label fusion step decreased the segmentation performance. This behavior may due to the fact that there were either not enough candidate segmentations or that some of them were irrelevant for decision making and misguided the process.

The best segmentation performance was obtained using the PWMV label fusion strategy with 50% of the candidate segmentations. Analyzing the results locally, it could be observed that the part of the aorta with the lowest performance is the PAAO. The performance difference between abdominal aorta and thoracic aorta may be related to the fact that the signal-to-noise ratio in this area decreases due to the distance to the center of the phase-array thoracic coil during acquisition. In addition, it was found that the differences between the manual and automatic segmentation of PAAO were mainly related to the length of the segments of PAAO covered by each one. In contrast, the error of the automatic segmentation at the level of the TDAO was the lowest, due to the fact that in this region the borders of the aorta are better defined than in the other regions, which can better guide the registration. The errors in segmentation at the level of the AAO+AA were generally related to the low definition of the aorta border at the level of the aortic valve and the inclusion of voxels belonging to the brachiocephalic, carotid and subclavian arteries that were not included in the manual segmentation. Considering the importance of this part of the aorta in the hemodynamic analysis, the method could be further improved with pre-processing to differentiate these edges in the magnitude images.

5   Conclusion

The proposed multi-atlas based segmentation method allows the automatic segmentation of the aorta in the systolic phase of the magnitude images from 4D flow MRI. The global and local analysis of the results lead to a detailed evaluation of the performance of the method and the identification of challenging regions. The results found with the proposed method show that such segmentation could be a first reliable step to further propagation to the other 4D flow MRI time frames and for the analysis of aortic hemodynamics.

References

1. Berhane, H., et al.: Fully automated 3D aortic segmentation of 4D flow MRI for hemodynamic analysis using deep learning. Magn. Reson. Med. **84**(4), 2204–2218 (2020)
2. Bustamante, M., Gupta, V., Forsberg, D., Carlhäll, C.J., Engvall, J., Ebbers, T.: Automated multi-atlas segmentation of cardiac 4D flow MRI. Med. Image Anal. **49**, 128–140 (2018)
3. Casciaro, M.E., et al.: 4D flow MRI: impact of region of interest size, angulation and spatial resolution on aortic flow assessment. Physiol. Meas. **42**(3), 035004 (2021)
4. Köhler, B., Preim, U., Grothoff, M., Gutberlet, M., Fischbach, K., Preim, B.: Guided analysis of cardiac 4D PC-MRI blood flow data. In: Eurographics (Dirk Bartz Prize), pp. 2–5 (2015)
5. Markl, M., Frydrychowicz, A., Kozerke, S., Hope, M., Wieben, O.: 4D flow MRI. J. Magn. Reson. Imaging **36**(5), 1015–1036 (2012)
6. van Pelt, R., Nguyen, H., ter Haar Romeny, B., Vilanova, A.: Automated segmentation of blood-flow regions in large thoracic arteries using 3D-cine PC-MRI measurements. Int. J. Comput. Assist. Radiol. Surg. **7**(2), 217–224 (2012)
7. Stankovic, Z., Allen, B.D., Garcia, J., Jarvis, K.B., Markl, M.: 4D flow imaging with MRI. Cardiovasc. Diagn. Therapy **4**(2), 173 (2014)
8. Wang, Z., Bovik, A.C., Sheikh, H.R., Simoncelli, E.P.: Image quality assessment: from error visibility to structural similarity. IEEE Trans. Image Process. **13**(4), 600–612 (2004)

Quality-Aware Cine Cardiac MRI Reconstruction and Analysis from Undersampled K-Space Data

Inês Machado[1]([✉]), Esther Puyol-Antón[1], Kerstin Hammernik[2,3],
Gastão Cruz[1], Devran Ugurlu[1], Bram Ruijsink[1,4], Miguel Castelo-Branco[5],
Alistair Young[1], Claudia Prieto[1], Julia A. Schnabel[1,2,6], and Andrew P. King[1]

[1] School of Biomedical Engineering and Imaging Sciences, King's College London, London, UK
ines.machado@kcl.ac.uk
[2] Technical University of Munich, Munich, Germany
[3] Biomedical Image Analysis Group, Imperial College London, London, UK
[4] Department of Adult and Paediatric Cardiology, Guy's and St Thomas' NHS Foundation Trust, London, UK
[5] Coimbra Institute for Biomedical Imaging and Translational Research, University of Coimbra, Coimbra, Portugal
[6] Helmholtz Center Munich, Munich, Germany

Abstract. Cine cardiac MRI is routinely acquired for the assessment of cardiac health, but the imaging process is slow and typically requires several breath-holds to acquire sufficient k-space profiles to ensure good image quality. Several undersampling-based reconstruction techniques have been proposed during the last decades to speed up cine cardiac MRI acquisition. However, the undersampling factor is commonly fixed to conservative values before acquisition to ensure diagnostic image quality, potentially leading to unnecessarily long scan times. In this paper, we propose an end-to-end quality-aware cine short-axis cardiac MRI framework that combines image acquisition and reconstruction with downstream tasks such as segmentation, volume curve analysis and estimation of cardiac functional parameters. The goal is to reduce scan time by acquiring only a fraction of k-space data to enable the reconstruction of images that can pass quality control checks and produce reliable estimates of cardiac functional parameters. The framework consists of a deep learning model for the reconstruction of 2D+t cardiac cine MRI images from undersampled data, an image quality-control step to detect good quality reconstructions, followed by a deep learning model for biventricular segmentation, a quality-control step to detect good quality segmentations and automated calculation of cardiac functional parameters. To demonstrate the feasibility of the proposed approach, we perform simulations using a cohort of selected participants from the UK Biobank (n = 270), 200 healthy subjects and 70 patients with cardiomyopathies. Our results show that we can produce quality-controlled images in a scan time reduced from 12 to 4 s per slice, enabling reliable estimates of cardiac functional parameters such as ejection fraction within 5% mean absolute error.

E. Puyol Antón et al. (Eds.): STACOM 2021, LNCS 13131, pp. 12–20, 2022.
https://doi.org/10.1007/978-3-030-93722-5_2

Keywords: Cardiac MRI · Deep learning reconstruction · Accelerated MRI · Image segmentation · Quality assessment

1 Introduction

Cardiac MRI is a common imaging modality for assessing cardiovascular diseases, which is the leading cause of death globally. Cine cardiac MRI enables imaging of the heart throughout the cardiac cycle, and is especially useful for quantifying left and right ventricular function by measuring parameters such as ejection fraction (EF), end-diastolic and end-systolic cardiac chamber volumes. However, cine cardiac MRI acquisition is slow and there has been much research interest in accelerating the scan without compromising the high resolution and image quality requirements. One approach that has been used to speed up the scan is to reduce the amount of acquired k-space data. However, reconstructing cine cardiac MRI from undersampled k-space data is a challenging problem, and approaches typically exploit some type of redundancy or assumption in the underlying data to resolve the aliasing caused by sub-Nyquist sampling [1]. Considerable efforts have been devoted to accelerate the reconstruction of cardiac MRI from undersampled k-space including parallel imaging and compressed sensing [2]. More recently, machine learning reconstruction approaches have been proposed to learn the non-linear optimization process employed in cardiac MRI undersampled reconstruction. In particular, deep learning (DL) techniques have been proposed to learn the reconstruction process from existing data sets in advance, providing a fast and efficient reconstruction that can be applied to all newly acquired data [3,4]. In this paper, we assess image quality from reconstructions of undersampled k-space data during acquisition, creating an 'active' acquisition process in which only a fraction of k-space data are acquired to enable the reconstruction of an image that can pass automated quality control (QC) checks and produce reliable estimates of cardiac functional parameters. The major contributions of this work are three-fold: 1) to the best of our knowledge, this is the first paper that combines cine cardiac MRI undersampled reconstruction with QC in downstream tasks such as segmentation in a unified framework; 2) our pipeline includes robust pre- and post-analysis QC mechanisms to detect good quality image reconstructions (QC1) and good quality image segmentations (QC2) during active acquisition and 3) we show that quality-controlled cine cardiac MRI images can be reconstructed in a scan time reduced from 12 to 4 s per slice, and that image quality is sufficient to allow clinically relevant parameters (EF and left- and right-ventricle chamber volumes) to be automatically estimated within 5% mean absolute error.

2 Materials

We evaluate our proposed reconstruction and analysis framework using a cohort of selected healthy (n = 200) and cardiomyopathy (n = 70) cases from the UK

Biobank obtained on a 1.5 T MRI scanner (MAGNETOM Aera, Siemens Health-care, Erlangen, Germany). The short-axis (SAX) image acquisition typically consists of 10 image slices with a matrix size of 208 × 187 and a slice thickness of 8 mm, covering both the ventricles from the base to the apex. The in-plane image resolution is $1.8 \times 1.8\,\text{mm}^2$, the slice gap is 2 mm, with a repetition time of 2.6 ms and an echo time of 1.10 ms. Each cardiac cycle consists of 50 time frames. More details of the image acquisition protocol can be found in [5]. The image reconstruction model and the segmentation model were trained using an additional set of 3,975 cine cardiac MRI datasets from the UK Biobank. For these subjects, pixel-wise segmentations of three structures (left-ventricle (LV) blood pool, right-ventricle (RV) blood pool and LV myocardium) for both end-diastolic (ED) frames and end-systolic (ES) frames were manually performed to act as ground truth segmentations [6]. The segmentations were performed by a group of eight observers and each subject was annotated only once by one observer. Visual QC was performed on a subset of the data to ensure acceptable inter-observer agreement. The segmentation model was evaluated using 600 different subjects from the UK Biobank for intra-domain testing and two other datasets for cross-domain testing: the Automated Cardiac Diagnosis Challenge (ACDC) dataset (100 subjects, 1 site, 2 scanners) and the British Society of Cardiovascular Magnetic Resonance Aortic Stenosis (BSCMR-AS) dataset (599 subjects, 6 sites, 9 scanners).

3   Methods

The developed image analysis pipeline consists of a DL model for accelerated reconstruction of SAX cine cardiac MRI acquisitions, a DL model for automatic segmentation of the LV blood pool, RV blood pool and LV myocardium, automated calculation of cardiac functional parameters such as EF and LV and RV chamber volumes, and two QC steps: a pre-analysis image QC step during the undersampling and reconstruction process (QC1) and a segmentation QC step to detect good quality segmentations (QC2). For an illustration of the pipeline see Fig. 1.

Undersampling and Reconstruction: We simulated an active acquisition process by first creating k-space data from all slices of cine SAX cardiac MRI images. We utilised a similar strategy to [7] to generate synthetic phase and a radial golden angle sampling pattern to simulate undersampled k-space data containing increasing numbers of profiles corresponding to scan times between 1 to 30 s, in steps of 1 s. These were then reconstructed using two reconstruction algorithms for comparison: the non-uniform Fast Fourier Transform (nuFFT) and the Deep Cascade of Convolutional Neural Networks (DCCNN) [3,4] which features alternating data consistency layers and regularisation layers within an unrolled end-to-end framework. Undersampled k-space data, along with the sampling trajectory and density compensation function, are provided as input to this unrolled model for DL reconstruction, and high-quality MRI images are obtained as an output in an end-to-end fashion. The regularisation layers of this network were implemented as a 5-layer CNN according to [3], and the data consistency layers follow a gradient descent scheme according to [4].

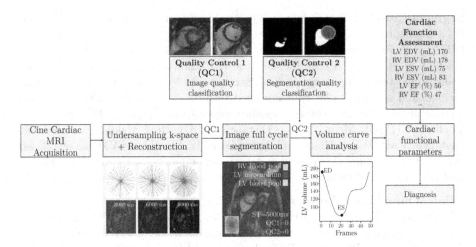

Fig. 1. Overview of the image analysis pipeline for fully-automated cine cardiac MRI undersampled reconstruction and analysis including comprehensive QC algorithms to detect erroneous output. As k-space profiles are acquired, images are continually reconstructed and passed through QC checks. The simulated acquisition terminates when the reconstructed image passes all QC checks.

Image Quality Control (QC1): QC1 was framed as a binary classification problem and addressed using a ResNet classification network [8]. Binary image quality labels (analyzable/non-analyzable) from 225 images of different levels of undersampling from UK Biobank subjects were generated by visual inspection and validated by an expert cardiologist. 80% were used for training and validation of the classification network and 20% were used for testing. The ResNet network was trained for 200 epochs with a binary cross entropy loss function. During training, data augmentation was performed on-the-fly including rotation, shifts and image intensity transformations. The probability of augmentation for each of the parameters was 50%. The training/testing images for QC1 were randomly selected from the UK Biobank dataset and were not used for training or evaluating the reconstruction/analysis framework. The training/testing dataset for QC1 consisted of 50% healthy subjects and 50% patients with cardiomyopathies.

Image Full Cycle Segmentation: We used a pre-trained U-net based architecture [9] for automatic segmentation of the LV blood pool, LV myocardium and RV blood pool from all SAX slices and all frames through the cardiac cycle. The UK Biobank dataset was split into three subsets, containing 3975, 300 and 600 subjects for training, validation and testing respectively. All images were resampled to 1.25 × 1.25 mm. The training dataset was augmented in order to cover a wide range of geometrical variations in terms of the heart pose and size. All images were cropped to the same size of 256×256 before being fed into the network.

Segmentation Quality Control (QC2): The segmentation process is followed by a segmentation QC step (QC2) based on Reverse Classification Accuracy (RCA) [10]. First, image registration is performed between the test image and a set of 20 pre-selected template images with known segmentations. Next, the transformed test image segmentation is compared to those of the atlas segmentations, and a high similarity is assumed to indicate a good quality test image segmentation. The segmentation quality metrics used were the Dice Similarity Coefficient (DSC), Mean Surface Distance (MSD), Root-Mean-Square Surface Distance (RMSD) and Hausdorff Distance (HD). Finally, a SVM binary classifier was trained using the quality metrics independently for LV and RV, and ED and ES frames to discriminate between poor and good quality segmentations.

Clinical Functional Parameters: The volumes were calculated by multiplying the number of voxels by the voxel volume for each of the LV/RV classes. The maximum volume over the cardiac cycle was used for (LV/RV)EDV and the minimum for (LV/RV)ESV. EF (for both LV and RV) was calculated as (EDV-ESV)/EDV.

4   Results

We validated our method in two ways. First, we evaluated the ability of the full pipeline to detect good quality image reconstructions (QC1) and good quality image segmentations (QC2) during simulated active acquisition of 270 cases (200 healthy subjects and 70 patients with cardiomyopathies) randomly selected from the UK Biobank cohort (Validation 1). Second, we compared the estimates of cardiac functional parameters obtained via our pipeline to those obtained from the fully-sampled data (Validation 2).

Validation 1. Image quality was evaluated with the Mean Absolute Error (MAE), Peak Signal to Noise Ratio (PSNR) and Structural Similarity Index (SSIM), calculated between the fully-sampled image and the undersampled image that passed all QC checks with the lowest scan time. Segmentation quality was quantified using Dice coefficients between the segmentations from the fully-sampled image (obtained using the U-net) and the segmentations derived from our pipeline. Image and segmentation quality results are shown in Table 1.

Validation 2. The performance of the method was also evaluated using clinically relevant measures: LVEDV, LVESV, LVEF, RVEDV, RVESV and RVEF. Our DCCNN-based approach resulted in a closer match to the fully-sampled data measures than the nuFFT method and also resulted in a lower scan time as shown in Table 1. There was a good correlation between estimations obtained from fully-sampled data and via our pipeline (Pearson's correlations: LVEDV: r = 0.98; LVESV: r = 0.97; LVEF: r = 0.98; RVEDV: r = 0.98; RVESV:

r = 0.95 and RVEF: r = 0.96). Figure 2 illustrates image reconstructions and undersampling trajectories as a function of the scan time using nuFFT and DCCNN. A Bland-Altman analysis between the volumes estimated from fully-sampled reconstructions and using our DCCNN-based pipeline is shown in Fig. 3.

Table 1. Top: Sensitivity, specificity, and average balanced accuracy for QC1. Middle: Image quality of 270 subjects from the UK Biobank after passing QC1. Bottom: Dice scores between the segmentations from the fully-sampled images and the segmentations derived from our pipeline. The mean and standard deviation of scan times at which the reconstructed images passed all QC checks are also reported.

	nuFFT	DCCNN
QC1 training model		
Average balanced accuracy	0.93	0.95
Sensitivity	0.86	0.87
Specificity	0.99	0.99
QC1 image quality metrics		
MAE	0.03 ± 0.02	0.04 ± 0.02
PSNR	29.82 ± 0.08	32.14 ± 0.06
SSIM	0.87 ± 0.04	0.90 ± 0.03
QC2 dice scores		
LV blood pool	0.91 ± 0.06	0.93 ± 0.04
LV myocardium	0.91 ± 0.06	0.94 ± 0.05
RV blood pool	0.89 ± 0.04	0.90 ± 0.06
Scan time (s)	11.82 ± 3.29	3.72 ± 0.54

5 Discussion

This work has demonstrated the feasibility of a DL-based framework for automated quality-controlled "active" acquisition of undersampled cine cardiac MRI data without a previously defined undersampling factor. The proposed pipeline results in a reduced scan time for 2D cardiac cine MRI, which takes ~12 s in our clinical protocol (spatial resolution = $1.8 \times 1.8 \times 8.0 \, mm^3$, temporal resolution = 31.56 ms and undersampling factor = 2). Our results show that by using a DCCNN for cine cardiac MRI reconstruction, we can pass QC checks after approximately 4 s of simulated acquisition (i.e. an undersampling factor of 4.5). One limitation of our approach is that in the current version of our pipeline, QC2 takes ~1 min. Therefore, this would not allow immediate quality feedback and true 'active' acquisition. Future investigations will focus on developing a real-time approach for segmentation quality control. For example, in [11], a convolutional autoencoder was trained to quantify segmentation quality without a

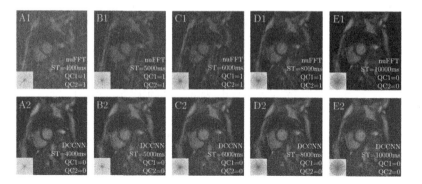

Fig. 2. Illustration of image reconstructions and undersampling trajectories as a function of the scan time using nuFFT and DCCNN. For this subject, the two QC steps were passed at a scan time of 10 s and 4 s with the nuFFT and DCCNN reconstruction models respectively. (QC = 0 means that the QC check was passed.)

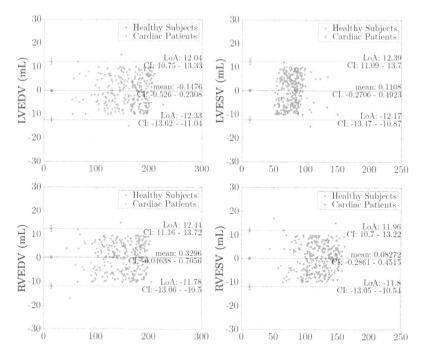

Fig. 3. Bland-Altman plots for cardiac volumes between U-net segmentations from fully sampled reconstructions and our DCCNN reconstructions with smallest scan time that passed all QC checks. The black solid line represents the mean bias and the black dotted lines the limits of agreement. The limits of agreement are defined as the mean difference ± 1.96 SD of differences.

ground truth at inference time. In [12], a 3D convolutional neural network was trained in order to predict the DSC values of 3D segmentations. We will investigate these and similar approaches to implement a segmentation quality control step that is efficient enough to facilitate real-time implementation on the MRI scanner. We believe that an approach such as the one we have proposed could have great clinical utility, reducing redundancies in the cardiac MRI acquisition process whilst still providing diagnostic quality images and robust estimates of functional parameters.

Acknowledgement. This work was funded by the Engineering and Physical Sciences Research Council (EPSRC) programme grant 'SmartHeart' (EP/P001009/1) and supported by the Wellcome/EPSRC Centre for Medical Engineering [WT 203148/Z/16/Z]. The research was supported by the National Institute for Health Research (NIHR) Biomedical Research Centre based at Guy's and St Thomas' NHS Foundation Trust and King's College London. The views expressed are those of the author(s) and not necessarily those of the NHS, the NIHR or the Department of Health. This work was also supported by Health Data Research UK, an initiative funded by UK Research and Innovation, Department of Health and Social Care (England) and the devolved administrations, and leading medical research charities. This research has been conducted using the UK Biobank Resource under Application Number 17806.

References

1. Bustin, A., Fuin, N., Botnar, R.M., Prieto, C.: From compressed-sensing to artificial intelligence-based cardiac MRI reconstruction. Front. Cardiovasc. Med. **7**, 17 (2020)
2. Menchón-Lara, R.M., Simmross-Wattenberg, F., Casaseca-de-la Higuera, P., Martín-Fernández, M., Alberola-López, C.: Reconstruction techniques for cardiac cine MRI. Insights Imaging **10**(1), 1–16 (2019)
3. Schlemper, J., Caballero, J., Hajnal, J.V., Price, A.N., Rueckert, D.: A deep cascade of convolutional neural networks for dynamic MR image reconstruction. IEEE Trans. Med. Imaging **37**(2), 491–503 (2017)
4. Hammernik, K., et al.: Learning a variational network for reconstruction of accelerated MRI data. Magn. Reson. Med. **79**(6), 3055–3071 (2018)
5. Petersen, S.E., et al.: UK biobank's cardiovascular magnetic resonance protocol. J. Cardiovasc. Magn. Reson. **18**(1), 8 (2015)
6. Petersen, S.E., et al.: Reference ranges for cardiac structure and function using cardiovascular magnetic resonance (CMR) in caucasians from the UK biobank population cohort. J. Cardiovasc. Magn. Reson. **19**(1), 1–19 (2017)
7. Haldar, J.P.: Low-rank modeling of local k-space neighborhoods (loraks) for constrained MRI. IEEE Trans. Med. Imaging **33**(3), 668–681 (2013)
8. He, K., Zhang, X., Ren, S., Sun, J.: Deep residual learning for image recognition. In: Proceedings of the IEEE Conference on Computer Vision and Pattern Recognition, pp. 770–778 (2016)
9. Chen, C., et al.: Improving the generalizability of convolutional neural network-based segmentation on CMR images. Front. Cardiovasc. Med. **7**, 105 (2020)
10. Robinson, R., et al.: Automated quality control in image segmentation: application to the UK biobank cardiovascular magnetic resonance imaging study. J. Cardiovasc. Magn. Reson. **21**(1), 1–14 (2019)

11. Galati, F., Zuluaga, M.A.: Efficient model monitoring for quality control in cardiac image segmentation. In: Ennis, D.B., Perotti, L.E., Wang, V.Y. (eds.) FIMH 2021. LNCS, vol. 12738, pp. 101–111. Springer, Cham (2021). https://doi.org/10.1007/978-3-030-78710-3_11
12. Robinson, R., et al.: Real-time prediction of segmentation quality. In: Frangi, A.F., Schnabel, J.A., Davatzikos, C., Alberola-López, C., Fichtinger, G. (eds.) MICCAI 2018. LNCS, vol. 11073, pp. 578–585. Springer, Cham (2018). https://doi.org/10.1007/978-3-030-00937-3_66

Coronary Artery Centerline Refinement Using GCN Trained with Synthetic Data

Zhanqiang Guo[1,2], Yifan Zhang[1,2], Jianjiang Feng[1,2(✉)], Eddy Yang[3],
Lan Qin[3], and Jie Zhou[1,2]

[1] Department of Automation, Tsinghua University, Beijing, China
jfeng@tsinghua.edu.cn
[2] Beijing National Research Center for Information Science and Technology,
Beijing, China
[3] UnionStrong (Beijing) Technology Co. Ltd., Beijing, China

Abstract. Coronary artery centerlines extraction from cardiac CT angiography (CCTA) is an important but challenging task. The popular U-net based coronary artery segmentation and thinning approaches rely on large number of labeled data and tend to produce noisy results. We proposed a graph convolutional network (GCN) for refining noisy centerlines outputted by U-net and developed a coronary artery tree synthesis approach for GCN pretraining. Experiments demonstrate that both modules led to improved performance.

Keywords: Coronary artery extraction · Graph convolutional network · Synthesis data · Coronary CT angiography

1 Introduction

Coronary artery disease (CAD) is one of the leading cause of death in the world. Coronary CT angiography (CCTA) is often applied to obtain coronary artery information because of its non-invasion and sensitivity [1]. In medical image analysis, extracting coronary artery centerlines from CT is the first step to evaluate the extent of plaque and stenosis area [2]. Many automatic (semi-automatic) centerlines extraction methods have been proposed.

Existing coronary artery centerlines extraction methods can be coarsely divided into three categories: shortest path, tracking based, and segmentation based. The first type of method computes the shortest path between the starting and ending points in vessel maps. The key to this method is to define an appropriate cost function, which is smaller for points on centerlines than those on other locations [1,3]. In recent years, algorithms based on deep learning are used to extract features of each point on centerlines [4]. However, these methods generally require numerous manual interactions, and there may be shortcuts off

This work was supported in part by the National Natural Science Foundation of China under Grants 61976121 and 82071921.

E. Puyol Antón et al. (Eds.): STACOM 2021, LNCS 13131, pp. 21–28, 2022.
https://doi.org/10.1007/978-3-030-93722-5_3

the true centerlines. The second type of method depends on iterative tracking, which means the location of the next point is determined according to the characteristics (the direction and radius of the blood vessel, etc.) of the points on the centerlines that have already been obtained [5,6]. These methods generally need additional processing for bifurcation points and they are sensitive to lesions in coronary arteries.

The third type of method segments the coronary artery first, and then extracts the centerline from the segmentation [7]. CNN, especially U-Net [8,9], is the mainstream method of blood vessel extraction in recent years, but its success depends on a large number of data with annotations. What's more, it is a time-consuming and expensive task to label vascular trees in CTs. A possible solution is applying synthetic data with annotations to training [7,10], but it is very challenging to generate realistic and diverse CCTA with groundtruth vessel trees.

In order to solve problems mentioned above, especially the massive demand for data, we propose a framework that can reduce the dependence on segmentation network, which mainly consists of two parts. We firstly use 3D U-net to obtain over-segmented results of coronary artery, and then refine the results to get centerlines via graph convolutional network (GCN). The task of coronary artery over-segmentation is relatively easy with a small amount of data. Compared with generating CCTA, it is more feasible to generate centerlines of coronary artery trees, and the synthetic centerline data can be used to improve the performance of GCN. An effective method of tree structure generation is to interpolate two existing tree structures via geometric and structural blending [11]. So we can complete the whole framework of extracting the coronary artery centerlines with a small amount of data.

The main contributions of this paper include: (1) we propose a GCN model to post-process the centerlines obtained by an over-segmentation network. (2) we propose an approach to generating pairs of training data by adding noise and fracture to synthetic centerlines, which can be regarded as labels. In this way, we augment the training data effectively which can be used to pretrain our GCN, so as to improve the network performance.

Fig. 1. Flowchart: CT image (a) is fed into the over-segmentation network to obtain segmented arteries (b); thinning image (c) is used to construct graph (d); (e) is the coronary artery centerlines after post-processing with GCN.

2   Method

The flowchart of the proposed algorithm is shown in the Fig. 1. The CT image is processed into the mask of the coronary artery by an over-segmentation network. Then, the result is refined to be a centerline image with noise and fracture, which is used to construct a graph. Finally, we can get the precise result through a GCN network. In this step, the synthetic data is applied to improve the performance of the GCN model. The detailed information about our framework is described as follows.

2.1   Over-Segmentation Network

Illustrated as Fig. 2, our segmentation network used in this framework is customized from the basic 3D U-Net network [8], which is generated with the help of nnUNet framework [12]. Similar to the standard version, our segmentation network consists of 3D convolution, max pooling, up sampling, and the shortcut connection from the down-sampling path layer to the up-sampling layer path. Considering that the resolution on each axis of the image differs, the step size $2 \times 2 \times 2$ is changed to $1 \times 2 \times 2$ in the first down-sampling, so that the resolution of each axis can be processed to approximately the same. In order to improve the performance of the network, loss is added for supervision in each step of up-sampling.

2.2   Graph Building

Consider an over-complete undirected graph $G_{in} = \{V, E\}$, which contains subgraphs of the coronary tree and some redundant nodes and edges, where V represents the set of nodes, and $|V| = N$ represents the number of nodes. For each node $i \in V$, its features are expressed as $x_i \in R^{F \times 1}$, and $X \in R^{F \times N}$ is the matrix of node features. E is the set of edges, and $(i, j) \in E$ indicates that there is an edge connection between nodes i and j. $y_{(i,j)} \in R^{H \times 1}$ stands for the features of edge (i, j). The task of extracting the coronary artery centerlines from

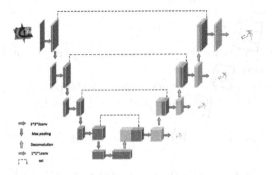

Fig. 2. 3D U-net based coronary artery over-segmentation

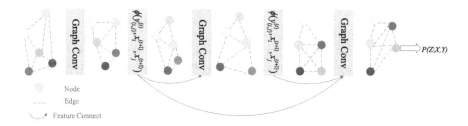

Fig. 3. GCN architecture: convolution of node features and full connection of edge features.

the image with noise and fracture is to train the model $f(\cdot)$, which can identify the points that belong to the centerline in the complete graph, $f : G_{in} \to G$, where G represents the graph of centerline.

The segmentation results in Sect. 3.1 can be refined to centerline images with noise and fracture, the points of which will be built to graphs. We select spatial features and pixel features for each point. The spatial features include the direction $[d_x, d_y, d_z]$, spatial position $[x, y, z]$ and two of the ten closest points with maximum and minimum distances $[d_{max}, d_{min}]$. The pixel features include the original image pixel value, the probability value of segmentation prediction and the value in the blood vessel enhancement image. For each point, it is connected with its nearest 10 points. The feature of each edge is defined as the reciprocal of the distance between involved two nodes.

2.3   GCN Architecture

The Graph Convolutional Network is shown in Fig. 3, which applies an alternate mode of updating node features and edge features. The updating process of node features is defined as:

$$x_v^{(t+1)} = \varphi(x_v^{(t)}, \{x_u^{(t)} : (u,v) \in E\}) \tag{1}$$

where $x_u^{(t)}$ is the features of all nodes connected to node v. $x_v^{(t+1)}$ and $x_v^{(t)}$ represent the features of node v at time $t+1$ and t respectively. The updating process of edge features is defined as:

$$y_{(i,j)}^{(t+1)} = \phi(y_{(i,j)}^{(t)}, x_i^{(t+1)}, x_j^{(t+1)}) \tag{2}$$

where $y_{(i,j)}^{(t)}$ and $y_{(i,j)}^{(t)}$ indicate the features of edge (i,j) at $t+1$ and t, respectively. $\phi(\cdot)$ is a fully connected network.

The output $P(z|i)$ of final network represents the probability that the node i belongs to the centerline. The loss is the cross-entropy loss, expressed as:

$$L = -\sum_{i=1}^{N}[z_i log(P(\hat{z}_i|X,Y,W)) + (1-z_i)log(1-P(\hat{z}_i|X,Y,W))] \tag{3}$$

where X, Y correspond to the features of nodes and edges. W represents the parameters of the network, and z represents the label of the node i.

2.4   Data Synthesis

We consider that centerlines and trees resemble the same structure. For a coronary artery tree, we classify its branches into three categories. We define the longest branch as the main tree, the branches connected to the main tree as primary branches, and the branches connected to primary branches as secondary branches (if existing). For two centerlines T and S in our dataset, we match their main trees, primary branches and secondary branches respectively. Empty branches will be added if one branch doesn't have pairing branch. In this way, we construct the pairing relationship between all the branches of T and S. For each pair of corresponding branches, we uniformly sample the same number of points and interpolate these points by B-spline respectively. After B-spline interpolation, we obtain a tuple B, containing the vector of B-spline coefficients. Thus, for a certain pair of branches t and s, two tuples B_t and B_s of the same size are obtained. Linear combination of B_t and B_s can generate intermediate state between t and s.

$$B_{new} = \alpha \cdot B_t + (1 - \alpha) \cdot B_s \quad (0 < \alpha < 1) \tag{4}$$

As α goes from 0 to 1, it illustrates a process that B_t converts to B_s. By adjusting the value of α, we get different B_{new}, which can be restored to different branches. Applying the above operation to each branch, finally we generate a complete coronary artery tree based on a certain α. The synthetic centerlines can be used as training data after adding noise and fracture, with the original centerlines as its pairing labels.

3   Experiments and Results

3.1   Evaluation Metrics

We evaluate the results based on the following indicators.

(1) The classification accuracy rate of GCN: the ratio of the number of nodes classified correctly to the number of total nodes, which shows the classification ability of GCN.

(2) The mean distance between centerlines:

$$d_{err} = \frac{\sum_{i=1}^{N_{seg}} min[d_E(c_i, C_{ref})]}{2N_{seg}} + \frac{\sum_{j=1}^{N_{ref}} min[d_E(c_j, C_{seg})]}{2N_{ref}} = \frac{d_{FP} + d_{FN}}{2} \tag{5}$$

where C_{ref} is the reference centerline, C_{seg} is the prediction centerline. N_{seg} and N_{ref} represent the number of points on the prediction centerline and the reference centerline respectively.

(3) Coverage percentage: a point on centerline A is covered by centerline B, if the distance between this point and the closest point on B does not exceed the threshold. We calculate the coverage percentage of the labels and the predicted centerlines respectively.

Given that the predicted results of some branches may be longer than the marked results, which should not affect the evaluation in practice, these points do not participate in the calculation when computing the above indicators.

3.2   Results on Our Dataset

In the public CAT08 dataset [13], only four main centerline labels of the coronary arteries are given, which cannot reflect the performance of our method on small branches. So we conducted experiments on a dataset from a local hospital, which includes 47 CT volumes with segmentation labels. The label of centerline is obtained by thinning manually marked coronary artery label image. We randomly selected 32 data for training and 15 data for testing. Half of training data was used to train the segmentation network and all training data was used to train graph network.

Table 1. Performance comparison of four methods: GNN (the results on the airway dataset [14]), refining the segmented image (3D U-net), GCN with spatial features (S), GCN with spatial and pixel features (S+V). The accuracy of node classification (Acc), the distance of predicted centerline and label (d_{PF}, d_{FN}, d_{err}), and the coverage ($C_{predict}, C_{label}$) are reported.

	Acc(%)	d_{PF}(mm)	d_{FN}(mm)	d_{err}(mm)	$C_{predict}$(%)	C_{label}(%)
GNN	–	2.21 ± 0.46	2.87 ± 0.50	2.54 ± 0.58	–	–
3D U-net	–	3.21	1.57	2.39	89.37	72.74
GCN(S)	80.41 ± 0.01	1.57 ± 0.05	1.21 ± 0.04	1.39 ± 0.03	92.41 ± 0.05	77.61 ± 0.11
GCN(S+V)	80.34 ± 0.18	1.58 ± 0.07	1.29 ± 0.04	1.44 ± 0.05	90.58 ± 0.02	75.02 ± 0.12

As shown in Table 1, our approach has better performance in reducing fracture and noise on the centerlines. And the results show that the spatial feature plays a key role in the GCN model, because the pixel feature has been used in the previous segmentation network.

3.3   Results with Synthesis Data

We access the usefulness of pretraining with synthetic data by comparing various scores mentioned above. As shown in Table 2, when using pretraining, the performance on each indicator improves or remains basically unchanged. Since the synthetic data is mainly used for training GCN, the improvement of its classification accuracy can directly prove the effectiveness of synthetic data in this task. We achieve the accuracy rate of 80.41% for training from scratch without pretraining and 81.32% when pretraining on synthetic data.

Figure 4 shows that compared to the method of refining the segmentation result, our approach perfoms better in avoiding noise and fracture on the centerline. So it can deal with stenosis and plaques that may exist on CT images.

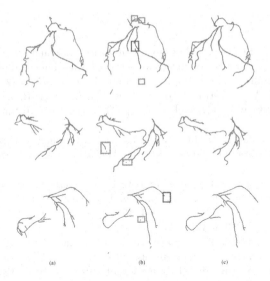

Fig. 4. Qualitative comparison of different algorithms on some examples. (a) annotated, (b) thinning image of over-segmentation result, (c) refining result of the proposed method. The green boxes and blue boxes respectively reflect the improvement of the noise suppression and fracture connection capabilities of our proposed method by using syntheic data. (Color figure online)

Table 2. Results of different data: the performance on synthesis data, the performance on real data and the results with pretraining.

	Acc(%)	d_{PF}(mm)	d_{FN}(mm)	d_{err}(mm)	$C_{predict}$(%)	C_{label}(%)
Synthesis data	96.10 ± 0.19	0.72 ± 0.07	0.17 ± 0.01	0.36 ± 0.04	96.30 ± 0.16	92.58 ± 0.33
Real data	80.41 ± 0.01	1.57 ± 0.05	1.21 ± 0.04	1.39 ± 0.03	92.41 ± 0.05	77.61 ± 0.11
Pretraining	81.32 ± 0.07	1.45 ± 0.04	1.19 ± 0.01	1.32 ± 0.2	92.83 ± 0.09	77.47 ± 0.10

4   Conclusion

In this paper, we propose a GCN-based centerline extraction framework. It consists of a segmentation network and post-processing steps. Since GCN performs better in the task of processing centerline images with noise, we just need an over-segmentation result, thereby reducing the dependence on segmentation performance. The experiments show that our method achieves good performance. In addition, synthetic data is used to further improve the effect of network. In our framework, the requirements for annotating data can be significantly reduced, which is also very helpful in practice.

References

1. Leipsic, J., et al.: SCCT guidelines for the interpretation and reporting of coronary CT angiography: a report of the society of cardiovascular computed tomography guidelines committee. J. Cardiovasc. Comput. Tomogr. **8**(5), 342–358 (2014)
2. Marquering, H.A., Dijkstra, J., de Koning, P.J.H., Stoel, B.C., Reiber, J.H.C.: Towards quantitative analysis of coronary CTA. Int. J. Card. Imaging **21**(1), 73–84 (2005)
3. Krissian, K., Bogunovic, H., Pozo, J., Villa-Uriol, M., Frangi, A.: Minimally interactive knowledge-based coronary tracking in CTA using a minimal cost path. Insight J. **1** (2008)
4. Guo, Z., et al.: DeepCenterline: a multi-task fully convolutional network for centerline extraction. In: Chung, A.C.S., Gee, J.C., Yushkevich, P.A., Bao, S. (eds.) IPMI 2019. LNCS, vol. 11492, pp. 441–453. Springer, Cham (2019). https://doi.org/10.1007/978-3-030-20351-1_34
5. Zhou, C., et al.: Automated coronary artery tree extraction in coronary CT angiography using a multiscale enhancement and dynamic balloon tracking (MSCAR-DBT) method. Comput. Med. Imaging Graph. **36**(1), 1–10 (2012)
6. Lesage, D., Angelini, E.D., Funka-Lea, G., Bloch, I.: Adaptive particle filtering for coronary artery segmentation from 3D CT angiograms. Comput. Vis. Image Underst. **151**, 29–46 (2016)
7. Tetteh, G., et al.: DeepVesselNet: vessel segmentation, centerline prediction, and bifurcation detection in 3-D angiographic volumes. Front. Neurosci. **14**, 1285 (2020)
8. Çiçek, Ö., Abdulkadir, A., Lienkamp, S.S., Brox, T., Ronneberger, O.: 3D U-net: learning dense volumetric segmentation from sparse annotation. In: Ourselin, S., Joskowicz, L., Sabuncu, M.R., Unal, G., Wells, W. (eds.) MICCAI 2016. LNCS, vol. 9901, pp. 424–432. Springer, Cham (2016). https://doi.org/10.1007/978-3-319-46723-8_49
9. Wang, Y., et al.: Deep distance transform for tubular structure segmentation in CT scans. In: Proceedings of the IEEE/CVF Conference on Computer Vision and Pattern Recognition, pp. 3833–3842 (2020)
10. Zhu, J.-Y., Park, T., Isola, P., Efros, A.A.: Unpaired image-to-image translation using cycle-consistent adversarial networks. In: Proceedings of the IEEE International Conference on Computer Vision, pp. 2223–2232 (2017)
11. Wang, G., Laga, H., Xie, N., Jia, J., Tabia, H.: The shape space of 3D botanical tree models. ACM Trans. Graph. (TOG) **37**(1), 1–18 (2018)
12. Isensee, F., Jäger, P.F., Kohl, S.A.A., Petersen, J., Maier-Hein, K.H.: Automated design of deep learning methods for biomedical image segmentation. arXiv preprint arXiv:1904.08128 (2019)
13. Schaap, M., et al.: Standardized evaluation methodology and reference database for evaluating coronary artery centerline extraction algorithms. Med. Image Anal. **13**(5), 701–714 (2009)
14. Selvan, R., et al.: Graph refinement based airway extraction using mean-field networks and graph neural networks. Med. Image Anal. **64**, 101751 (2020)

Novel Imaging Biomarkers to Evaluate Heart Dysfunction Post-chemotherapy: A Preclinical MRI Feasibility Study

Peter Lin[1,2], Terenz Escartin[2,3], Matthew Ng[3], Mengyuan Li[4], Melissa Larsen[3], Jennifer Barry[3], Idan Roifman[3], and Mihaela Pop[3(✉)]

[1] University Health Network, Toronto, Canada
[2] Medical Biophysics, University of Toronto, Toronto, Canada
[3] Sunnybrook Research Institute, Toronto, Canada
mihaela.pop@utoronto.ca
[4] Institute of Biomedical Engineering, University of Toronto, Toronto, Canada

Abstract. Cardiotoxicity is a major complication of chemotherapy, which leads in time to progressive electro-mechanical dysfunction, irreversible myocardial tissue remodeling and eventual heart failure. In this pilot study, we seek to find key biomarkers of early cardiotoxic effects. For this, a weekly dose of doxorubicin, DXO was injected in swine during a four-week treatment. Longitudinal MR imaging was performed pre- and post-DXO (at: 1, 5 and 9 weeks after ending the DXO-therapy) using a scanning protocol at 3T that included: functional CINE imaging; T2 and T1 mapping, and 3D late gadolinium enhancement (LGE) for tissue characterization. Our results showed that, compared to the mean baseline values pre-DXO, ejection fraction gradually decreased from a mean 47% to ~34% post-DXO, indicating that the cardiac biomechanical function started to deteriorate within weeks post-DXO. The initially increased T2-derived edema appeared to resolve by week 5. Furthermore, a gradual deposition of diffuse fibrosis post-DXO was identified by T1 values and LGE, and confirmed by collagen-sensitive histological stains. Our preliminary results will help establish MR imaging protocols and models that predict early DXO-induced cardiotoxicity, which could be used by clinicians to develop cardioprotective strategies for preventing heart failure progression post-chemotherapy.

Keywords: Cardiac MR imaging · Chemotherapy · Cardiotoxicity

1 Introduction

Cardio-oncology is a new research field with a focus on detection and monitoring of reversible and irreversible heart damage (at cellular, tissue or organ level) caused by chemotherapeutic agents and radiotherapy in cancer patients [1, 2]. In recent years, it has been observed in animal and clinical studies that progressive electro-mechanical heart dysfunction is a significant adverse consequence [3–5], limiting the effectiveness of chemotherapy using anthracyclines (e.g., doxorubicin, DXO) in the cancer patients.

© Springer Nature Switzerland AG 2022
E. Puyol Antón et al. (Eds.): STACOM 2021, LNCS 13131, pp. 29–37, 2022.
https://doi.org/10.1007/978-3-030-93722-5_4

Notably, while apoptosis and edema often occur in the acute/subacute phases of DXO therapy, the irreversible myocardial damage such as deposition of collagenous fibrosis is suspected to occur in early chronic and late-onset phases after treatment [2, 6].

Current clinical diagnostic tools (such as: invasive biopsy, echocardiography) often lack appropriate protocols to monitor and detect DXO-mediated myocardial injury, particularly soon after the therapy ended [7]. As a consequence, the irreversible injury evolves and worsens in time, leading within months and years to cardiomyopathy and heart failure, respectively (Fig. 1). Thus, there has been an unmet need to develop better non-invasive imaging tools for early detection of DXO-induced cardiotoxicity.

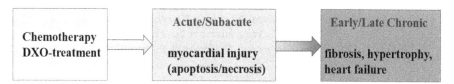

Fig. 1. Cardiotoxic effects and heart remodeling following chemotherapy are first detected as myocardial injury (apoptosis and cellular necrosis), which gradually degenerates into fibrosis, hypertrophy and heart failure.

With this respect, several clinical groups have employed MRI methods to characterize late cardiotoxic effects [8, 9]; however, most MR acquisitions are limited to a large slice thickness, missing subtle myocardial changes [10, 11]. Other groups have focused on developing animal models of chemotherapy-induced cardiotoxicity; however, these either employed small animals [12] or the DXO-delivery was performed intracoronary [13] which is invasive and does not replicate a clinical chemotherapy treatment.

The specific aims of this paper are: 1) to develop a translational swine model of DXO-therapy, mimicking precisely the chemotherapeutic dosage plan in cancer patients; and, 2) in a pilot study, to establish subtle MR imaging biomarkers of DXO-induced cardiotoxic effects in the early phases post-DXO (i.e., subacute and early chronic, weeks after DXO-therapy). More precisely, in this preclinical work we seek to employ our 2D MR imaging sequences (e.g. Cine, T1 and T2 mapping) to monitor the temporal evolution of biomarkers such as left ventricle (LV) function (i.e., ejection fraction index) and edema (as in [14]), as well as a high resolution free-breathing 3D contrast-enhanced MR method to detect gradual collagen deposition. We hypothesize that these biomarkers will provide a better mechanistic understanding regarding the irreversible alterations in structural and functional tissue characteristics, which could trigger LV remodelling and lead in time to heart failure.

2   Materials and Methods

2.1   Preclinical Model of Doxorubicin Chemotherapy

All animal studies received ethical approval from Sunnybrook Research Institute and were conducted in accordance with protocols instated by the Animal Care Committee.

In this feasibility pilot study, we used 3 juvenile Yorkshire pigs of approximative 30 kg each. Each animal underwent a total of 4 weekly intravenous injections of 1mg/kg DXO (i.e., the typical chemotherapeutic doses given to cancer patients). Specifically, the total DXO dosage per treatment was calculated based on the pig weight and administered slowly, intravenously (i.v.), over a 20–30-min period.

The study design is summarized in Fig. 2. MR imaging was performed prior and after chemotherapy. MR studies were conducted at baseline, 1-week post, 5-week post, and 9-week post-doxorubicin therapy. The hearts were explanted at the study endpoint (i.e., at the completion of the experiment). Select myocardial tissue samples were prepared for histopathological staining and visualization.

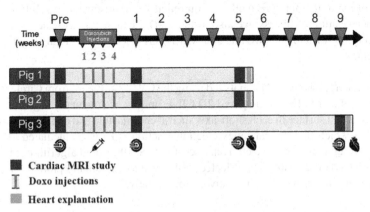

Fig. 2. Diagram of the study design for the control pig and the pigs undergoing chemotherapy (for the latter, *in vivo* MRI was performed for baseline measurements prior to the weekly DXO injections, and at 1-week, 5-week and/or 9-week timepoints following the completion of DXO injections).

2.2 In Vivo MR Imaging Studies and Image Analysis

Cardiac MRI was performed *in vivo* on a 3T whole body scanner (MR 750, General Electric Healthcare, Waukesha, Wisconsin, USA), using a cardiac phase-array coil. The imaging protocol developed and proposed for this study included: short-axis 2D CINE sequences, 2D T2 and T2* mapping; as well as 2D T1 mapping and 3D late gadolinium enhancement (LGE) for fibrosis quantification. CMR imaging was performed as per the diagram in Fig. 2: at baseline healthy state (pre-DXO injections) and at 1 week following the completion of DXO therapy in all three pigs; then at 5 weeks following the completion of DXO therapy in pigs #1 and #2, respectively; as well as at 9 weeks post-DXO injections in pig #3.

The following MR scan protocol was tested and implemented:

a) *Cardiac function* was evaluated using a steady-state-free-precession (SSFP) CINE sequence: 16–20 short-axis slices, TR/TE = 4.2/1.8 ms; flip angle = 45°; 20

phases/slice; matrix $= 224 \times 160$; in-plane spatial resolution $\sim 1 \times 1$ mm; and, slice thickness 5 mm.

b) *Edema* was quantified using a T2 mapping sequence with spiral readout, with T2-prep in 5 short-axis slices; flip angle $= 90°$, TR $= 3.93$ ms; variable echo times TE; in-plane resolution $\sim 1 \times 1$ mm; slice thickness 5 mm (note that T2 mapping was performed only in DXO-treated pig #1 and pig #2, due to limited availability of the MR sequence).

c) *Fibrosis* (i.e., collagen deposition) was assessed post-DXO using a recently implemented high resolution free breathing 3D late gadolinium enhancement (LGE) MR sequence with navigator (as in [15]), initiated at approximative 5 min post-injection of a bolus of contrast agent Gd-DTPA (0.2 mmol/kg; Magnevist, Bayer Healthcare). In addition to the 3D LGE method acquired at 1.4 mm isotropic resolution, for the characterization of fibrosis we also tested a 2D T1 mapping sequence (employing imaging for T1 native as well as after Gd-DTPA injection), using a slice thickness of 5 mm.

All MR images were further analyzed using the CVI 42 software from Circle (https://www.circlecvi.com/). For the acquired 2D CINE images, we manually segmented each heart endocardial surface in each short-axis slice at end-diastole and end-systole, respectively, from which the clinical index ejection fraction (EF) was calculated. 3D LGE images were segmented using a clinically accepted 5SD threshold algorithm. Mean values $\pm$ SD (standard deviation) on selected ROIs were longitudinally plotted for various parameters of interest (i.e., EF; T1 native; and, T2 native).

2.3   Histology

Explanted hearts were fixed and preserved in formaldehyde. Myocardial tissue samples were collected from each heart. Furthermore, select cross-sections corresponding to the short-axis MR images (as guided by anatomical landmarks), were cut at 4 μm thickness. These were stained with collagen-specific Mason Trichrome stain to detect and clearly visualize collagen deposition. Following staining, the histopathology samples were mounted on large glass slides and digitally scanned using a TissueScope (Huron Technologies). Lastly, the digital images were visualized using the Aperio ImageScope open-source software (https://aperio-imagescope.software.informer.com/), which is specifically designed for digital pathology evaluations. Select regions of interest ROIs from histology-defined fibrosis and from MR-derived fibrosis were then qualitatively compared.

3   Results

Figure 3 shows results from the functional analysis of the MR CINE images obtained in the left ventricle, LV. Figures 3-A and 3-B illustrate an example from the segmentation step in Pig #3. The red contours correspond to the endocardium while the green contours correspond to the epicardium of the LV, at end-diastole and end-systole of the cardiac cycle, respectively. Figure 3-C shows examples of Cine MR images in the short-axis planes selected for analysis (from apex to base).

The temporal evolution of the ejection fraction (EF) parameter plotted in Fig. 3-D for all three pigs, suggests that there is a gradual decline in the overall cardiac function in the weeks following DXO-therapy. In particular, we observed an important decrease in EF values within 9 weeks post-DXO (i.e., from 47% to ~34%).

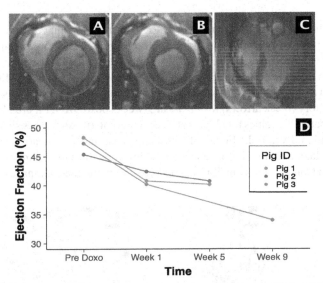

Fig. 3. Characterization of left ventricular function through analysis of MR CINE images. The images captured complete cardiac cycles and timepoints representing end-systole and end-diastole were manually annotated. ED = end-diastole, ES = end-systole, SAX = short-axis image.

Figures 4 shows illustrative results from myocardial T2 analysis. Each T2 map was constructed by performing pixel-wise fitting of MR signal intensities with an exponential model (as exemplified in Fig. 4A). Moreover, as demonstrated in Fig. 4B, the initial increase of native T2 values (from 55 ± 11 ms to 68 ± 11 ms) at 1-week after DXO-therapy was significant, indicating edema development (i.e., fluid accumulation and increased interstitial space). This was followed by a T2 decrease (to 61 ± 14 ms), suggestive of edema resorption within the following month.

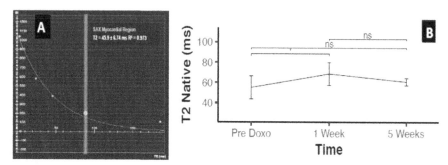

Fig. 4. Results from the T2 mapping analysis: (A) MR signal intensity fitting of T2 relaxation time for a selected ROI; (B) evolution of mean T2 native across pig 1 and 2 during the study (*p < 0.05 denotes significant differences; ns = not significant).

Figure 5 illustrates results from the analysis of T1 mapping images, before and after the gadolinium-based contrast agent injection. Figures 5 (A-B) show exemplary pixel-wise native T1 maps (before Gd-DTPA injection) constructed from Pig #3 at baseline and at 9 weeks following the last injection of DXO, respectively. Figure 5C illustrates the T1 map 2-min after Gd injection, indicating (qualitatively) evidence of diffuse fibrosis in the septal territory (black arrows) for Pig #3 at 9-weeks post-DXO. Figure 5D shows the plot of mean native T1 values for all pigs at three representative short-axis slice locations: apex, base, and mid-ventricle. Figure 5E shows the quantitative comparison (*p < 0.05) between T1 relaxation times before and after Gd injection, on a ROI comprised of septal tissue selected from the apparent fibrotic area (note: from the same animal as in Fig. 5C).

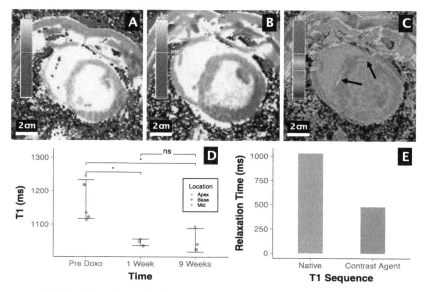

Fig. 5. Example results from T1 mapping analysis (see text for details)

Figure 6 shows data from high resolution 3D LGE images and histology in Pig #3, to highlight the evidence of diffuse fibrosis deposition. In the segmented short-axis image selected as an example, the pixels with diffuse fibrosis (in green in Fig. 6B) were identified in the raw image (shown in Fig. 6A) using a clinical threshold of signal intensity SI = 5 SD above healthy tissue (pixels in white). Figures 6C and 6D show representative histological images stained with collagen-sensitive Masson Trichrome selected from pathological regions with clear collagen deposition and from normal tissue (i.e., lacking diffuse fibrosis), respectively. As qualitatively observed in Fig. 6C, patches of interstitial diffuse fibrosis (blue stain) are scattered throughout the myocardium (red stain).

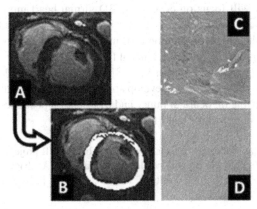

Fig. 6. Identification of fibrosis by high resolution 3D LGE (at 1.4 mm isotropic spatial resolution) and collagen-sensitive histological stain (see text for details).

4 Discussion and Conclusion

Functional and structural MRI biomarkers may be critical in predicting the recovery of cardiac function, potentially allowing the experimental testing of combined therapies targeted to prevent or alleviate DXO-related injury.

Overall, in this feasibility pilot study, our high-resolution MR imaging methods were able to identify key biomarkers of reversible and irreversible DXO-mediated cardiotoxic effects along with histological validation. Firstly, the initially increased edema (supported by T2-derived values) appeared to reverse by week 5, in agreement with other preclinical studies [13]. Secondly, the gradual decline of EF within the first couple of months post-DXO suggests an early (i.e., sub-acute and early chronic phases) triggering of biomechanical dysfunction. Thirdly, a gradual deposition of diffuse fibrosis post-DXO therapy was identified by both T1 maps and LGE images (confirmed by histology). Collagen deposition in the myocardium has been observed in preclinical models of structural heart disease with histopathological validation (e.g., in chronic infarction with focal fibrosis [16, 17], and in diffuse fibrosis cases [18]). Thus, we believe that the irreversible collagen accumulation substantially contributed to the observed gradual

EF decline post-DXO. This is in agreement with other studies suggesting that DXO-induced fibrosis is an important cause of the heart biomechanical function deterioration and evolution towards heart failure [19–21].

We acknowledge that our pilot study was limited to a small number of animals. However, we successfully addressed our objectives, while respecting the '3Rs principle' in animal research (replacement, reduction and refinement). In particular, we demonstrated the feasibility of a novel preclinical pig model of cardiotoxicity using a noninvasive i.v. delivery of anthracyclines mimicking the clinical delivery of chemotherapy treatment (in contrast to the invasive intracoronary approach used in [13]). We also successfully implemented an MR protocol using higher spatial resolution compared to clinical scans that use 8–10 mm slice thickness (producing partial volume effects).

Our future work will focus on developing 3D virtual heart models to predict late effects of cardiotoxicity based on early image-based features of cardiac structure and function post-DXO therapy. Computational 3D models are powerful tools used in cardiovascular research to predict electro-mechanical function in normal or pathological state, and to design better delivery of therapies [22, 23]. Thus, we envision that such predictive models will be able to help clinicians develop appropriate cardio-protective strategies to minimize the irreversible myocardial injury and dysfunction post-chemotherapy, and to limit further progression towards heart failure stage.

Acknowledgment. This work was financially supported by a CIHR project grant (PJT 153212) awarded to Dr. Mihaela Pop. The authors would like to thank Mr. Adebayo Adeeko (Biomarker Imaging Lab, Sunnybrook Research Institute, Toronto, CA) for processing and staining the histopathological slices.

References

1. Kostakou, P.M., Kouris, N.T., Kostopoulos, V.S., Damaskos, D.S., Olympios, C.D.: Cardio-oncology: a new and developing sector of research and therapy in the field of cardiology. Heart Fail. Rev. **24**(1), 91–100 (2018). https://doi.org/10.1007/s10741-018-9731-y
2. Von Hoff, D.D., Rozencweig, M., Piccart, M.: The cardiotoxicity of anticancer agents. Semin. Oncol. **9**, 23–33 (1982)
3. Christiansen, S., Autschbach, R.: Doxorubicin in experimental and clinical heart failure. Eur. J. Cardiothorac. Surg. **30**, 611–616 (2006)
4. Podyacheva, E.Y., Kushnareva, E.A., Karpov, A.A., Toropova, Y.G.: Analysis of models of doxorubicin-induced cardiomyopathy in rats and mice. a modern view from the perspective of the pathophysiologist and the clinician. Front. Pharmacol. **12**, 670479 (2021)
5. Zeiss, C.J., et al.: Doxorubicin-induced cardiotoxicity in collaborative cross (CC) mice recapitulates individual cardiotoxicity in humans. G3 Genes|Genomes|Genetics **9**(8), 2637–2646 (2019)
6. Angsutararux, P., Luanpitpong, S., Issaragrisil, S.: Chemotherapy-induced cardiotoxicity: overview of the roles of oxidative stress. Oxid. Med. Cell Longev. **2015**, 795602 (2015)
7. Silbiger, J.J.: Pathophysiology and echocardiographic diagnosis of left ventricular diastolic dysfunction. J. Am. Soc. Echocard. **32**(2), 216-232.e2 (2019)
8. Song, L., et al.: Serial measurements of left ventricular systolic and diastolic function by cardiac magnetic resonance imaging in patients with early stage breast cancer on Trastuzumab. Am. J. Cardiol. **123**, 1173–1179 (2019)

9. Safaei, A.M., et al.: Detection of the early cardiotoxic effects of doxorubicin-containing chemotherapy regimens in patients with breast cancer through novel cardiac magnetic resonance imaging: a short-term follow-up. J. Clin. Imag. Sci. **11**, 33 (2021)
10. Thavendiranathan, P., Wintersperger, B.J., Flamm, S.D., Marwick, T.H.: Cardiac MRI in the assessment of cardiac injury and toxicity from cancer chemotherapy: a systematic review. Circul. Cardiovas. Imag. **6**(6), 1080–1091 (2013)
11. Gong, I.Y., et al.: Early diastolic strain rate measurements by cardiac MRI in breast cancer patients treated with trastuzumab: longitudinal study. Int. J. Card. Imag. **35**, 653–662 (2019)
12. Manno, R.A., Grassetti, A., Oberto, G., Nyska, A., Ramot, Y.: The minipig as a new model for the evaluation of doxorubicin-induced chronic toxicity. J. Appl. Toxico. l **36**, 1060–1072 (2016)
13. Galan-Arriola, C., et al.: Serial magnetic resonance imaging to identify early stages of anthracycline-induced cardiotoxicity. J. Am. Coll. Cardiol. **73**, 779–791 (2019)
14. Ghugre, N.R., Ramanan, V., Pop, M., et al.: Quantitative tracking of edema, hemorrhage, and microvascular obstruction in subacute myocardial infarction in a porcine model by MRI. Magn. Reson. Med. **66**(4), 1129–1141 (2011)
15. Zhang, L., Lai, P., Pop, M., Wright, G.A.: Accelerated multicontrast volumetric imaging with isotropic resolution for improved peri-infarct characterization using parallel imaging, low-rank and spatially varying edge-preserving sparse modeling. Magn. Reson. Med. **79**(6), 3018–3031 (2018)
16. Pop, M., Ramanan, V., Yang, F., et al.: High-resolution 3-D T1*-mapping and quantitative image analysis of gray zone in chronic fibrosis. IEEE Trans. Biomed. Eng. **61**(12), 2930–2938 (2014)
17. Oduneye, S.O., et al.: Postinfarction ventricular tachycardia substrate characterization: a comparison between late enhancement magnetic resonance imaging and voltage mapping using an MR-guided electrophysiology system. IEEE Trans. Biomed. Eng. **60**(9), 2442–2449 (2013)
18. Siu, A.G., et al.: Characterization of the ultrashort-TE (UTE) MR collagen signal: UTE MRI of collagen. NMR Biomed. **28**(10), 1236–1244 (2015). https://doi.org/10.1002/nbm.3372
19. Meléndez, G.C., Gregory Hundley, W.: Is myocardial fibrosis a new frontier for discovery in cardiotoxicity related to the administration of anthracyclines? Circul. Cardiovas. Imag. **9**, 12 (2016)
20. Jordan, J.H., et al.: Anthracycline-associated t1 mapping characteristics are elevated independent of the presence of cardiovascular comorbidities in cancer survivors. Circul. Cardiovas. Imag. **9**, 8 (2016)
21. Octavia, Y., et al.: Doxorubicin-induced cardiomyopathy: from molecular mechanisms to therapeutic strategies. J. Mol. Cell. Cardiol. **52**, 1213–1225 (2012)
22. Yuan, Y., Bai, X., Luo, C., Wang, K., Zhang, H.: The virtual heart as a platform for screening drug cardiotoxicity. Br. J. Pharmacol. **172**, 5531–5547 (2015)
23. Marchesseau, S., Delingette, H., Sermesant, M., et al.: Fast parameter calibration of a cardiac electromechanical model from medical images based on the unscented transform. Biomech. Model Mechanobiol. **12**, 815–831 (2013)

A Bi-atrial Statistical Shape Model as a Basis to Classify Left Atrial Enlargement from Simulated and Clinical 12-Lead ECGs

Claudia Nagel[✉] ⓘ, Matthias Schaufelberger ⓘ, Olaf Dössel ⓘ,
and Axel Loewe ⓘ

Institute of Biomedical Engineering, Karlsruhe Institute of Technology (KIT),
76131 Karlsruhe, Germany
publications@ibt.kit.edu

Abstract. Left atrial enlargement (LAE) is one of the risk factors for atrial fibrillation (AF). A non-invasive and automated detection of LAE with the 12-lead electrocardiogram (ECG) could therefore contribute to an improved AF risk stratification and an early detection of new-onset AF incidents. However, one major challenge when applying machine learning techniques to identify and classify cardiac diseases usually lies in the lack of large, reliably labeled and balanced clinical datasets. We therefore examined if the extension of clinical training data by simulated ECGs derived from a novel bi-atrial shape model could improve the automated detection of LAE based on P waves of the 12-lead ECG.

We derived 95 volumetric geometries from the bi-atrial statistical shape model with continuously increasing left atrial volumes in the range of 30 ml to 65 ml. Electrophysiological simulations with 10 different conduction velocity settings and 2 different torso models were conducted. Extracting the P waves of the 12-lead ECG thus yielded a synthetic dataset of 1,900 signals. Besides the simulated data, 7,168 healthy and 309 LAE ECGs from a public clinical ECG database were available for training and testing of an LSTM network to identify LAE.

The class imbalance of the training data could be reduced from 1:23 to 1:6 when adding simulated data to the training set. The accuracy evaluated on the test dataset comprising a subset of the clinical ECG recordings improved from 0.91 to 0.95 if simulated ECGs were included as an additional input for the training of the classifier.

Our results suggest that using a bi-atrial statistical shape model as a basis for ECG simulations can help to overcome the drawbacks of clinical ECG recordings and can thus lead to an improved performance of machine learning classifiers to detect LAE based on the 12-lead ECG.

Keywords: Left atrial enlargement · Atrial fibrillation · ECG simulation

© Springer Nature Switzerland AG 2022
E. Puyol Antón et al. (Eds.): STACOM 2021, LNCS 13131, pp. 38–47, 2022.
https://doi.org/10.1007/978-3-030-93722-5_5

1 Introduction

Left atrial enlargement (LAE) contributes to the arrhythmogenesis and maintenance of atrial fibrillation [1]. Thus, an established LAE diagnosis could serve as a valuable risk marker for the early detection of atrial fibrillation which is crucial for choosing the appropriate patient-specific treatment therapy.

As an alternative to cardiac chamber size measurements with imaging techniques, a clinical diagnosis of LAE could initially also be informed by an evaluation of the 12-lead electrocardiogram (ECG) [2]. In contrast to imaging techniques applied to quantify the atrial chamber sizes, the ECG features an inexpensive, easily accessible and widely available tool in clinical practice. An automated analysis of ECG signals with e.g. machine learning techniques could facilitate the early diagnosis of LAE and in turn contribute to a proper risk stratification for atrial fibrillation.

However, the application of machine learning algorithms to clinical ECGs for an automated disease classification entails several challenges. On the one hand, large clinical datasets are rarely available and also the diagnostic classes of interest are usually not balanced. Moreover, ground truth labels marking the underlying cardiac pathology have to be set manually by cardiologists. However, expert annotations are inevitably subject to inter- and intra-observer variabilities contributing to unreliably and inconsistently labeled signals that can severely impair the performance of a machine learning classifier [3].

These limitations call for simulated ECG signals as an additional data source with precisely known ground truth labels and arbitrarily selectable class distributions to overcome the drawbacks of clinical signals. Especially for the use case of identifying LAE, a bi-atrial statistical shape model carries the potential to produce a vast amount of atrial geometries with predefined and equally distributed left atrial volumes that can be employed for electrophysiological simulations to obtain simulated P waves of the 12-lead ECG. We therefore investigated if synthetic P waves resulting from simulations carried out on anatomical models derived as instances from a bi-atrial statistical shape model can improve the detection of clinical LAE ECGs.

2 Methods

2.1 Database

Simulated Data. We used the bi-atrial statistical shape model previously developed in our group [4] to generate 95 anatomical models of the atria and augmented them with fiber orientation, inter-atrial connections and tags for anatomical structures and made them publicly available [5]. We chose the eigenmode coefficients of the statistical shape model such that the left atrial volumes of the 95 geometries were uniformly distributed in a range of 30–65 ml, covering the LA volume range reported in a large clinical cohort study including a healthy control group and LAE patients [6]. For our computer models, left atrial volumes (LAV) were calculated as the volume bounded by the surface of the left atrial body as shown in Fig. 1. The left atrial appendage as well as the pulmonary vein ostia were excluded for the volume assessment. The atrial statistical shape model was built based on magnetic resonance (MR) and computed tomography (CT) segmentations but clinical LAV reference values are usually specified based on 2D echocardiography data. We therefore multiplied each initially calculated LAV value of the virtual cohort with a correction factor of 0.75 to account for the systematic underestimation of the atrial size using echocardiography compared to MR and CT measurements [7]. Examples of three atrial instances with a left atrial volume of 31 ml, 47 ml and 65 ml are shown in Fig. 1.

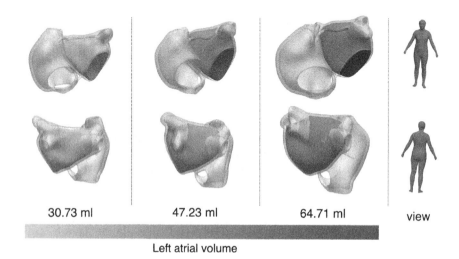

| 30.73 ml | 47.23 ml | 64.71 ml | view |

Left atrial volume

Fig. 1. Examples of three atrial geometries (endo- and epicardium) derived from the statistical shape model from the anterior view (top row) and the posterior view (bottom row). The left atrial volume (LAV) is marked in blue and was calculated as the volume enclosed by the surface of the left atrial body excluding the left atrial appendage area and the pulmonary vein ostia. (Color figure online)

For each atrial model, a baseline conduction velocity (CV) was assigned based on the values reported in [8] to 4 different atrial regions as listed in Table 1 and depicted in Fig. 2. Nine variants of the baseline CV setup were defined by modifying the velocities by ±20% in each region. For each model and each CV setup, we solved the Eikonal equation using a mesh resolution of 1.22 mm with the Fast Iterative Method and obtained local activation times (LAT). Excitation was initiated at a sinus node exit site located at the junction of the right atrial appendage and the superior caval vein. Shifting a precomputed Courtemanche action potential in time according to the resulting LATs yielded the spatio-temporal distribution of the transmembrane voltage in the atria [9]. We rotated each atrial geometry by a random rotation angle drawn from a uniform distribution in a range of [−15, 15]° around the x-, y- and z-axis [10] and placed them in two different torso geometries (see Fig. 4) derived from a human body statistical shape model [11]. Torso 1 represents a male subject (BSA = $1.7\,\mathrm{m}^2$) and torso 2 a female subject (BSA = $1.2\,\mathrm{m}^2$). The ECG forward problem was solved by means of the boundary element method as implemented in [12] and 12-lead ECGs were extracted at the standardized electrode positions. In this way, we generated 1900 (95 atrial geometries × 2 torso geometries × 10 CV settings) simulated ECGs.

Table 1. Conduction velocities in transversal fiber direction ($CV_\perp$) and anisotropy ratios (AR) for the baseline CV setup.

Atrial region	$CV_\perp$ in m/s	AR = $CV_\parallel/CV_\perp$
Bulk tissue	0.59	2.11
Bachmann's bundle	0.64	3.33
Crista terminalis	0.59	2.84
Pectinate muscles	0.46	3.78

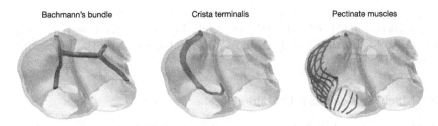

Bachmann's bundle   Crista terminalis   Pectinate muscles

Fig. 2. Example atrial geometry with three regions that are assigned different conduction velocities and anisotropy ratios. The bulk tissue makes up for the remaining areas in the left and right atrium.

Ground truth labels for the simulated dataset were set based on the left atrial volume indexed to the body surface area (LAV/BSA). Applying the cutoff value for LAV/BSA of $34\,\mathrm{ml/m}^2$ as recommended in the cardiac chamber size quantification guidelines [13] for 2D echocardiography derived data resulted in 1050 healthy and 850 LAE signals in the simulated database. Left and right atrial volumes indexed to the body surface area are visualized in Fig. 4 for both torso models and all 95 atrial geometries.

Clinical Data. Clinical ECGs of 10 s length from 9485 healthy subjects and 421 LAE patients sampled at 500 Hz were extracted from the publicly available PTB-XL ECG database [14]. For 7168 healthy and 309 LAE signals in this database, the certainty with which the expert cardiologist labeled the respective pathology

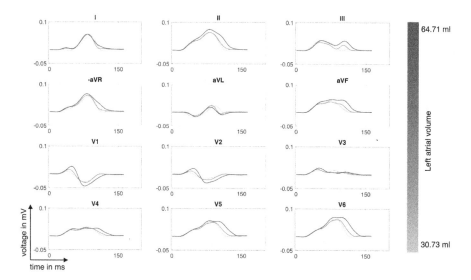

Fig. 3. Examples for the P waves of the 12-lead ECG resulting from electrophysiological simulations conducted on the three models with different left atrial volumes shown in Fig. 1 and constant rotation angles.

was specified as 100%. ECGdeli [15] was applied to extract the P waves from the time series of all healthy and LAE signals and to build a mean P wave template for each subject.

2.2   Machine Learning Classifier

We trained a long short-term memory (LSTM) network to perform the binary classification between the healthy control group (NORM) and the left atrial enlargement cohort (LAE). The simulated P waves and the clinical P wave templates were filtered (0.5 Hz highpass cutoff frequency, 60 Hz lowpass cutoff frequency) and cut 40 samples ($\widehat{=}$ 80 ms) before and 40 samples after the P wave peak in lead II occurred. In this way, we ensured that the P waves are robustly and consistently extracted throughout the simulated and clinical dataset. The resulting time series signals of all 12 leads served as an input to the network. We considered three different training scenarios. For scenario 1, only clinical signals with ground truth labels annotated with a certainty of 100% were considered. 70%, 15% and 15% of the ECGs in both classes were split into a training, validation and test set, respectively. For training scenario 2, the 1900 simulated P waves were additionally used as an input data source during training. The validation and test sets remained unchanged compared to scenario 1. In scenario 3, also the clinical NORM and LAE signals annotated with a certainty <100% by the expert cardiologist were added to the training split. The train, validation and test compositions for all training scenarios are summarized in Table 2. To avoid a bias of the network due to the larger number of samples in the NORM

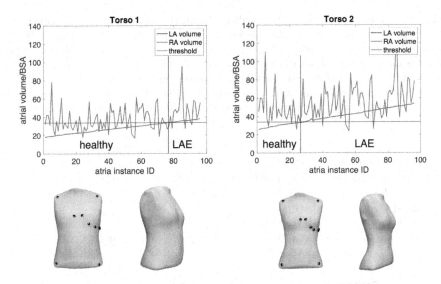

Fig. 4. Left (blue) and right (red) atrial volumes indexed to the respective body surface area (BSA) for both torso models. Torso model 1 and 2 had a BSA of $1.7\,\mathrm{m}^2$ and $1.2\,\mathrm{m}^2$, respectively, and are depicted together with the positions of the attached electrodes from the anterior and lateral view. Applying a threshold value of $34\,\mathrm{ml/m}^2$ to the LAV/BSA values yields the ground truth labels for the healthy (NORM) and the left atrial enlargement (LAE) cohorts. (Color figure online)

class in all training scenarios, we weighted the cross-entropy loss function by the inverse of the respective class support.

3   Results

Figure 5 depicts the confusion matrix for all training scenarios evaluated on the test set. Sensitivity, specificity and accuracy obtained for training scenario 1 were 0.83, 0.92 and 0.91, respectively. When additionally including simulated ECGs during training, as in scenario 2, the results for sensitivity, specificity and accuracy were 0.87, 0.95 and 0.95 respectively. Thus, all three performance metrics improved when including simulated data for the training of the classifier. When extending the clinical training dataset of scenario 1 with additional clinical signals labeled with a certainty <100% in scenario 3, sensitivity, specificity and accuracy decreased to 0.78, 0.84 and 0.83 compared to the training scenario comprising only reliably labeled data.

4   Discussion

We investigated whether the application of a bi-atrial statistical shape model for generating a large synthetic ECG database can yield valuable additional input

44 C. Nagel et al.

Table 2. Number of samples in the healthy (NORM) and the left atrial enlargement (LAE) classes for different training scenarios. In scenario 1, a subset of the reliably labeled clinical data is used for training. For training scenario 2, simulated ECGs served as an additional input to the clinical training data from scenario 1. In scenario 3, clinical ECGs for which ground truth labels were specified with a certainty <100% by the expert cardiologist were added to the training data. For all three cases, the same validation set and a test set comprising each 15% of the reliably labeled clinical data was used.

	Training scenario 1	Training scenario 2	Training scenario 3	Validation	Test
Clinical	NORM: 5018 LAE: 216	NORM: 5018 LAE: 216	NORM: 7335 LAE: 328	NORM: 1075 LAE: 47	NORM: 1075 LAE: 46
Simulated	–	NORM: 1050 LAE: 850	–	–	–
Total	NORM: 5018 LAE: 216	NORM: 6068 LAE: 1066	NORM: 7335 LAE: 328	NORM: 1075 LAE: 47	NORM: 1075 LAE: 46

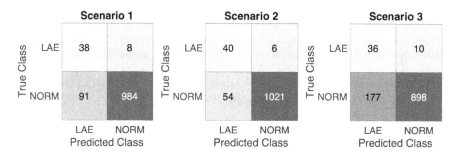

Fig. 5. Confusion matrices resulting from predicting the labels of the test set with the network trained with each of the three training sets summarized in Table 2.

data for the training of a machine learning classifier to distinguish between clinical healthy and LAE ECGs. We have shown that the performance metrics of the network increased if simulated ECGs were added to reliably labeled clinical training data. The lowest scores for sensitivity, specificity and accuracy were achieved when additional clinical data with uncertain ground truth labels were included during training. Thus, expanding a clinical training dataset with simulated data can be preferable to either resigning oneself to a smaller clinical training split or extending the latter with additional unreliably tagged clinical data. By drawing on a statistical shape model for deducing the simulated data, were able to derive an arbitrary number of atrial geometries with an a priori definable class distribution for the virtual healthy and LAE cohort, so that the class imbalance for the training set could be reduced from 1:23 to 1:6.

One key aspect for a successful and meaningful application of simulated ECGs to pathology classification tasks with clinical signals is assessing fidelity and comparability of the simulated data to clinical ECGs and previously published simulation studies. In our study, the absolute amplitude and the duration of both, the positive and the negative deflection of the P wave in lead V1 increased with the left atrial volume (see Fig. 3). Andlauer et al. [16] report that left atrial concentric hypertrophy causes an increase in the absolute P wave amplitude of the negative deflection in V1, whereas left atrial dilation with a constant myocardial volume does not have a marked effect on the ECG morphology in V1. In contrast to our work, the shape of the right atrium remained constant which might explain the additional alterations in the positive deflection in V1 for our simulation results. Furthermore, for any stage of left atrial dilation, the myocardial volume was kept constant in [16], implying that the wall thickness decreased with an increasing left atrial volume. In our study, we ensured a constant wall thickness of 3 mm for all atrial geometries, i.e. the myocardial volume increases with the left atrial size. This could thus explain the decrease in amplitude in V1 for increasing LAV values as seen in Fig. 3 and reported for left atrial hypertrophy in [16].

The network used for the classification was not explicitly optimized. We used a standard setup for a binary LSTM classifier. Tuning the network's hyperparameters or using a different network architecture might further improve the results. However, in this work, we did not focus on developing a robust classification method for healthy and LAE patients in the first place. We instead wanted to test the hypothesis that applying a bi-atrial statistical shape model as a basis to generate a large database of synthetic simulation-derived electrophysiological signals with a predefined class distribution and precisely known ground truth labels can improve real-world classifier performance. We showed that this approach can help to overcome the drawbacks of only using clinically recorded ECGs as the classifier performance improved when a combination of clinical and simulated data was employed during training.

The class imbalance during training was addressed by multiplying the cross-entropy loss function with the inverse of the class support. Considering sensitivity and specificity metrics in addition to evaluating the network's accuracy performance demonstrated that the class-wise accuracy was above 0.8 in all scenarios and thus, no notable overfitting towards the overrepresented NORM class occurred.

One key aspect for a successful and meaningful application of simulated ECGs to classification problems with clinical signals is assessing fidelity and comparability of the simulated data to the clinical ECGs. In [4] we have shown that the P wave duration distribution originating from simulations on instances of the statistical shape model with Gaussian distributed eigenvector coefficients is in accurate accordance with P wave durations reported in large clinical cohort studies of healthy subjects. The altered ECG signal characteristics in LAE patients reported in clinical studies and comprising an increased duration and absolute amplitude in the negative deflection of the P wave in lead V1 as well as an increased overall duration of P wave in all leads are also reflected in our model (see Fig. 3). However, a data-driven comparison between the synthetic and the clinical dataset for the healthy and LAE cases would further help to assess fidelity of the simulated ECGs.

Using the Eikonal model and the boundary element method as simplified model solutions for normal sinus rhythm simulations has been shown to yield P waves of the 12-lead ECG resembling the full bidomain simulation results to a high degree [17]. In our study, we used two torso models for the generation of the in silico database. As shown previously [18], the torso geometry mainly has an influence on the ECG amplitudes, whereas the P wave morphology is primarily dependent on the atrial geometry and functional variability. Therefore, we considered 95 atrial geometries with 10 different conduction velocity settings and only 2 different torso models representing a male and a female thoracic geometry. Since the torso models were derived from a statistical shape model [11], the location of the electrodes could be consistently defined on both models and no electrode placement variability is reflected in the simulated dataset.

Future directions could include the consideration of further patient characteristics such as the age, sex, weight and accompanying pathologies for an improved classification result. Furthermore, adding more simulated data covering a larger torso variability and electrode placement variations to the training dataset could also yield improved results.

Acknowledgement. This work was supported by the EMPIR programme co-financed by the participating states and from the European Union's Horizon 2020 research and innovation programme under grant MedalCare 18HLT07.

References

1. Hamatani, Y., et al.: Left atrial enlargement is an independent predictor of stroke and systemic embolism in patients with non-valvular atrial fibrillation. Sci. Rep. **6**, 31042 (2016)
2. Hazen, M.S., Marwick, T.H., Underwood, D.A.: Diagnostic accuracy of the resting electrocardiogram in detection and estimation of left atrial enlargement: an echocardiographic correlation in 551 patients. Am. Heart J. **122**(3), 823–828 (1991)
3. Hannun, A.Y., et al.: Cardiologist-level arrhythmia detection and classification in ambulatory electrocardiograms using a deep neural network. Nat. Med. **25**(1), 65–69 (2019)
4. Nagel, C., Schuler, S., Dössel, O., Loewe, A.: A bi-atrial statistical shape model for large-scale in silico studies of human atria: model development and application to ECG simulations. Med. Image Anal. 102210 (2021)
5. Nagel, C., et al.: A bi-atrial statistical shape model and 100 volumetric anatomical models of the atria. Zenodo (2021). https://doi.org/10.5281/zenodo.5095379
6. Pritchett, A.M., Jacobsen, S.J., Mahoney, D.W., Rodeheffer, R.J., Bailey, K.R., Redfield, M.M.: Left atrial volume as an index of left atrial size: a population-based study. J. Am. Coll. Cardiol. **41**(6), 1036–1043 (2003)
7. Rodevan, O., Bjornerheim, R., Ljosland, M., Maehle, J., Smith, H.J., Ihlen, H.: Left atrial volumes assessed by three- and two-dimensional echocardiography compared to MRI estimates. Int. J. Card. Imaging **15**(5), 397–410 (1999)

8. Loewe, A., Krueger, M.W., Holmqvist, F., Dössel, O., Seemann, G., Platonov, P.G.: Influence of the earliest right atrial activation site and its proximity to interatrial connections on P-wave morphology. Europace **18**(Suppl. 4), iv35–iv43 (2016)
9. Kahlmann, W., Poremba, E., Potyagaylo, D., Dössel, O., Loewe, A.: Modelling of patient-specific Purkinje activation based on measured ECGs. CDBME **3**(2), 171–174 (2017)
10. Odille, F., et al.: Statistical variations of heart orientation in healthy adults. In: 2017 Computing in Cardiology Conference (CinC), vol. 44 (2017)
11. Pishchulin, L., Wuhrer, S., Helten, T., Theobalt, C., Schiele, B.: Building statistical shape spaces for 3D human modeling. Pattern Recogn. **67**, 276–286 (2017)
12. Stenroos, M., Mäntynen, V., Nenonen, J.: A Matlab library for solving quasi-static volume conduction problems using the boundary element method. Comput. Methods Programs Biomed. **88**(3), 256–263 (2007)
13. Lang, R.M., et al.: Recommendations for cardiac chamber quantification by echocardiography in adults: an update from the American Society of Echocardiography and the European Association of Cardiovascular Imaging. Eur. Heart J. Cardiovasc. Imaging **16**(3), 233–70 (2015)
14. Wagner, P., et al.: PTB-XL, a large publicly available electrocardiography dataset. Sci. Data **7**(1), 154 (2020)
15. Pilia, N., Nagel, C., Lenis, G., Becker, S., Dössel, O., Loewe, A.: ECGdeli - an open source ECG delineation toolbox for MATLAB. SoftwareX **13**, 100639 (2020)
16. Andlauer, R., et al.: Influence of left atrial size on P-wave morphology: differential effects of dilation and hypertrophy. Europace **20**(S3), iii36–iii44 (2018)
17. Nagel, C., et al.: Comparison of propagation drivers and forward calculation in atrial electrophysiology regarding activation times and electrocardiograms. In: iHEART Congress - Modelling the Cardiac Function (2021)
18. Nagel, C., Luongo, G., Azzolin, L., Schuler, S., Dössel, O., Loewe, A.: Non-invasive and quantitative estimation of left atrial fibrosis based on P waves of the 12-lead ECG-a large-scale computational study covering anatomical variability. J. Clin. Med. **10**(8), 1797 (2021)

Vessel Extraction and Analysis of Aortic Dissection

Hui Fang[1,2], Zhanqiang Guo[1,2], Guozhu Shao[3], Zimeng Tan[1,2], Jinyang Yu[1,2], Jia Liu[3], Yukun Cao[3], Jie Zhou[1,2], Heshui Shi[3], and Jianjiang Feng[1,2(✉)]

[1] Department of Automation, Tsinghua University, Beijing, China
jfeng@tsinghua.edu.cn
[2] Beijing National Research Center for Information Science and Technology, Beijing, China
[3] Department of Radiology, Union Hospital, Tongji Medical College, Huazhong University of Science and Technology, Wuhan, China

Abstract. Aortic dissection (AD) is a dangerous disease usually diagnosed by computed tomography angiography. Segmentation of true and false lumens of aortic trunk and major branches is very important for the diagnosis and treatment of this disease. In this paper, we proposed a fully automatic vessel analysis algorithm for dissected aorta, which can output centerlines, true lumen, and false lumen of trunk and major branches, and perfusion source of branches. In our experiment, the mean dice similarity coefficient (DSC) of true lumen segmentation was 0.939 for trunk and 0.912 for branch while the mean DSC of whole lumen segmentation was 0.974 for trunk and 0.937 for branch, and the classification accuracy of branch perfusion source was 0.863.

Keywords: Aortic dissection · Centerline extraction · Lumen segmentation · Branch perfusion source

1 Introduction

Aortic dissection (AD) is a life-threatening vascular disease, which occurs by disruption of the intima along the aortic wall, resulting in separation of a true lumen (TL) and a false lumen (FL) by the intimal flap. During AD surgical planning, it is crucial to select and place the stent-graft based on the geometrical characteristics of the patient's aorta, such as lumen diameters [1,2]. CT is the standard reference for AD imaging. Manual segmentation of dissected aorta and branches from CT volumes is tedious and time-consuming [1], making automatic algorithms very desirable. However, vessel extraction of AD is a quiet challenging task because of the thin intimal flap and irregular shape changes and large intensity variations between true and false lumen [1].

Earlier studies empirically designed segmentation algorithms by considering special shape and intensity of dissected aorta. Kovács et al. [3] proposed a multi-stage method for AD segmentation, which consisted of segmentation of the aorta,

This work was supported in part by the National Natural Science Foundation of China under Grants 61976121 and 82071921.

detection of the intimal flap on the 2D cross-sections and recognition of the true and false lumens. Lee et al. [4] proposed a multi-stage method using wavelet analysis, whose core was to find the intimal flap through edge detection on the 2D cross-sections. These earlier studies have some limitations: (1) extracted 3D model is incomplete (lack of semantic branches), (2) the algorithm is not robust to noise and challenging cases, (3) evaluation is usually based on a very small number of samples, and (4) some methods are not fully automatic.

In recent years, deep CNNs have achieved the state-of-the-art results on many medical image segmentation tasks [5,6]. Researchers have started developing deep learning based segmentation algorithms for aorta dissection. Li et al. [7] used two cascaded convolutional networks to extract the adventitia and intima contours of aortic trunk. Hahn et al. [8] developed a five-step segmentation pipeline that segments the true and false lumens of aortic trunk based on centerline. Cao et al. [9] designed two U-net: the first was used for coarse aortic trunk segmentation while the second was used for fine segmentation of aortic trunk. However, none of these recent studies [7–11] tackled the problem of branch segmentation, semantic labeling, and perfusion source analysis.

We proposed a fully automatic and comprehensive vessel analysis algorithm for aortic dissection, which can perform centerline extraction, lumen segmentation of aortic trunk and nine main branches, and branch perfusion source classification. Experiments showed that the mean dice similarity coefficient (DSC) of true lumen segmentation was 0.939 for trunk and 0.912 for branch while the mean DSC of whole lumen was 0.974 for trunk and 0.937 for branch, and the classification accuracy of branch perfusion source was 0.863. Besides, as far as we know, this is the first work for lumen segmentation and perfusion source classification of AD branches using deep learning.

2   Methodology

Our scheme (see Fig. 1) is divided into three stages. The first stage is to extract centerline of aortic trunk and branches, while the second stage is to obtain true and false lumens of aortic trunk and branches with the help of the centerline, and the last stage is to determine the branch perfusion source. These steps are introduced in detail in the following three sections respectively.

2.1   Centerline Extraction

Nine main branches are considered in our study, including right common carotid artery, left common carotid artery, left subclavian artery, celiac trunk, superior mesenteric artery, two renal arteries and two iliac arteries. We take several points to represent centerline to simplify centerline extraction. This simplification is not only resistant to noise, but also beneficial to aorta image reconstruction.

For the aortic trunk, we define the centerline as 40 equidistant points. We implement landmark detection method [12] to output heat map with 40 channels, then obtain the centerline by connecting detected points (see Fig. 2).

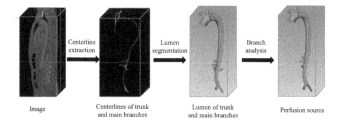

Fig. 1. Flowchart of the proposed system. The centerlines of different branches are marked by different colors and green represents true lumen while dark red represents false lumen. In the fourth image, green, yellow and blue indicates that the branch perfusion source is true lumen, false lumen and both of them respectively. (Color figure online)

Detection of branch centerline is more challenging because of the different number and length of branches for different people. To simplify the problem, nine main branches are selected and cut into fixed length, and then nine mean centerlines are calculated relative to their starting point. The mean centerline on the training dataset is defined in Eq. (1), where i represents the branch label and the branch centerline is defined as 5 equidistant points.

$$M^i = \frac{1}{N} \sum_{dataset} [0, c_2 - c_1, c_3 - c_1, c_4 - c_1, c_5 - c_1]^i \tag{1}$$

Our branch centerline extraction algorithm is divided into three steps: (1) Jointly extract the branch label and the starting point of nine branches by implementing landmark detection method [12]. (2) Calculate the initial branch centerline ($initC^i = c_0^i + M^i$) based on the predicted starting point and the mean centerline corresponding to the predicted label. (3) Fine-tune the branch centerline based on local information through the finetuning network (see Fig. 2), whose output is the centerline coordinate deviation ($\Delta x^i = C_{GT}^i - initC^i$).

2.2 Lumen Segmentation

Our lumen segmentation method takes a local volume along the centerline as input and outputs the whole lumen mask and the true lumen mask. False lumen can be simply obtained by subtracting the true lumen from the whole lumen.

Data Preprocessing. Firstly, all image resolutions are resized to 1 mm. Based on a centerline composed of several key points, rotation and cropping are used to reconstruct aortic vessel image orthogonal to local centerline, which is similar to the multiplanar reformation (MPR). The dimension of reconstructed image is $64 \times 64 \times L$ for trunk and $32 \times 32 \times L$ for branch, where L represents the distance between two adjacent points. Then 3D lumen segmentation takes these reconstructed image blocks as input.

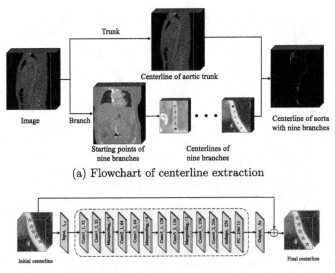

(a) Flowchart of centerline extraction

(b) Branch centerline finetuning network

Fig. 2. Flowchart of centerline extraction. (a) The trunk centerline is directly obtained by landmark detection method. The branches starting points are obtained by landmark detection, and then fine-tune the branch centerline by network. (b) The input of fine-tuning network is a 5-channel image with a size of $32 \times 32 \times 32$, where each channel image is centered on the corresponding point of the initial centerline. The output of the network is a 15-dimensional vector represents the centerline coordinate deviation.

Network. Considering the anisotropy of the reconstructed image, that is, the amount of information between Z-axis and XY-axis is different, 3D flat U-net for lumen segmentation called FU-net is designed (see Fig. 3). Different from traditional U-net, FU-net retains the Z-axis resolution during down-sampling, whose advantage is to improve the segmentation accuracy in XY plane while considering the continuity of the Z-axis.

Loss Function. Dice loss function and deep supervision are used in the lumen segmentation experiment. The final loss function is defined as follows, where the subscript of loss represents the resolution.

$$LOSS = loss_1 + 0.6 \times loss_{\frac{1}{2}} + 0.3 \times loss_{\frac{1}{4}} + 0.1 \times loss_{\frac{1}{8}}. \tag{2}$$

2.3 Branch Perfusion Source Classification

The application of endovascular repair of aortic dissection is different for patients with different branch perfusion sources. Therefore, the classification of perfusion source is of great significance in clinical practice. There are three types of branch perfusion sources, true lumen, false lumen, or both of them, the latter two of which are abnormal branches. The ground truth of category labels can be simply

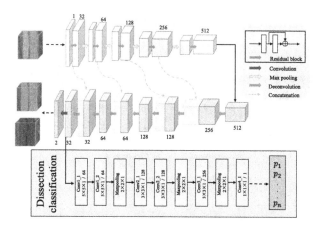

Fig. 3. Lumen segmentation and dissection classification network structure. The resolution of Z-axis remains the same in the segmentation part, which means after maxpooling layer, the size changes from (x, y, z) to $(x/2, y/2, z)$. The output of dissection classification network is a column vector whose length is equal to the length of the Z-axis of the input image.

obtained by comparison of true lumen and whole lumen masks. Our classification algorithm is divided into two steps. First, a dissection classification network is used to determine whether the branch comes from both lumens. It is built based on segmentation network since they are quite related tasks (see Fig. 3). In the second stage, we exploit the distance difference to determine whether the branch comes from false lumen. If the distance of a branch from true lumen of trunk is 3 mm greater than the distance from the whole lumen of trunk, its perfusion source is attributed to false lumen.

3   Experiment

Data. We collected 255 contrast-enhanced AD CT from 255 different patients with type A dissection (97 patients) or type B dissection (158 patients). The size of volume is $512 \times 512 \times Z(389 - 1479)$. Firstly, we discarded 36 CT volumes with low contrast. Then the branch perfusion source of the remaining 219 CT volumes was labeled by an experienced doctor. 100 of them (49 type A and 51 type B) were manually labeled with vessel centerlines, and the whole lumen and true lumen of the aorta and nine main branches with 3D slicer (https://www.slicer.org/). The voxel spacing of all images was resized to 1 mm. 100 CT volumes with manually labeled vascular lumens were randomly divided into training set (60) and test set (40). The test set was increased to 159 when evaluating the branch perfusion source classification performance. To improve the robustness of our model, training images are augmented by flip transform, rotation transform and the combination of them.

Table 1. Comparison of aortic lumen segmentation performance in the reconstruction space, where the evaluation standard is dice similarity coefficient (DSC), and the number in brackets represents the standard deviation.

		Method	DSC	Precision	Recall
Trunk	Whole lumen	Li et al. [7]	0.960(0.009)	0.952(0.017)	0.967(0.012)
		FU-net	**0.974(0.009)**	**0.978(0.012)**	**0.970(0.019)**
	True lumen	Li et al. [7]	0.891(0.026)	0.895(0.040)	0.887(0.026)
		FU-net	**0.939(0.032)**	**0.942(0.034)**	**0.937(0.035)**
Branch	Whole lumen	FU-net	0.937(0.024)	0.929(0.029)	0.946(0.026)
	True lumen	FU-net	0.912(0.051)	0.920(0.038)	0.907(0.074)

Table 2. Comparison of the results of true lumen segmentation of the entire aorta with branches in the original coordinate space. In other methods, it is difficult to separate the trunk and branches. In addition, true lumen segmentation is more challenging and more clinically meaningful.

Method	DSC	Precision	Recall
2D U-net [5]	0.812(0.103)	0.847(0.066)	0.790(0.127)
3D U-net [6]	0.833(0.051)	0.832(0.060)	0.840(0.077)
3D V-net [13]	0.809(0.099)	0.810(0.091)	0.821(0.126)
FU-net	**0.907(0.027)**	**0.906(0.028)**	**0.908(0.034)**

Results. We reproduced the algorithm in Li et al. [7] which can segment only the aortic trunk. In addition, we compared our algorithm with 2D U-net [5], 3D U-net [6], and 3D V-net [13], because (1) these networks have good generalization, (2) recent AD segmentation algorithms [8–10] are based on these networks, and (3) open source implementations are available.

Firstly, we compared our method with Li et al. [7] in the reconstruction space, namely, aligned to local centerline (see Table 1). Furthermore, we mapped the result of our segmentation algorithm back to the original coordinate space and compared with segmentation algorithms without utilizing centerline information (see Table 2).

In Table 1, the average DSC for FU-net was 0.939 for trunk and 0.912 for branch on the task of true lumen segmentation. The score is higher than [7], which demonstrated that it is helpful to add 3D information in lumen segmentation. In the original coordinate space (see Table 2), the average DSC of our proposed method was at least 7% higher than those algorithms without utilizing centerline information, which proved that the mutil-stage method we designed is effective.

Figure 4 shows the 3D meshes of the segmentation results of four algorithms and the comparison between our method and 3D U-net in sagittal view in the original coordinate space. Compared to our method, other algorithms have some problems such as the disconnection of blood vessels and the confusion of true and false lumen.

Table 3. Confusion matrix of the classification task of branch perfusion sources.

Label	Predict				
	True lumen	False lumen	Both	Total	Recall
True lumen	970	122	38	1130	0.858
False lumen	15	95	2	112	0.848
Both	12	7	170	189	0.899

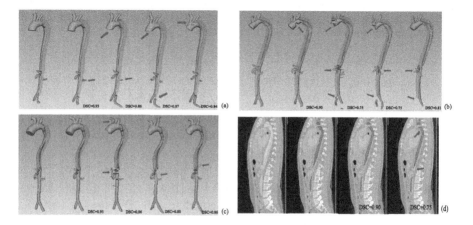

Fig. 4. In (a) (b) (c), which represent three different samples, from left to right are 3D meshes of the ground truth, ours, 2D U-net, 3D U-net and 3D V-net. In sagittal view (d), from left to right are original image, ground truth, ours and 3D U-net. Note that green represents true lumen while dark red represents false lumen, and the red arrow marks the poor segmentation part. Dice scores are shown below. (Color figure online)

Table 3 shows the confusion matrix of the classification task of branch perfusion sources. The average classification accuracy is 0.863. Misclassification usually occurs in difficult samples such as calcification and thrombosis. In the task of judging whether a branch arises from false lumen, the performance of our algorithm is not ideal, indicating that other criteria need to be added in addition to distance.

4   Conclusion

In this paper, we proposed an automatic vascular extraction algorithm for CT of aortic dissection, which can obtain centerlines, true lumen, and false lumen of trunk and major branches, and can estimate the perfusion source of branches. To our knowledge, it is the most comprehensive vessel analysis algorithm for AD in the public literature. However, our current method still has two shortcomings. One is that the variation of branch topology is not considered (although uncommon), and the other is that the lumen segmentation result at the bifurcation point is not accurate enough. We intend to overcome them in the future.

References

1. Pepe, A., et al.: Detection, segmentation, simulation and visualization of aortic dissections: a review. Med. Image Anal. **65**, 101773 (2020)
2. Chiesa, R., Melissano, G., Zangrillo, A., Coselli, J.S.: Thoraco-Abdominal Aorta: Surgical and Anesthetic Management, vol. 783. Springer, Milano (2011). https://doi.org/10.1007/978-88-470-1857-0
3. Kovács, T., Cattin, P., Alkadhi, H., Wildermuth, S., Székely, G.: Automatic segmentation of the aortic dissection membrane from 3D CTA images. In: Yang, G.-Z., Jiang, T.Z., Shen, D., Gu, L., Yang, J. (eds.) MIAR 2006. LNCS, vol. 4091, pp. 317–324. Springer, Heidelberg (2006). https://doi.org/10.1007/11812715_40
4. Lee, N., Tek, H., Laine, A.F.: True-false lumen segmentation of aortic dissection using multi-scale wavelet analysis and generative-discriminative model matching. In: Medical Imaging 2008: Computer-Aided Diagnosis, vol. 6915, p. 69152V. International Society for Optics and Photonics (2008)
5. Ronneberger, O., Fischer, P., Brox, T.: U-net: convolutional networks for biomedical image segmentation. In: Navab, N., Hornegger, J., Wells, W.M., Frangi, A.F. (eds.) MICCAI 2015. LNCS, vol. 9351, pp. 234–241. Springer, Cham (2015). https://doi.org/10.1007/978-3-319-24574-4_28
6. Çiçek, Ö., Abdulkadir, A., Lienkamp, S.S., Brox, T., Ronneberger, O.: 3D U-net: learning dense volumetric segmentation from sparse annotation. In: Ourselin, S., Joskowicz, L., Sabuncu, M.R., Unal, G., Wells, W. (eds.) MICCAI 2016. LNCS, vol. 9901, pp. 424–432. Springer, Cham (2016). https://doi.org/10.1007/978-3-319-46723-8_49
7. Li, Z., et al.: Lumen segmentation of aortic dissection with cascaded convolutional network. In: Pop, M., et al. (eds.) STACOM 2018. LNCS, vol. 11395, pp. 122–130. Springer, Cham (2019). https://doi.org/10.1007/978-3-030-12029-0_14
8. Hahn, L.D., et al.: CT-based true-and false-lumen segmentation in type B aortic dissection using machine learning. Radiol. Cardiothorac. Imaging **2**(3), e190179 (2020)
9. Cao, L., et al.: Fully automatic segmentation of type B aortic dissection from CTA images enabled by deep learning. Eur. J. Radiol. **121**, 108713 (2019)
10. Fantazzini, A., et al.: 3D automatic segmentation of aortic computed tomography angiography combining multi-view 2D convolutional neural networks. Cardiovasc. Eng. Technol. **11**(5), 576–586 (2020)
11. Xu, X., He, Z., Niu, K., Zhang, Y., Tang, H., Tan, L.: An automatic detection scheme of acute stanford type A aortic dissection based on DCNNs in CTA images. In: Proceedings of the 2019 4th International Conference on Multimedia Systems and Signal Processing, pp. 16–20 (2019)

12. Tan, Z., Duan, Y., Wu, Z., Feng, J., Zhou, J.: A cascade regression model for anatomical landmark detection. In: STACOM 2019. LNCS, vol. 12009, pp. 43–51. Springer, Cham (2020). https://doi.org/10.1007/978-3-030-39074-7_5
13. Milletari, F., Navab, N., Ahmadi, S.-A.: V-net: fully convolutional neural networks for volumetric medical image segmentation. In: 2016 Fourth International Conference on 3D Vision (3DV), pp. 565–571. IEEE (2016)

The Impact of Domain Shift on Left and Right Ventricle Segmentation in Short Axis Cardiac MR Images

Devran Ugurlu[1]([✉]), Esther Puyol-Antón[1], Bram Ruijsink[1,4], Alistair Young[1], Inês Machado[1], Kerstin Hammernik[2,3], Andrew P. King[1], and Julia A. Schnabel[1,2,5]

[1] School of Biomedical Engineering and Imaging Sciences,
King's College London, London, UK
[2] Technical University of Munich, Munich, Germany
[3] Biomedical Image Analysis Group, Imperial College London, London, UK
[4] Department of Adult and Paediatric Cardiology,
Guy's and St Thomas' NHS Foundation Trust, London, UK
[5] Helmholtz Center Munich, Munich, Germany

Abstract. Domain shift refers to the difference in the data distribution of two datasets, normally between the training set and the test set for machine learning algorithms. Domain shift is a serious problem for generalization of machine learning models and it is well-established that a domain shift between the training and test sets may cause a drastic drop in the model's performance. In medical imaging, there can be many sources of domain shift such as different scanners or scan protocols, different pathologies in the patient population, anatomical differences in the patient population (e.g. men vs women) etc. Therefore, in order to train models that have good generalization performance, it is important to be aware of the domain shift problem, its potential causes and to devise ways to address it. In this paper, we study the effect of domain shift on left and right ventricle blood pool segmentation in short axis cardiac MR images. Our dataset contains short axis images from 4 different MR scanners and 3 different pathology groups. The training is performed with nnUNet. The results show that scanner differences cause a greater drop in performance compared to changing the pathology group, and that the impact of domain shift is greater on right ventricle segmentation compared to left ventricle segmentation. Increasing the number of training subjects increased cross-scanner performance more than in-scanner performance at small training set sizes, but this difference in improvement decreased with larger training set sizes. Training models using data from multiple scanners improved cross-domain performance.

Keywords: Domain shift · Cardiac MR · Short axis · Segmentation

© Springer Nature Switzerland AG 2022
E. Puyol Antón et al. (Eds.): STACOM 2021, LNCS 13131, pp. 57–65, 2022.
https://doi.org/10.1007/978-3-030-93722-5_7

1   Introduction

Cardiac magnetic resonance (CMR) allows non-invasive and radiation-free imaging of the heart and is the recommended imaging modality for many clinical scenarios [6]. Cine CMR involves acquiring images at several time steps through the cardiac cycle. These images can be used to measure diagnostic metrics such as left and right ventricle (LV and RV) volumes and ejection fraction (EF) [9,10]. In order to calculate these metrics, segmentation of the LV and RV from the cine images is required. Manual segmentation requires expertise, is very time-consuming and is prone to operator bias and error. Hence, automatic segmentation methods are highly desirable. Several challenges have been organized to tackle this problem and many methods have been proposed. Recent challenges such as ACDC [1] and M&Ms [2] have been dominated by U-Net-based [8] deep learning methods, with nnUNet [5] being the best overall performer in the most recent M&Ms challenge.

A significant problem with deep learning methods is that their performance can drastically drop if the data distribution of the training set and test set are different, and this phenomenon is known as domain shift or dataset shift [7]. Medical images might have several sources of domain shift such as different scanners or scan protocols, different pathologies in the patient population and anatomical differences in the patient population (e.g. men vs women). Hence, it is recommended for clinicians using AI assistance to be aware of potential sources of domain shift that could cause the AI system to fail [3].

With intra-domain performance reaching comparable quality to manual expert segmentations [1], analysis of domain shift and ways to improve cross-domain generalization are becoming popular topics in recent studies. Significant cross-scanner performance drop has indeed been observed on the M&Ms challenge data [2] and several studies have proposed various methods to improve cross-domain generalization. On the M&Ms dataset, an aggressive data augmentation scheme used with nnUNet [4], and an approach based on domain-adversarial learning [11] were shown to improve cross-scanner performance. In another study, an adverserial domain adaptation approach was utilized on a multi-modality dataset showing improved cross-modality performance [12]. However, there has been relatively little work on investigating and quantifying the relative impact of different sources of domain shift.

In this paper, we study the effect of domain shift on LV and RV blood pool segmentation in end-diastolic (ED) and end-systolic (ES) frames of short axis cine CMR images. Our dataset contains short axis images from 4 different MR scanners and 3 different pathology groups. The training is performed using the nnUNet model [5]. The contributions of our study are as follows: (1) We separate domains not only by scanner but also by pathology group in order to systematically analyse the pathology-based and scanner-based domain shifts. (2) Using a large dataset of 1373 subjects with manual segmentations, we analyse the impact of training set size on domain shift both on intra-scanner and cross-scanner test sets.

2   Materials and Methods

2.1   Dataset

We employ a dataset of 1373 cine CMR examinations acquired for clinical purposes from the clinical imaging system of Guy's and St Thomas' NHS Foundation Trust Hospital. All patients gave informed consent for research use of their imaging data. For each subject, an experienced cardiologist manually segmented the LV and RV blood pools in the ES and ED frames of the short axis sequence. The scanners and numbers of subjects from each scanner are as follows: Siemens Aera 1.5 T (555 subjects), Philips Ingenia 1.5 T (324 subjects), Philips Achieva 3 T (452 subjects) and Siemens Biograph mMR 3 T (42 subjects). For brevity, we name these domains as A, B, C and D, respectively.

The dataset is partially annotated with the following pathology groups: healthy, congenital heart disease, dilated cardiomyopathy (DCM), hypertrophic cardiomyopathy (HCM), hypertensive cardiomyopathy (CMP), ischaemic, mixed CMP, non-ischaemic CMP, other CMP, other (for example, myocarditis, infiltrative myocardial diseases, etc.). From these, the following groups were created as the pathology domains for our experiments: Pathology Group 1 (PG 1): Healthy, PG 2: DCM, hypertensive CMP, mixed CMP, non-ischaemic CMP. PG 3: HCM. Other groups were removed either due to very small numbers of subjects or high structural variance within the group. DCM and various CMPs were grouped together in PG 2 due to structural similarity. Using the scanners A, B, C and the 3 pathology domains, we created 9 domains containing different combinations of scanner/pathology. The domain names, descriptions and numbers of subjects are summarized in Table 1.

2.2   Experimental Setup and Training

The first set of experiments was aimed at analysing the pathology-based and scanner-based domain shifts separately. From the 9 scanner and pathology-specific domains, the ones that had more than 30 subjects, i.e. A1, A2, B1, C1 and C2 were separated into training and test sets, with 30 subjects in the training set for each domain. The rest of the domains were only used for testing. In addition, mixed domains were created by mixing the pathologies and scanners and named as follows: AM: A1 + A2, CM: C1 + C2, 1M: A1 + B1 + C1, 2M: A2 + C2. The mixed domain names and descriptions are summarized in Table 1. Training set sizes were also fixed to 30 for mixed domains and subjects were randomly selected with equal proportion from the domains that make up the mixed domains. For example, domain 1M's training set consisted of 10 subjects each from A1, B1 and C1 domains and domain AM's training set consisted of 15 subjects each from domains A1 and A2. This was done to keep the training set size fixed across all domains in order to separate the effect of domain shift on performance from increasing training set size.

The second set of experiments was aimed at measuring the effect of increasing training set size on intra-domain and cross-domain performance. For these

Table 1. Description of domains defined within our dataset.

Single-scanner domains	Scanner	Pathology group	No. subjects
A	Siemens Aera 1.5 T	All	554
B	Philips Ingenia 1.5 T	All	324
C	Philips Achieva 3 T	All	452
D	Siemens Biograph_mMR 3 T	All	42
A1	Siemens Aera 1.5 T	Healthy	74
A2	Siemens Aera 1.5 T	DCM, CMP	37
A3	Siemens Aera 1.5 T	HCM	15
B1	Philips Ingenia 1.5 T	Healthy	42
B2	Philips Ingenia 1.5 T	DCM, CMP	14
B3	Philips Ingenia 1.5 T	HCM	9
C1	Philips Achieva 3 T	Healthy	64
C2	Philips Achieva 3 T	DCM, CMP	36
C3	Philips Achieva 3 T	HCM	8
Mixed domains	Domains included		
AM	A1, A2		
CM	C1, C2		
1M	A1, B1, C1		
2M	A2, C2		
M	A, B, C		

experiments, we only analyse the effect of scanner-based domain shift on the A, B, C and D domains because our pathology-specific domains were not large enough for this purpose. Models were trained on A, B, C and M domains, where M is the union of A, B and C, and were trained with 30, 60, 120 and 240 training set sizes. A final model was trained with a training set size of 1064 for domain M. Domain D was only used as a test set for cross-domain testing.

For all experiments, 2D training was performed using the nnUNet [5] model with batch size 8, 250 max epochs and 4 fold cross-validation. We did not use 3D training because of training time constraints. All other parameters, including data augmentation settings, are default settings for the version 2 trainer of nnUNet (nnUNetTrainerV2). An ensemble automatically created by nnUNet from the 4 folds was used for testing.

3   Results and Discussion

The mean Dice scores between manual segmentations and nnUNet-produced segmentations over LV, RV, ED and ES are given in Table 2. An interesting observation is that the cross-domain performance is clearly not symmetric. For example, the model trained on domain C2 performs very poorly on domain A1 but the reverse is not true.

An in-domain dominance is not immediately obvious from Table 2. On closer inspection, however, it can be noticed that the best model for every test domain was trained on a domain that includes the same scanner. Furthermore, there are a number of very poor performances on cross-scanner tests (e.g. train C2, test A1) which is not the case for any in-scanner test. To test the hypothesis that in-domain performance is better than cross-domain performance, we computed means and standard deviations of Dice scores for intra-scanner, cross-scanner, intra-pathology and cross-pathology cases using the domains A1, A2, A3, B1, B2, B3, C1, C2 and C3, and statistically compared these using Mann-Whitney U tests. The results are given in Table 3, which shows a clear effect of domain shift across scanners for both LV and RV and for both ED and ES frames, whereas the domain shift effect is not significant across pathology groups for any segmented structure and frame combination. Furthermore, Table 3 also shows a more significant cross-domain performance drop on RV segmentations compared to LV segmentations across scanners.

Examining the significance results in Table 3, it can be hypothesized that mixing different scanner domains will improve generalization performance. To investigate this hypothesis, we produced a boxplot of Dice scores for different training domains (including mixed pathology domains AM and CM, and mixed scanner domains 1M and 2M) for a single pooled test domain that aggregates all of the individual test domains A1, A2, A3, B1, B2, B3, C1, C2 and C3 (see Fig. 1). Note that the numbers of subjects in the training sets were kept fixed when mixing domains in order to remove the effect of increasing training set size from the analysis. Indeed, the healthy group with mixed scanners (1M) had the best overall performance, especially due to its performance on the more difficult RV segmentation task.

In Fig. 2, we plot the mean Dice scores (over LV, RV, ED and ES) vs. the number of subjects in the training set (30, 60, 120, 240) for each model trained on domains A, B and C (i.e. different scanners) to each test domain A, B, C and D. A model trained on the mixed domain M and tested on domain M and D is also added to the plot. For the mixed domain experiment, an additional training size of 1064 that makes use of all available data is added. It can be seen that for every combination except C→A, that is, the model trained on domain C and tested on domain A, and M→D, performance increases with increasing number of subjects in the training set. On C→A and M→D, the performance kept increasing until 120 training subjects but dropped when the training size was increased further. These two might be outlier cases since a similar drop was not observed in any of the other tests. For all test set sizes, the model trained on multiple scanners (i.e. M) has comparable performance to that of single-scanner models on intra-scanner experiments and performed better than single scanner models on cross-scanner experiments. In addition, when trained with all available data (1064 subjects), this model's performance matched the best performance of any of the intra-domain models, demonstrating the utility of using large training sets from multiple scanners. It also generalised well to the unseen domain D, as can be seen from the right-hand plot in Fig. 2.

Table 2. Mean Dice scores between manual segmentations and nnUNet-produced segmentations over LV, RV, ED and ES. The mean Dice score for the best performing model for each test domain is displayed in bold and the second best model is displayed in italic. Multiple scores may be in bold or italic when there are ties.

Train domain	Test domain								
	A1	B1	C1	A2	C2	B2	A3	B3	C3
A1	0.866	0.859	0.873	0.901	0.872	0.890	0.866	0.843	0.869
B1	0.852	*0.889*	0.870	0.877	0.861	*0.912*	0.858	**0.889**	0.907
C1	0.822	0.875	**0.903**	0.893	**0.910**	0.889	0.828	0.865	*0.920*
A2	**0.881**	0.874	0.884	**0.903**	0.888	0.888	0.867	0.861	0.898
C2	0.677	0.764	*0.900*	0.688	*0.909*	0.794	0.656	0.731	**0.922**
1M	*0.879*	**0.890**	*0.900*	*0.902*	0.907	**0.914**	**0.882**	*0.876*	0.919
2M	0.870	0.879	0.896	0.896	0.907	0.906	0.856	0.871	0.918
AM	0.872	0.870	0.863	0.894	0.875	0.871	*0.871*	0.862	0.866
CM	0.745	0.835	0.899	0.784	0.904	0.849	0.695	0.760	**0.922**

Table 3. Intra vs. cross domain mean and standard deviation of Dice scores between manual and nnUNet-produced segmentations separately for LV and RV for end-systolic (ES) and end-diastolic (ED) frames on domains A1, A2, A3, B1, B2, B3, C1, C2, C3. In intra-scanner, the training and test sets are from the same scanner but do not have to be from the same pathology group. In intra-pathology, the training and test sets are from the same pathology group but do not have to be from the same scanner. p-values for Mann-Whitney U tests are given for each intra-vs-cross group.

	LV ED	LV ES	RV ED	RV ES
Intra-scanner	0.944 (0.025)	0.887 (0.072)	0.888 (0.057)	0.838 (0.102)
Cross-scanner	0.937 (0.030)	0.873 (0.105)	0.790 (0.214)	0.726 (0.253)
In vs cross-scanner p-val	0.0003	0.0285	0.0000	0.0000
Intra-pathology	0.939 (0.027)	0.880 (0.090)	0.837 (0.137)	0.784 (0.181)
Cross-pathology	0.940 (0.029)	0.877 (0.098)	0.815 (0.208)	0.751 (0.244)
In vs cross-path. p-val	0.2716	0.4126	0.0959	0.4778

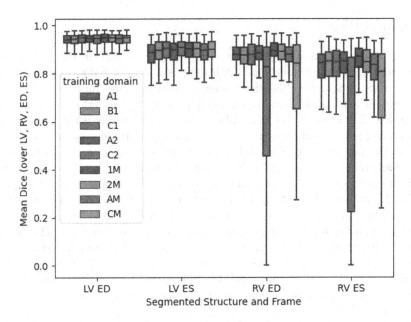

Fig. 1. Boxplot of Dice scores for different training domains (including mixed pathology domains AM and CM, and mixed scanner domains 1M and 2M) for a single aggregated test domain that combines all of A1, A2, A3, B1, B2, B3, C1, C2, C3. See Table 1 for domain descriptions.

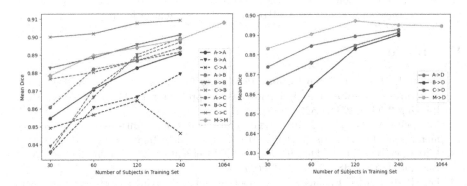

Fig. 2. Lineplots of mean Dice scores (over LV, RV, ED and ES) for each domain model to each domain test set with increasing training set size. A→B means the model was trained on domain A and tested on domain B. On the left lineplot, intra-scanner tests are shown as solid lines and cross-scanner tests as dotted lines.

4   Conclusion

We have presented an analysis of the effect of domain shift on RV and LV segmentations on a large clinical cine CMR dataset and observed several important results. Firstly, scanner differences seem to play a more important role than pathology group differences on cross-domain segmentation performance and this also means that mixing data from different scanners for training is more helpful than mixing pathologies. Secondly, cross-domain performance is not symmetric, that is, the performance of a model trained on some domain A and tested on some domain B cannot be used to anticipate the performance of a model trained on B and tested on A. This phenomenon was also previously observed in [4] on the M&Ms dataset. Thirdly, in-scanner vs. cross-scanner performance difference is much larger on RV compared to LV. Lastly, increasing the number of training subjects improves both in-domain and cross-domain performance. The performance improvement is faster for cross-domain on smaller training sets but the cross-domain improvement slows faster than in-domain improvement as training sets get larger. We expect our work to be helpful to researchers studying the effect of domain shift and working on improving generalization of segmentation networks in CMR.

Acknowledgements. This work was funded by the Engineering and Physical Sciences Research Council (EPSRC) programme grant 'SmartHeart' (EP/P001009/1) and supported by the Wellcome/EPSRC Centre for Medical Engineering [WT 203148/Z/16/Z]. The research was supported by the National Institute for Health Research (NIHR) Biomedical Research Centre based at Guy's and St Thomas' NHS Foundation Trust and King's College London. The views expressed are those of the authors and not necessarily those of the NHS, the NIHR or the Department of Health.

References

1. Bernard, O., et al.: Deep learning techniques for automatic MRI cardiac multi-structures segmentation and diagnosis: is the problem solved?, IEEE Trans. Med. Imaging **37**(11), 2514–2525 (2018)
2. Campello, V.M., et al.: Multi-centre, multi-vendor and multi-disease cardiac segmentation: the MandMs challenge. IEEE Trans. Med. Imaging, pp. 1–1 (2021)
3. Finlayson, S.G., et al.: The clinician and dataset shift in artificial intelligence. New England J. Med. **385**(3), 283–286 (2021)
4. Full, P.M., Isensee, F., Jäger, P.F., Maier-Hein, K.: Studying robustness of semantic segmentation under domain shift in cardiac MRI. In: Puyol Anton, E., et al. (eds.) Statistical Atlases and Computational Models of the Heart. M&Ms and EMIDEC Challenges, pp. 238–249. Springer International Publishing, Cham (2021). https://doi.org/10.1007/978-3-030-68107-4_24
5. Isensee, F., Jaeger, P.F., Kohl, S.A.A., Petersen, J., Maier-Hein, K.H.: nnu-net: a self-configuring method for deep learning-based biomedical image segmentation. Nat. Methods **18**(2), 203–211 (2021)
6. von Knobelsdorff-Brenkenhoff, F., Pilz, G., Schulz-Menger, J.: Representation of cardiovascular magnetic resonance in the AHA/ACC guidelines. J. Cardiovascular Magn. Reson. **19**(1), 70 (2017)

7. Quionero-Candela, J., Sugiyama, M., Schwaighofer, A., Lawrence, N.D.: Dataset shift in machine learning. The MIT Press (2009)

8. Ronneberger, O., Fischer, P., Brox, T.: U-net: Convolutional networks for biomedical image segmentation. In: Navab, N., Hornegger, J., Wells, W.M., Frangi, A.F. (eds.) Medical Image Computing and Computer-Assisted Intervention – MICCAI 2015, pp. 234–241. Springer International Publishing, Cham (2015). https://doi.org/10.1007/978-3-319-24574-4_28

9. Ruijsink, B.,et al.: Quality-aware semi-supervised learning for CMR segmentation. In: Puyol Anton, E., et al. (eds.) Statistical Atlases and Computational Models of the Heart. M&Ms and EMIDEC Challenges, pp. 97–107. Springer International Publishing, Cham (2021). https://doi.org/10.1007/978-3-030-68107-4_10

10. Ruijsink, B., et al.: Fully automated, quality-controlled cardiac analysis from CMR: Validation and large-scale application to characterize cardiac function. JACC: Cardiov. Imaging **13**(3), 684–695 (2020)

11. Scannell, C.M., Chiribiri, A., Veta, M.: Domain-adversarial learning for multicentre, multi-vendor, and multi-disease cardiac MR image segmentation. In: Puyol Anton, E., et al. (eds.) Statistical Atlases and Computational Models of the Heart. M&Ms and EMIDEC Challenges, pp. 228–237. Springer International Publishing, Cham (2021). https://doi.org/10.1007/978-3-030-68107-4_23

12. Wang, J., Huang, H., Chen, C., Ma, W., Huang, Y., Ding, X.: Multi-sequence cardiac MR segmentation with adversarial domain adaptation network. In: Pop, M., et al. (eds.) Statistical Atlases and Computational Models of the Heart. Multi-Sequence CMR Segmentation, CRT-EPiggy and LV Full Quantification Challenges, pp. 254–262. Springer International Publishing, Cham (2020). https://doi.org/10.1007/978-3-030-39074-7_27

Characterizing Myocardial Ischemia and Reperfusion Patterns with Hierarchical Manifold Learning

Benoît Freiche[1(✉)], Patrick Clarysse[1], Magalie Viallon[1,2], Pierre Croisille[1,2], and Nicolas Duchateau[1]

[1] Univ Lyon, Université Claude Bernard Lyon 1, INSA-Lyon, CNRS, Inserm, CREATIS UMR 5220, U1294, F-69621 Lyon, France
`benoit.freiche@creatis.insa-lyon.fr`
[2] Department of Radiology, Hôpital Nord, University Hospital of Saint-Étienne, Saint-Étienne, France

Abstract. We aim at better understanding the mechanisms of ischemia and reperfusion, in the context of acute myocardial infarction. For this purpose, imaging and in particular magnetic resonance imaging are of great value in the clinic, but the richness of the images is currently under exploited. In this paper, we propose to characterize myocardial ischemia and reperfusion patterns across a population beyond the scalar measurements used in the clinic. Specifically, we adapted representation learning techniques to not only characterize the population distribution in terms of scar and microvascular obstruction patterns, but also regarding the appearance of late gadolinium images which reflects tissue heterogeneity. To do so, we implemented a hierarchical manifold learning approach where the embedding from a higher-level content (the images) is guided by one from a lower-level content (the infarct and microvascular obstruction segmentations). We demonstrate its relevance on 1711 late gadolinium enhancement slices from 123 patients with acute ST-elevation myocardial infarction. We designed ways to balance the contribution of each level in the hierarchy, and quantify its impact on the overall distribution and on sample neighborhoods. We notably observe that the obtained latent space is a balanced contribution between the two levels of the hierarchy, and is more robust to challenging images subjected to artifacts or specific lesion patterns.

Keywords: Manifold learning · Decision hierarchy · Information fusion · Cardiac imaging · Myocardial infarction

1 Introduction

The ischemic mechanims following the obstruction of a coronary artery can lead to both structural and functional myocardial damage. For acute myocardial infarction, the benefits of treatments that restore the coronary circulation are counterbalanced by potential reperfusion injuries (microvascular obstruction, MVO) due to a sudden blood reflow in areas that were deprived of it [2].

© Springer Nature Switzerland AG 2022
E. Puyol Antón et al. (Eds.): STACOM 2021, LNCS 13131, pp. 66–74, 2022.
https://doi.org/10.1007/978-3-030-93722-5_8

In this context, imaging and in particular cardiac magnetic resonance imaging plays a crucial role to understand the mechanisms of ischemia-reperfusion [5]. However, due to limited analysis tools, the richness of the acquired images is under-exploited in the clinical analysis. The lesion characteristics are limited to simple scalar descriptors (extent and transmurality, mainly) [1], and the image contents are not exploited (e.g. the heterogeneity of pixel values within the segmented lesions).

The field of representation learning offers very efficient tools to better characterize the lesion patterns within a population. It allows mapping the high-dimensional data (e.g. images) to a simplified latent representation that facilitates the analysis of individual or subgroup trends. Within this field, manifold learning offers a sound framework where statistical distances in the latent space can be exploited for such a mapping. It assumes that the input samples lie on a (non-linear) mathematical manifold that is unknown but can be learnt from data.

However, the analysis of the imaging content within the myocardium is not straightforward. For instance, in late gadolinium enhancement (LGE) images, MVO is displayed as dark areas within bright and larger regions indicating the infarct. The gray levels of MVO and healthy tissues may be close (see Fig. 1), which can fool the metric used to compare images (often in a pixel-wise fashion), in particular for large MVO. Other critical issues can happen in case of image artifacts. This analysis could be substantially more robust by using additional imaging information, up to the local appearance of the acquired images.

To merge the information from different imaging descriptors, several fusion strategies exist within the field of manifold learning. For instance, Multiple Manifold Learning (MKL) [8] finds the best linear combination between the affinity matrices associated to each descriptor. A non-linear fusion process called Similarity Network Fusion (SNF) [11] has also been proposed to iteratively merge several descriptors. Nonetheless, these methods perform the early fusion of all the descriptors at the same time, which may be suboptimal in our context. A better integration scheme could consist in a hierarchical learning process, to guide the embedding from a given descriptor by a previous embedding from a lower-level descriptor. This approach is already part of standard clinical reasoning through decision trees [10] and of computational methods through random forests [7]. However, it is hardly scalable to multiple high-dimensional descriptors from images. *Bhatia et al.* [4] proposed an interesting hierarchical manifold learning scheme that could overcome this data integration problem, but it has only been exploited to study a single medical imaging modality at different resolutions. In contrast, we want to incorporate the data from several medical imaging modalities in a hierarchical way and estimate a single representation for a population of patients.

In this paper, we aim at improving the analysis of tissue heterogeneity in LGE images by prior knowledge from the segmented lesions, which can be seen as a hierarchical way of estimating a latent space. We propose to use the hierarchical manifold learning framework such that the embedding from a higher level content (the LGE images) is guided by the one from a lower level content (the infarct and microvascular obstruction zones segmented on these images). Here, we keep the problem unsupervised and examine the distribution of ischemia and reperfusion patterns across a population. As there is no groundtruth for such an unsupervised problem, we design ways of selecting relevant hyperparameters a posteriori. We demonstrate its relevance on a population of 123 patients with acute myocardial infarct, to improve the analysis of potentially subtle tissue heterogeneities beyond the prior segmentation masks.

2   Method

2.1   Hierarchical Representation Problem Statement

We build a two-level hierarchical model, where $\mathbf{x}_i^0$ stands for the i-th sample from the lower/parent level (the image of the segmented lesions, where pixel values lie in the interval $[0,2]$, with 0, 1 and 2 respectively standing for the healthy myocardium, the infarct, and the MVO), and $\mathbf{x}_i^1$ corresponds to the same sample from the higher level (the LGE image). We aim at estimating the higher/child level latent space $\mathbf{Y}^1 = [\mathbf{y}_i^1]_{i \in [1,K]}$ guided by the lower level latent space $\mathbf{Y}^0 = [\mathbf{y}_i^0]_{i \in [1,K]}$, K being the number of samples.

2.2   Spectral Embedding

In this work, manifold learning is achieved within the diffusion maps framework [6], which served as a basis for fusion [8,11] and hierarchical [4] algorithms.

For each level $m = \{0,1\}$ in the hierarchy, pairwise affinities between individuals are encoded within the matrix $\mathbf{W}^m = [W_{ij}^m] \in \mathbb{R}^{K \times K}$ defined as:

$$W_{ij}^m = \begin{cases} \exp(-\frac{\|\mathbf{x}_i^m - \mathbf{x}_j^m\|^2}{2\sigma^2}) & \text{if } j \in \mathcal{N}_k(i), \\ 0 & \text{otherwise,} \end{cases} \quad (1)$$

where $\mathcal{N}_k(i)$ stands for the neighborhood of the i-th sample, based on the k closest samples. The graph Laplacian is defined from this matrix as $\mathbf{L}^m = \mathbf{D}^m - \mathbf{W}^m$, where $\mathbf{D}^m$ is a diagonal matrix such that $D_{ii}^m = \sum_j W_{ij}^m$.

Diffusion maps consists in performing the spectral decomposition of the graph Laplacian to estimate $\mathbf{Y}^m$. In practice, this is achieved by diagonalizing the matrix $\tilde{\mathbf{W}}^m = (\mathbf{D}^m)^{-\frac{1}{2}} \mathbf{W}^m (\mathbf{D}^m)^{-\frac{1}{2}}$, which corresponds to working with the normalized graph Laplacian. It can be seen as a Markov chain matrix, where $\tilde{W}_{ij}^m$ represents the probability of moving from sample i to j in one step of a random walk on the graph [6]. The embedding corresponds to the eigenvectors associated to the first higher eigenvalues of $\tilde{\mathbf{W}}^m$, after removing the trivial case associated to the eigenvalue 1. It stands for the principal directions of diffusion across the data manifold, approximated by the graph made of the available samples.

2.3   Hierarchical Manifold Learning

The hierarchical embedding proposed in *Bhatia et al.* [4] builds upon this framework and minimizes the following cost function:

$$\underset{\mathbf{Y}^1}{\arg\min}\ (1-\mu)\sum_i\sum_j\|\mathbf{y}_i^1-\mathbf{y}_j^1\|^2\,W_{ij}^1+\mu\sum_i\|\mathbf{y}_i^1-\mathbf{y}_i^0\|^2, \qquad (2)$$

where $\mathbf{Y}^0 = [\mathbf{y}_i^0]$ was previously estimated by applying diffusion maps to the lower-level data, and $\mu \in [0, 1]$ balances the contributions of the higher and lower levels to the hierarchical embedding. If $\mu = 1$, the optimal solution is $\mathbf{Y}^1 = \mathbf{Y}^0$, so the hierarchical embedding is in this case the lower-level embedding. If $\mu = 0$, it corresponds to performing diffusion maps on the higher level only. In their paper, *Bhatia et al.* showed that there is an analytic solution to this cost function minimization for $\mu \neq 0$:

$$\mathbf{Y}^1 = (\mu\mathbf{I} + 2(1-\mu)\mathbf{L}^1)^{-1}\mu\mathbf{Y}^0, \qquad (3)$$

where $\mathbf{I}$ stands for the identity matrix.

2.4   Hyperparameters Optimization

Bhatia et al. arbitrarily set the weighting parameter μ. In contrast, we propose two complementary strategies to find the best embedding for our application.

First, we computed a-posteriori each term in the energy function (Eq. 2), and defined the optimal μ as the value for which the two terms are balanced (Fig. 2b).

In addition, we quantified the point-to-point distances between the estimated hierarchical embedding $\mathbf{Y}^1$ and the embeddings estimated for the higher and lower levels considered independently. To reduce bias in the distances, we rescaled the embeddings globally so that the standard deviations along their first dimension match, and determined the sign of the eigenvectors that produced the best match. The optimal μ using this second strategy corresponds to the embedding at equal distance from the high and low levels (Fig. 2c).

We implemented the method in Python 3.7.6. The whole algorithm was computed on an standard laptop within a few seconds. The limiting part is the computation of the affinity matrices for the whole set of images. The hierarchical part of the algorithm (Eq. 2) is really quick as it doesn't require any optimization step.

3   Experiments and Results

3.1   Data and Preprocessing

We analyzed the data from 123 patients with acute myocardial infarction recruited in the MIMI study [3]. The myocardial content of the LGE images was resampled to a reference anatomy using atlas-based techniques, as done in

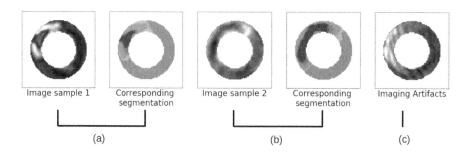

Fig. 1. Representative pairs of LGE images and their corresponding segmentation. The MVO and the infarct respectively correspond to the dark/red and the bright/yellow regions in the grayscale/segmented images. Some challenging samples are displayed: a very large MVO that covers most of the lesion and could be counfounded with healthy myocardium (b), and a slice with typical MRI artifacts (c). Note that during the transport of imaging data to a common reference, imaging contents may have been interpolated within a slice and across slices, leading to non-categorical values for the segmentated images (not completely blue/yellow/red). (Color figure online)

[9]. In brief, this consists in a pixel-wise parameterization based on the local radial, circumferential, and long-axis coordinates, which only requires labelling the LV-RV junction on each slice, and the identification of the apex and base levels in the stack of slices.

In this paper, we analyzed the 2D slices independently to benefit from a larger dataset. Besides, the image stacks were artificially reoriented for patients with LCX or RCA infarcts to match the location of the LAD infarct subgroup and therefore prevent the infarct location from confounding the analysis. As a result, we analyzed 1711 samples of segmented images as the parent, and their corresponding grayscale images as the child (see representative examples in Fig. 1).

3.2   Latent Space Organization

We first applied diffusion maps to the segmented images in which the infarct and MVO are labeled, leading to the lower-level embedding $\mathbf{Y}^0$ (Fig. 2a-b-c, left column). Then, we computed the affinity matrix associated to the LGE image contents to define the graph Laplacian $\mathbf{L}^1$, and estimated the higher-level embedding $\mathbf{Y}^1$ from Eq. 3 for several values of μ spanning the interval $[0, 1]$. For comparison purposes, diffusion maps were also directly applied to the LGE images (Fig. 2a-b-c, right column).

The bandwidth σ of the kernel involved in the affinity matrices $\mathbf{W}^m$ was set experimentally for each latent space. In the literature, it is commonly set as the average distance between a sample and its k-th nearest neighbor. However, this choice was not relevant in our case, in particular with MVO patterns. For instance, if σ is too small, the principal directions of diffusion can be strongly affected by one specific sample. Conversely, if σ is too large, it can lead to a

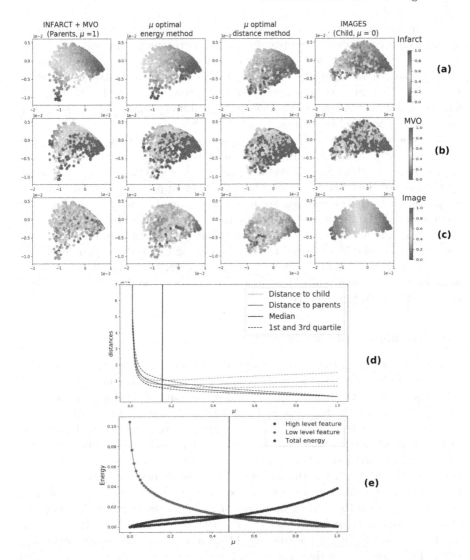

Fig. 2. First three rows (a-b-c): Latent spaces obtained for the parent (left column) and child (right column) data independently, and for the optimal hierarchical embedding (middle columns). The first two dimensions are displayed, and colored according to the infarct (a), MVO size (purple dots correspond to slices without MVO) (b) or the average pixel value in the images (c). Last two rows (d-e): The two optimal μ (vertical black line) were obtained from the crossing of the distance curves (d) or energy curves (e), as explained in Sect. 2.4.

shrinkage of the latent spaces to zero. We therefore heuristically set σ to the lowest value allowing a meaningful data spread in the latent space (in our case $\sigma = 35$ for the segmentation embedding, and $\sigma = 25$ for the image-based one).

As visible in the two central columns of Fig. 2a-b, the optimal values for μ lead to intermediate latent spaces to which both the parent and child levels contribute. Samples are no longer fully organized according to the segmentation characteristics (the amount of infarct and MVO encoded in the colorscale of Fig. 2a-b). They are neither strongly disorganized as obtained with the child data only, but contain part of the information on image appearance (the average pixel value is encoded in the colorscale of Fig. 2c).

The energy and distance curves in Fig. 2d-e confirm that $\mu = 1$ leads to matching the parent data embedding. In contrast, a jump is observed when approaching $\mu = 0$, as Eq. 3 would amount to $\mathbf{Y}^1 = 0$. This might be explained by the sizes of the embeddings we obtained when approaching $\mu = 0$ (typically for $\mu \leq 0.1$). As the embeddings are really small (due to numerical issues), their rescaling might be less accurate, leading to the jump in the distance curves. On this population, a value of $\mu = 0.47$ leads to balanced energies in Eq. 3, although a smaller value of 0.16 is needed to get a hierarchical embedding equally distant from the child and parent embeddings. In our case, the energy-based solution is closer to the parent embedding structure, whereas the distance-based one is closer to the child latent space.

3.3   Neighborhood Consistency

Figure 3 complements these observations by showing representative cases picked from the latent spaces. Row (a) shows a slice with a standard infarct pattern containing a small MVO. The four images and segmentations displayed on the left of the figure stand for the four nearest neighbors in the parent embedding, whereas the ones on the right stand for the four neighbors from the hierarchical embedding (balanced energies solution). We observe that both the neighbors from the parent and hierarchical embedding have segmentations close to the sample slice. However, the images from the hierarchical method are closer (compared with the parent embedding) to the original subject. It means that the embedding is more faithful to the MR image appearance. Row (b) displays the neighbors of a slice with a large MVO that covers most of the infarct. In this case, we display on the left the four nearest neighbors from the image-based embedding, while the images on the right correspond to the neighbors from the hierarchical embedding. We observe here that the image-based neighbors have very different MVO patterns, despite close image appearance. In contrast, the hierarchy-based neighbors are much more consistent with the original pattern. This highlights the interest of guiding the hierarchical embedding by the segmented data, which leads to embeddings more robust to challenging image contents for unusual samples.

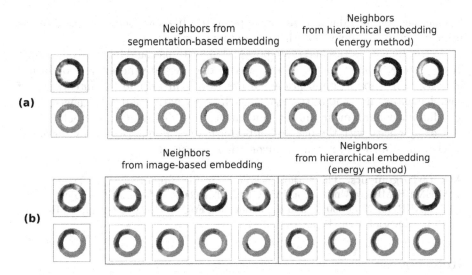

Fig. 3. Illustration of the robustness to challenging image contents. The first column represents the segmentation and image contents for two representative samples: a standard infarct pattern with small MVO (a), and a large MVO covering most of the infarct (b). The remaining columns show their four nearest neighbors for these two cases, either picked from the segmentation- or image-based embeddings, compared to the ones from the hierarchical embedding.

4 Discussion

We have demonstrated the relevance of a hierarchical approach for the analysis of tissue heterogeneities in LGE images. Our hierarchy consisted in guiding the representation of a higher-level (and more challenging) content, by a lower-level (and easier to represent) content corresponding to the segmented images. We also introduced two ways of selecting a relevant weighting parameter μ to balance the contribution of each level in the hierarchy.

Our approach comes from the hierarchical way physicians integrate several data from different sources, for more robustness and confidence in their diagnosis. In our case, the hierarchy allows the integration of prior knowledge corresponding to the lesion location. It can be seen as a way to circumvent the simplicity of a pixelwise distance metric between samples, which may be counfounded by specific lesion patterns or image artifacts. We used only simple metrics (pixelwise distances) and a well-known manifold learning framework (diffusion maps), which may reveal limited for complex datasets. Our experiments demonstrated that meaningful embeddings can be achieved even for the challenging cases included in our database.

This hierarchical manifold learning algorithm could be extended to other imaging protocols featuring several levels of data complexity (e.g. typical echocardiography or MRI examinations). The latent space can serve as an intermediate representation to feed a classification algorithm, or as a simplified way to examine complex data, as in our application. It allows the extraction of the principal diffusion dimensions in such data spaces.

Better image metrics and more powerful algorithms could improve our results in the future, but our purpose here was to demonstrate the interest of such a hierarchy on a sound application. Future work will include the integration of complementary imaging of the lesions such as T1 native and T2* images, early gadolinium enhancement images, and myocardial deformation extracted from image sequences, with the purpose of better understanding ischemia-reperfusion mechanisms in the case of acute myocardial infarction.

Acknowledgements. The authors acknowledge the support from the French ANR (LABEX PRIMES of Univ. Lyon [ANR-11-LABX-0063] and the JCJC project "MIC-MAC" [ANR-19-CE45-0005]), and the Fédération Francaise de Cardiologie ("MI-MIX" project, Allocation René Foudon).

References

1. Alexandre, J., et al.: Scar extent evaluated by late gadolinium enhancement CMR: a powerful predictor of long term appropriate ICD therapy in patients with coronary artery disease. J. Cardiovasc. Magn. Reson. **15**, 12 (2013)
2. Bekkers, S., et al.: Microvascular obstruction: underlying pathophysiology and clinical diagnosis. J. Am. Coll. Cardiol. **55**, 1649–60 (2010)
3. Belle, L., et al.: Comparison of immediate with delayed stenting using the Minimalist Immediate Mechanical Intervention approach in acute ST-segment-elevation myocardial infarction: the MIMI study. Circ. Cardiovasc. Interv. **9**, e003388 (2016)
4. Bhatia, K., et al.: Hierarchical manifold learning for regional image analysis. IEEE Trans. Med. Imaging **33**, 444–61 (2014)
5. Bulluck, H., et al.: Cardiovascular magnetic resonance in acute st-segment-elevation myocardial infarction: recent advances, controversies, and future directions. Circulation **137**, 1949–64 (2018)
6. Coifman, R.S.L.: Diffusion maps. Appl. Comput. Harm. Anal. **21**, 5–38 (2006)
7. Criminisi, A., Shotton, J.: Decision forests for computer vision and medical image analysis. Springer Publishing Company (2013)
8. Lin, Y., et al.: Multiple kernel learning for dimensionality reduction. IEEE Trans. Pattern Anal. Mach. Intell. **33**, 1147–60 (2011)
9. Mom, K., et al.: Population-based personalization of geometric models of myocardial infarction. Proc. FIMH **12738**, 3–11 (2021)
10. Shekelle, P.: Clinical practice guidelines: what's next? JAMA **320**, 757–8 (2018)
11. Wang, B., et al.: Unsupervised metric fusion by cross diffusion. In: Proceedings CVPR, pp. 2997-3004 (2012)

Generating Subpopulation-Specific Biventricular Anatomy Models Using Conditional Point Cloud Variational Autoencoders

Marcel Beetz[1]([✉]), Abhirup Banerjee[1,2][iD], and Vicente Grau[1][iD]

[1] Institute of Biomedical Engineering, Department of Engineering Science, University of Oxford, Oxford OX3 7DQ, UK
marcel.beetz@eng.ox.ac.uk
[2] Division of Cardiovascular Medicine, Radcliffe Department of Medicine, University of Oxford, Oxford OX3 9DU, UK
abhirup.banerjee@cardiov.ox.ac.uk

Abstract. Generative statistical models have a wide variety of applications in modelling of cardiac anatomy and function, including disease diagnosis and prediction, personalized shape analysis, and generation of population cohorts for electrophysiological and mechanical computer simulations. In this work, we propose a novel geometric deep learning method based on the variational autoencoder (VAE) framework capable of accurately encoding, reconstructing, and synthesizing 3D surface models of the biventricular anatomy. Our non-linear approach works directly with memory-efficient point clouds and is able to process multiple substructures of the cardiac anatomy at the same time in a multi-class setting. Furthermore, we introduce subpopulation-specific characteristics as additional conditional inputs to allow the generation of new personalized anatomies. Our method achieves high reconstruction quality on a dataset derived from the UK Biobank study with average Chamfer distances between reconstructed and gold standard point clouds below the underlying image pixel resolution, for all anatomical substructures and combinations of conditional inputs. We investigate our method's generative capabilities and show that it is able to synthesize virtual populations of realistic hearts with volumetric measurements in line with established clinical precedent. We also analyse the effects of variations in the latent space of the autoencoder on the generated anatomies and find interpretable changes in cardiac shapes and sizes.

Keywords: Cardiac anatomy synthesis · Point cloud generation · Beta-VAE · Cardiac anatomy reconstruction · Conditional generative models · Geometric deep learning · Cardiac MRI

1 Introduction

The human heart exhibits considerable inter-person variability both in terms of its shape and function, which significantly impacts the effectiveness of cardiac

© Springer Nature Switzerland AG 2022
E. Puyol Antón et al. (Eds.): STACOM 2021, LNCS 13131, pp. 75–83, 2022.
https://doi.org/10.1007/978-3-030-93722-5_9

disease prevention, diagnosis, and treatment. The ability to capture this variability with data-driven methods is highly beneficial for clinical practice and therefore a key objective of the cardiac image analysis community, as it allows population-specific shape analysis, disease and outcome prediction, dimensionality reduction, and computer modelling of cardiac function [14]. While traditional statistical models such as principal component analysis (PCA) have been widely used for this purpose [1,11,14], recent research efforts focus increasingly on deep learning methods [5,6,9,13]. In this paper, we propose a novel variational autoencoder (VAE) [8] architecture acting directly on memory-efficient point clouds to generate subpopulation-specifc 3D biventricular anatomy models. To the best of our knowledge, this is the first geometric deep learning approach for cardiac anatomy generation. Our point cloud surface representations avoid the sparsity issues of 3D voxelgrids leading to quick execution and high resolution. Compared to PCA and other traditional shape modelling techniques, our method can capture non-linear relations in the data and does not require any prior landmark detection or registration, making its application significantly simpler and less error-prone. The choice of VAE framework enables stable training and a compact but also interpretable latent space representation of population datasets. By additionally introducing multiple conditional inputs, we can generate arbitrarily large subpopulation-specific cohorts of artificial hearts, which allows us to visualize and better understand the effects of combinations of different subject characteristics on biventricular anatomy and function.

2 Methods

We first briefly describe the dataset used for method development, followed by the network architecture and training procedure.

2.1 Dataset

Our point cloud dataset is based on 3D reconstructions of cine MRI acquisitions obtained from volunteers of the UK Biobank study [10]. We randomly select ∼500 female and ∼500 male subjects and extract the end-diastolic (ED) and end-systolic (ES) slices from the temporal sequence for each case [2], allowing us to condition our method on two binary metadata variables (sex and cardiac phase). We follow the pipeline described in [3] to create the 3D point cloud reconstructions from each acquisition and split our dataset into ∼1700 and ∼300 point clouds for training and testing respectively with equal representation of all conditions.

2.2 Network Architecture

Our proposed model architecture consists of a point cloud-based geometric deep learning network embedded in a conditional β-VAE [7,8] framework (Fig. 1).

We choose the PointNet++ [12] and the Point Completion Network [15] as the baseline architectures of our encoder and decoder, respectively. We adapt

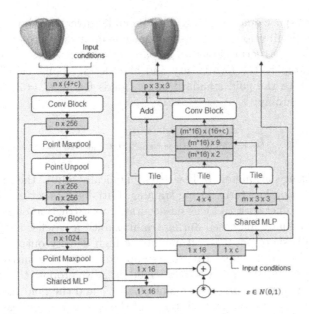

Fig. 1. Architecture of the proposed conditional Point Cloud β-VAE. The input (top left) is an unstructured point cloud with n points. Each point is represented by a 4-dimensional vector with three coordinate values (x,y,z) and a class label. A conditional vector c which contains additional information about the subject is concatenated to each input point vector as well as the latent space vector. The output consists of a coarse (top right) and a dense (top centre) point cloud generated from a random sample of the latent space distribution. Separate 3D coordinate values are used for each of the three classes in both output point clouds.

them to our multi-class setting by adding class information about the cardiac substructures (left ventricular (LV) endocardium, LV epicardium, right ventricular (RV) endocardium) to the encoder input and adjust the decoder architecture to output separate point clouds for each class. We enable conditional point cloud generation by concatenating our global input conditions to both encoder and decoder inputs. In order to effectively process high-density surface data and cope with the difficulty of latent space sampling, we also insert multiple fully connected layers to facilitate the exchange of spatial, class, and condition information. The standard reparameterization approach [8] is applied in the network's latent space. We choose a latent space size of 16, which we found to be sufficiently large to capture almost all of the variability in cardiac shapes and maintain good disentanglement.

2.3 Loss Function

Our loss function follows the design of the β-VAE [7] with a reconstruction loss and a latent space loss balanced by a weighting parameter β. We use a β value of 0.2, chosen empirically as a good trade-off between low reconstruction error and high latent space quality. The Kullback-Leibler divergence between the prior and posterior distributions of the latent space is used as a second loss term [8]. We split the reconstruction loss into a coarse and a dense loss term [15], which respectively compare the low-density and high-density point cloud predictions of our network to the gold standard point clouds for all $C = 3$ classes in the biventricular anatomy:

$$L_{recon} = \sum_{i=1}^{C} \left(L_{coarse,i} + \alpha * L_{dense,i} \right). \tag{1}$$

The weighting parameter α allows to dynamically adjust the importance of each reconstruction loss term during training. Initially, it is set to a low value of 0.01 to allow the network to focus on accurate reconstruction of global shapes, and is then gradually increased during training until it reaches the value 5.0 to put more emphasis on local structures in the high density output while maintaining a good overall shape. Due to its approximation of a surface-to-surface distance and its ability to process point cloud data, we propose the Chamfer distance (CD) between the predicted point cloud P_1 and the gold standard input point cloud P_2 as a metric for both terms of the reconstruction loss:

$$CD(P_1, P_2) = \frac{1}{2} \left(\frac{1}{|P_1|} \sum_{x \in P_1} \min_{y \in P_2} \|x - y\|_2 + \frac{1}{|P_2|} \sum_{y \in P_2} \min_{x \in P_1} \|y - x\|_2 \right). \tag{2}$$

3 Experiments

We evaluate our method in terms of both its point cloud reconstruction and generation performance. We also analyze its ability to correctly incorporate conditional inputs into the generation process and calculate commonly used clinical metrics over the generated heart shapes.

3.1 Reconstruction Quality

In order to assess the VAE's reconstruction ability, we select the point clouds of the unseen test dataset as our gold standard, input them into the network, and compare these inputs to the network's reconstructions using the Chamfer distance. We report the results separated by class and subpopulation in Table 1.

We find mean distance values to be consistently below the pixel resolution of the underlying MR images (1.8 × 1.8 × 8.0 mm) [10] and standard deviations all in the range of 0.19 mm to 0.32 mm.

For a qualitative evaluation of our method's reconstructions, we visualize the network input and output point clouds of five sample cases in Fig. 2. We observe that our method is able to reconstruct anatomical surfaces with high accuracy on both a global and local level for all biventricular substructures and can successfully cope with considerable variations.

Table 1. Reconstruction results of the proposed method on the test dataset.

Condition		Chamfer distance (mm)		
Sex	Phase	LV endocardium	LV epicardium	RV endocardium
Female	ED	1.06 (±0.23)	1.17 (±0.32)	1.37 (±0.29)
	ES	0.88 (±0.23)	1.00 (±0.21)	1.11 (±0.25)
Male	ED	1.16 (±0.23)	1.25 (±0.21)	1.38 (±0.24)
	ES	0.91 (±0.19)	1.08 (±0.21)	1.22 (±0.23)

Values represent mean (± standard deviation) in all cases.

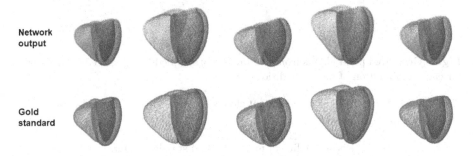

Fig. 2. Qualitative reconstruction results of our method on five sample cases.

3.2 Conditional Point Cloud Generation

In order to evaluate the generative performance of our method, we randomly sample from the latent space probability distribution and add either a 'male' or a 'female' label as well as either an 'ED' or an 'ES' label as conditional inputs to assess the ability of the method to generate specific subpopulations. We then pass the samples through the trained decoder part of our network. Figure 3 shows the generated point clouds from two such samples.

Comparing the point clouds in Fig. 3, we observe noticeable differences in sizes and shapes, indicating the decoder's ability to generate diverse point clouds. The effects of changing conditional inputs of each latent space vector on the reconstructed anatomy are also easily visible in a column-wise comparison and match well-known clinical expectations. For example, male hearts exhibit a larger size in both ED and ES phases than their female counterparts.

Next, we randomly sample 500 latent space vectors and use our trained decoder to generate random subpopulations for each combination of conditional inputs (ED female, ES female, ED male, ES male). We then convert both generated and test set point clouds into meshes using the Ball Pivoting algorithm [4]. This allows us to calculate common clinical metrics for each mesh and thereby quantify the clinical accuracy of our generated subpopulations compared to the meshes of the test dataset, that we consider to be our gold standard (Table 2).

ED female　　　**ED male**　　　**ES female**　　　**ES male**

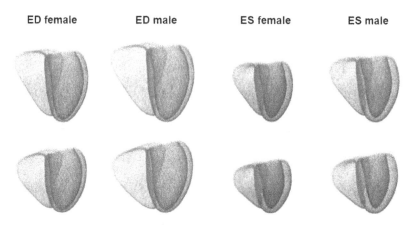

Fig. 3. Generated point clouds from two randomly sampled latent space vectors (rows) for each combination of input conditions (columns).

Table 2. Clinical metrics of meshed point clouds generated by our method with specific input conditions.

Sex	Phase	Clinical metric	Gold standard	Ours
Female	ED	LV volume (ml)	124 (±24)	122 (±24)
		RV volume (ml)	147 (±26)	146 (±27)
	ES	LV volume (ml)	50 (±13)	53 (±13)
		RV volume (ml)	64 (±15)	69 (±14)
	ED/ES	LV mass (g)	85 (±19)	85 (±21)
Male	ED	LV volume (ml)	153 (±28)	154 (±29)
		RV volume (ml)	190 (±27)	192 (±32)
	ES	LV volume (ml)	65 (±13)	73 (±15)
		RV volume (ml)	91 (±16)	98 (±16)
	ED/ES	LV mass (g)	121 (±23)	122 (±27)

Values represent mean (± standard deviation) in all cases.

We find comparable values across all clinical metrics and subpopulations in terms of both means and standard deviations. Slightly better scores are achieved for female hearts and the ED phase than for male hearts and the ES phase.

3.3 Latent Space Analysis

The quality of the latent space distribution plays an important role in the VAE's ability to synthesize artificial populations of realistic hearts that are also sufficiently diverse. We analyze the contributions of each part of the latent space to the generated point clouds by varying individual latent space components, while keeping the remaining latent space constant, and passing the resulting vectors through the decoder to obtain the respective outputs. Figure 4 shows the synthesized point clouds corresponding to variations in three sample latent space dimensions, similar to the most important modes of variation in a PCA analysis.

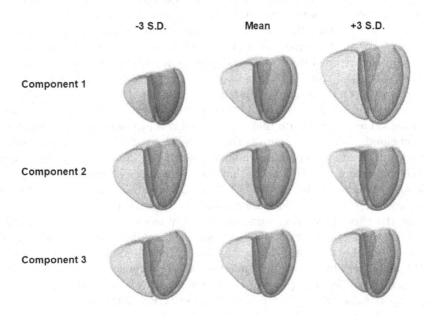

Fig. 4. Effect of different latent space components on point cloud reconstructions.

We observe gradual interpretable changes to the biventricular shapes and sizes without loss of a realistic appearance, while individual components encode different aspects of the biventricular anatomy. Among other things, component 1 is responsible for the overall heart size, component 2 changes the orientation angle of the basal plane of the heart, while component 3 transforms thin hearts with small mid-ventricular short-axis diameters into thicker ones.

4   Discussion

In this work, we have developed an efficient and easy-to-use method for synthesizing 3D biventricular anatomies conditioned on subject metadata. The method does not require any registration or point correspondence while maintaining high accuracy and diversity in its generation task. It is also capable of efficiently working with high-dimensional 3D MRI-based surface data due to its usage of point clouds instead of highly-sparse and memory-intensive voxelgrids. We achieve mean Chamfer distances considerably below the pixel resolution of the underlying images, demonstrating good reconstruction quality, while the small standard deviation values indicate that our method is highly robust and can successfully cope with a variety of different morphologies, both within and between subpopulations. Our approach is able to process multi-class point clouds which allows us to model different cardiac substructures with a single network. Despite no explicit constraint on the connectivity of the different substructures, we do not observe any sizeable disconnected or overlapping components between them. We therefore conclude that the low values in the general reconstruction loss were sufficient to implicitly impose correct inter-class connectivity. The closeness in mean clinical metrics of the synthesized subpopulation-specific distributions and the respective gold standard values show our method's good generative performance as well as its ability to accurately incorporate multiple conditional inputs into the generation process. In addition, the observed similarities in standard deviation values demonstrate that our method can produce a highly diverse set of point clouds that is representative of the real population. We find easily interpretable and gradual anatomical changes resulting from latent space variations of each component, which indicate that the latent space resembles a continuous unimodal probability distribution. This finding is also in line with other commonly used statistical approaches for population-based cardiac shape modelling, such as the effect of varying along the primary modes of variation in a PCA model. However, due to its non-linear design, our method is capable of capturing more complex relationships in the data while maintaining interpretability. Furthermore, we observe good latent space disentanglement with each component encoding different aspects of the biventricular anatomy. To this end, the weighting parameter β of the β-VAE framework was important for our high-dimensional dataset as it allowed for the right balance to be set between latent space and reconstruction quality.

5   Conclusion

In this work, we have presented an easy and efficient geometric deep learning method capable of generating arbitrarily-sized populations of realistic biventricular anatomies. We have shown how different subject metadata can be successfully incorporated into our approach to synthesize subpopulation-specific heart cohorts and how our method's compact latent space representation enables an interpretable shape analysis of cardiac anatomical variability.

Acknowledgments. This research has been conducted using the UK Biobank Resource under Application Number '40161'. The authors express no conflict of interest. The work of M. Beetz was supported by the Stiftung der Deutschen Wirtschaft (Foundation of German Business). The work of A. Banerjee was supported by the British Heart Foundation (BHF) Project under Grant HSR01230. The work of V. Grau was supported by the CompBioMed 2 Centre of Excellence in Computational Biomedicine (European Commission Horizon 2020 research and innovation programme, grant agreement No. 823712).

References

1. Bai, W., et al.: A bi-ventricular cardiac atlas built from 1000+ high resolution MR images of healthy subjects and an analysis of shape and motion. Med. Image Anal. **26**(1), 133–145 (2015)
2. Banerjee, A., et al.: A completely automated pipeline for 3D reconstruction of human heart from 2D cine magnetic resonance slices. Philosoph. Trans. Royal Soc. A., p. 20200257 (2021)
3. Beetz, M., Banerjee, A., Grau, V.: Biventricular surface reconstruction from cine MRI contours using point completion networks. In: 2021 IEEE 18th International Symposium on Biomedical Imaging (ISBI), pp. 105–109 (2021)
4. Bernardini, F., Mittleman, J., Rushmeier, H., Silva, C., Taubin, G.: The ball-pivoting algorithm for surface reconstruction. IEEE Trans. Visual. Comput. Graphics **5**(4), 349–359 (1999)
5. Biffi, C., et al.: Explainable anatomical shape analysis through deep hierarchical generative models. IEEE Trans. Med. Imaging **39**(6), 2088–2099 (2020)
6. Gilbert, K., Mauger, C., Young, A.A., Suinesiaputra, A.: Artificial intelligence in cardiac imaging with statistical atlases of cardiac anatomy. Front. Cardiovasc. Med. **7**, 102 (2020)
7. Higgins, I., et al.: beta-VAE: learning basic visual concepts with a constrained variational framework. In: 5th International Conference on Learning Representations (ICLR), pp. 1–13 (2017)
8. Kingma, D.P., Welling, M.: Auto-encoding variational Bayes. arXiv preprint arXiv:1312.6114 (2013)
9. Litjens, G., et al.: A survey on deep learning in medical image analysis. Med. Image Anal. **42**, 60–88 (2017)
10. Petersen, S.E., et al.: UK Biobank's cardiovascular magnetic resonance protocol. J. Cardiovasc. Magn. Reson. **18**(1), 1–7 (2015)
11. Piazzese, C., Carminati, M.C., Pepi, M., Caiani, E.G.: Statistical shape models of the heart: applications to cardiac imaging. In: Statistical Shape and Deformation Analysis, pp. 445–480. Elsevier (2017)
12. Qi, C.R., Yi, L., Su, H., Guibas, L.J.: Pointnet++: deep hierarchical feature learning on point sets in a metric space. In: Advances in neural information processing systems, pp. 5099–5108 (2017)
13. Rezaei, M.: Chapter 5 - Generative adversarial network for cardiovascular imaging. In: Al'Aref, S.J., Singh, G., Baskaran, L., Metaxas, D. (eds.) Machine Learning in Cardiovascular Medicine, pp. 95–121. Academic Press (2021)
14. Tavakoli, V., Amini, A.A.: A survey of shaped-based registration and segmentation techniques for cardiac images. Comput. Vision Image Understanding, **117**(9), 966–989 (2013)
15. Yuan, W., Khot, T., Held, D., Mertz, C., Hebert, M.: PCN: point completion network. In: 2018 International Conference on 3D Vision (3DV), pp. 728–737 (2018)

Improved AI-Based Segmentation of Apical and Basal Slices from Clinical Cine CMR

Jorge Mariscal-Harana[1]([✉]), Naomi Kifle[1], Reza Razavi[1,2], Andrew P. King[1], Bram Ruijsink[1,2,3], and Esther Puyol-Antón[1]([✉])

[1] School of Biomedical Engineering and Imaging Sciences, King's College London, London, UK
{jorge.mariscal_harana,esther.puyol_anton}@kcl.ac.uk
[2] Department of Adult and Paediatric Cardiology, Guy's and St Thomas' NHS Foundation Trust, London, UK
[3] Department of Cardiology, Heart and Lung Division, University Medical Center Utrecht, Utrecht, The Netherlands

Abstract. Current artificial intelligence (AI) algorithms for short-axis cardiac magnetic resonance (CMR) segmentation achieve human performance for slices situated in the middle of the heart. However, an often-overlooked fact is that segmentation of the basal and apical slices is more difficult. During manual analysis, differences in the basal segmentations have been reported as one of the major sources of disagreement in human interobserver variability. In this work, we aim to investigate the performance of AI algorithms in segmenting basal and apical slices and design strategies to improve their segmentation. We trained all our models on a large dataset of clinical CMR studies obtained from two NHS hospitals (n = 4,228) and evaluated them against two external datasets: ACDC (n = 100) and M&Ms (n = 321). Using manual segmentations as a reference, CMR slices were assigned to one of four regions: non-cardiac, base, middle, and apex. Using the 'nnU-Net' framework as a baseline, we investigated two different approaches to reduce the segmentation performance gap between cardiac regions: (1) non-uniform batch sampling, which allows us to choose how often images from different regions are seen during training; and (2) a cardiac-region classification model followed by three (i.e. base, middle, and apex) region-specific segmentation models. We show that the classification and segmentation approach was best at reducing the performance gap across all datasets. We also show that improvements in the classification performance can subsequently lead to a significantly better performance in the segmentation task.

Keywords: Cardiac magnetic resonance · Deep learning · Segmentation · Class imbalance

---

B. Ruijsink and E. Puyol-Antón—Shared last authors.

E. Puyol Antón et al. (Eds.): STACOM 2021, LNCS 13131, pp. 84–92, 2022.
https://doi.org/10.1007/978-3-030-93722-5_10

1   Introduction

Segmenting the ventricles and the myocardium from cardiac magnetic resonance (CMR) images allows cardiologists to assess cardiac function, thus guiding diagnosis, prognosis and treatment of cardiac disease [10]. However, for both ventricles, the selection and segmentation of the basal slice are major determinants of variability in the assessment of cardiac function [3]. For most of the slices, the ventricular blood pool is a circular shape whose border is well defined by the presence of a ring of myocardial muscle. However, in the basal slices, the myocardial muscle does not completely enclose the blood pool and its shape can be variable and non-circular due to the presence of the outflow tract and the atrio-ventricular and ventriculo-arterial valves. Moreover, through-plane motion of the cardiac structures over the cardiac cycle further challenges reproducible segmentations. In the apex, the myocardial rim is angulated with respect to the short-axis imaging plane, which means that the borders are less well defined. Incorrect segmentation of these two areas of the ventricle contributes to errors in measurement of cardiac function. Due to the longitudinal motion of the heart during contraction, the basal slices contribute greatly to estimates of stroke volume (i.e. the ejected volume of blood during ventricular contraction).

Deep learning (DL) models have achieved human-level performance for automatic segmentation of short-axis cine CMR images [1], and have been recently proposed as a means for the automated characterisation of cardiac function [11]. However, model performance is not uniform throughout the heart, with basal and apical regions posing a greater challenge than middle regions [1]. This is likely to be caused by the inherent difficulty in segmenting basal and apical regions, as well as the higher variability of manual segmentations.

In this paper, to the best of our knowledge, we perform the first statistical analysis of the performance of DL-based segmentation across the different sections of the heart. We discover significant differences between middle slices and both basal and apical slices after training a state-of-the-art segmentation model on a large clinical dataset ($n = 4{,}228$) and testing it on three independent test datasets. We propose two approaches to bridge the performance gap between cardiac regions, resulting in significantly improved segmentations for basal and apical slices with a reduced error spread.

2   Methods

To investigate whether DL-based automated CMR segmentation suffers from significant performance differences between cardiac regions, we trained a baseline segmentation model using a large dataset of clinical CMR images ($n = 4{,}228$). We subsequently investigated two potential approaches to reduce the performance gap based on (1) non-uniform batch sampling and (2) a combination of a cardiac-region classification model and three region-specific segmentation models. The baseline and the proposed approaches are illustrated in Fig. 1.

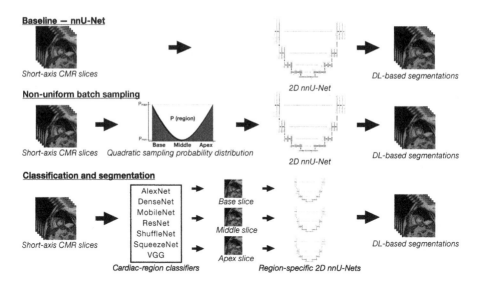

Fig. 1. Baseline and proposed approaches to reduce the performance gap between cardiac regions. The baseline corresponds to the original nnU-Net framework trained on the NHS dataset. (1) Non-uniform batch sampling approach using a quadratic probability distribution. (2) Cardiac-region classifier followed by region-specific segmentation models.

Segmentation Model: We used the 'nnU-Net' framework (2D U-Net) [7] to segment the left and right ventricular blood pools (LVBP and RVBP, respectively) and the left ventricular myocardium (LVM) from end-diastole (ED) and end-systole (ES) short-axis CMR stacks. Details of the training and test datasets and experimental setup are provided in Sect. 3.

Baseline: We trained the nnU-Net model on the NHS clinical CMR dataset. This approach is considered the baseline since its performance is comparable to that of other state-of-the-art CMR segmentation methods [1].

Approach (1) - Non-uniform Batch Sampling: We modified nnU-Net's data loader to allow for a "region-aware" non-uniform batch sampling. This means that, for each mini-batch during training, slices from the apex and the base are more likely to be selected than middle slices. This is achieved via a quadratic probability distribution whose lower and upper limits determine the minimum and maximum sampling probabilities, respectively. This approach does not affect the probability of non-cardiac slices being selected.

Approach (2) - Classification and Segmentation: We fine-tuned seven state-of-the-art DL-based classification models to determine which cardiac region each slice belonged to: AlexNet [8], DenseNet121 [5], MobileNetV2 [12], ResNet18 [4], ShuffleNet V2 [9], SqueezeNet [6], and VGG Net-A [13] (with batch normalisation). Since there was a class imbalance in terms of cardiac regions – with non-cardiac and middle slices being more common than basal and apical ones – we investigated the use of a weighted loss function for classification. However, this resulted in a decrease in performance across all models, thus we used a non-weighted loss function instead. We also trained three region-specific segmentation models using slices from the corresponding cardiac regions only. To ensure that the same number of images would be seen by each model, we used nnU-Net's default hyperparameters. Finally, we combined the best-performing classifier with the individual segmentation models for our predictions. Note that slices classified as non-cardiac were analysed using the baseline model.

Evaluation Metrics: Classification performance was evaluated using precision and recall for each slice. To evaluate segmentation performance, we used the Dice similarity coefficient (DSC) between manual segmentations and model predictions for the LVBP, RVBP, and LVM. Since the number of segmented slices was variable, DSC distributions from basal to apical slices were calculated via linear interpolation ($n = 101$) to evaluate regional segmentation performance. Finally, we assessed improvements in segmentation performance between the baseline and the proposed approaches via the differences in the mean and the standard deviation (SD) of the DSC.

3   Materials and Experiments

Datasets: The baseline and proposed approaches were trained and evaluated using manual segmentations from a dataset of clinical CMR scans ($n = 4{,}684$) acquired at two NHS hospitals and including the full spectrum of cardiac disease phenotypes. All approaches were evaluated on a randomly selected subset from the NHS ($n = 456$) dataset, and on two external datasets of clinical CMRs: ACDC [1] ($n = 100$) and M&Ms [2] ($n = 321$). All datasets consisted of ED and ES short-axis cine CMR images. For the NHS dataset, left-ventricular endocardial and epicardial borders and RV endocardial borders were manually segmented using the commercially available cvi42 CMR analysis software (Circle Cardiovascular Imaging Inc., Calgary, Alberta, Canada, version 5.10.1). Although working with data acquired at different spatial resolutions and MR parameters is challenging, nnU-Net's resampling during preprocessing helps to mitigate its impact on the results.

Training: For all approaches, we split the NHS dataset into random training, validation, and test splits with 3382, 846, and 456 cases, respectively. We found that the single fold performance of the baseline segmentation model was similar to that of the five-fold cross-validation, so all models used in this work were trained on a single random fold. All segmentation models were trained for 1,000 epochs with 250 random mini-batches of size 32 per epoch, and optimised using nnU-Net's default configuration: stochastic gradient descent (initial learning rate of 0.01 with a polynomial learning rate policy, Nesterov momentum of 0.99, and weight decay of 3e−5). Classification models were trained for 1,000 epochs with the Adam optimiser (initial learning rate of 5e−4). All models were trained on a NVIDIA TITAN RTX GPU. The approximate training times for each segmentation and classification models were 24 and 42 h, respectively.

We compared the nnU-Net baseline to the two proposed approaches using metrics of segmentation performance. For each short-axis CMR stack, slices containing segmentations were considered to belong to the base/middle/apex cardiac regions following a 20%/60%/20% split, respectively. The ratio of stratification was based on cardiac anatomical knowledge and clinical experience.

Experiment 1 - Performance Gap Assessment: We first looked at the regional differences in segmentation performance for the baseline model. Table 1 shows the mean and SD of the DSC for each cardiac region and segmentation label. Statistically significant differences between the base-middle and apex-middle regions were found using a Welch unequal variances t-test for all labels and test datasets, except for the ACDC dataset, which did not show significant differences in basal LVBP and RVBP.

Table 1. Mean (and SD) of the DSC (%) for each cardiac region and segmentation label. Asterisks indicate statistically significant differences in the mean DSC between the base or the apex and the middle region ($p < 0.01$).

Dataset	Label	Cardiac regions		
		Base	Middle	Apex
NHS	LVBP	87.53* (23.29)	94.93 (5.62)	81.08* (24.86)
	LVM	79.82* (19.34)	87.22 (6.04)	70.18* (24.96)
	RVBP	76.25* (28.90)	90.72 (9.57)	73.24* (27.03)
M&Ms	LVBP	84.41* (20.38)	93.63 (6.73)	89.86* (16.47)
	LVM	70.23* (29.13)	86.46 (8.08)	75.25* (25.84)
	RVBP	76.60* (26.56)	90.11 (11.84)	81.62* (25.36)
ACDC	LVBP	94.42 (7.76)	94.28 (6.87)	82.86* (23.47)
	LVM	84.01* (17.59)	87.66 (9.34)	64.99* (33.28)
	RVBP	90.23 (16.02)	88.54 (16.22)	67.42* (33.97)

Experiment 2 - Reducing the Performance Gap: Given the results from the first experiment, we focused on improving the segmentation performance for basal and apical slices with the two proposed approaches: (1) non-uniform batch sampling and (2) classification and segmentation. For approach (1), we tried two batch sampling levels - "low" and "high" - by setting the ratio of maximum to minimum sampling probabilities to 4 and 20, respectively. A higher level of batch sampling produced better results. For approach (2), we compared the seven classifiers according to the weighted-average precision and recall values for all cardiac regions in the NHS test cases (Table 2). The best-performing classifier, DenseNet121, was combined with the region-specific segmentation models.

Table 2. Mean precision and recall values for seven state-of-the-art classification models trained and tested on the NHS dataset.

	AlexNet	DenseNet	MobileNet	ResNet	ShuffleNet	SqueezeNet	VGG
Precision [%]	87.54	91.72	89.56	90.94	90.33	87.62	91.02
Recall [%]	87.56	91.50	89.50	90.94	90.25	87.56	91.06

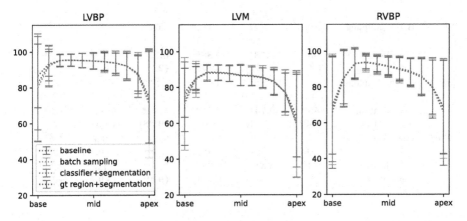

Fig. 2. Interpolated DSC distribution across the heart for the LVBP (left), LVM (middle), and RVBP (right) for the baseline and proposed methods for the NHS test cases. Blue: baseline, orange: approach (1); green: approach (2); red: upper-bound performance of approach (2). (Color figure online)

Additionally, the ground truth cardiac regions were combined with the region-specific segmentation models to determine the upper-bound performance (i.e. equivalent to a perfect classifier) of this approach. Figure 2 shows the interpolated DSC distributions for the baseline, the best-performing batch sampling approach, the classification and segmentation approach, and the ground truth region and segmentation. For the base and the apex, approach (2) produces the greatest performance improvements.

Finally, Table 3 shows the differences in the mean and the SD of the DSC between the baseline and the proposed approaches for all datasets. Paired Student's t-tests were performed to determine statistically significant differences between the means of the baseline and the proposed approaches. The results show that the classification and segmentation approach outperforms both the baseline and the non-uniform batch sampling approach across all segmentation labels and datasets. We observed that segmentation performance for the background label (i.e. not LVBP, LVM, or RVBP) was uniform across all slices and similar for all proposed methods.

Table 3. Differences in DSC mean (and SD) values between the baseline and the proposed approaches for the base and the apex. Positive values indicate improvements over the baseline (i.e. a higher mean and a lower SD). The largest improvement for each label and cardiac region is highlighted in bold. Asterisks indicate statistically significant differences between the means ($p < 0.01$).

Dataset	Approach	LVBP		LVM		RVBP	
		Base	Apex	Base	Apex	Base	Apex
NHS	Batch sampling	−0.06 (0.01)	−0.27 (−0.29)	−0.91* (−1.53)	0.09 (−0.22)	0.89* (0.67)	1.03* (1.82)
	Classification + segmentation	**1.33*** **(3.35)**	**0.78*** **(1.91)**	**0.88*** **(3.90)**	**0.97** **(2.89)**	**1.04*** **(1.74)**	**1.58*** **(3.21)**
M&Ms	Batch sampling	0.42 (1.39)	−0.21 (−0.13)	0.28 (0.34)	−1.33* (−1.37)	1.61* (3.30)	**0.76** **(2.38)**
	Classification + segmentation	**0.85*** **(2.59)**	**0.74*** **(2.83)**	**5.71*** **(8.02)**	**5.53*** **(9.38)**	**2.44*** **(4.57)**	0.02 (2.58)
ACDC	Batch sampling	−0.28 (−0.00)	−0.75 (−0.46)	−0.88 (−0.94)	0.46 (0.92)	−0.61 (−0.57)	3.03* (3.78)
	Classification + segmentation	−0.46 (−0.08)	**0.51** **(3.36)**	**1.88*** **(9.34)**	**7.60*** **(9.19)**	−1.37* (−0.78)	**5.74*** **(6.87)**

4   Discussion

To the best of our knowledge, this paper has presented the first statistical analysis of the performance of DL models for cardiac segmentation across different regions of the heart. We have shown that there are significant segmentation performance differences between the middle of the heart and both the apex and the base in three test datasets. In the ACDC dataset, however, the CMR volumes were truncated and did not include the full basal portion of the heart, which explains the lack of significant differences between the base and the middle for the LVBP and RVBP. These findings are in line with experience from clinicians in daily practice. Although human-level segmentation performance has been achieved by DL-based models overall, these regional differences are critical for clinical evaluation since errors in segmentation at the base and apex can impact subsequent estimates of functional metrics such as stroke volume and ejection fraction.

We have proposed two approaches to increase the DL-based segmentation performance for the base and the apex and compared them to a nnU-Net baseline. The first approach (non-uniform batch sampling) allowed us to modify how often CMR slices from different cardiac regions were seen by the model during training. The second approach (classification and segmentation) combined a cardiac-region classifier with region-specific segmentation models with the aim of having specialised models for each cardiac region. Our results show that both approaches improve performance in the basal and apical regions, although only the classification and segmentation approach produced significantly better results across all labels and datasets. This shows that the characteristics of basal and apical regions seem to differ enough from the middle slices to regard them as different tasks with respect to the segmentation model. Furthermore, the results for the upper-bound performance of approach (2) are promising, as they indicate that further segmentation improvements could be achieved by improving the performance of the cardiac-region classifiers.

In conclusion, in this work we have highlighted the uneven segmentation performance of DL-based models across the heart and proposed two approaches to bridge this performance gap. We have shown that combining a classifier with region-specific segmentation models significantly improves segmentation performance at the base and the apex. In the future, motivated by the current results, we aim to train deeper classification models to improve our current cardiac-region classification performance and to investigate a simultaneous classification and segmentation approach. Additionally, we will analyse how an improved segmentation performance affects the estimation of cardiac function metrics such as diastolic and systolic ventricular volumes, ejection fraction, and left ventricular mass.

Acknowledgements. This work was supported by the EPSRC (EP/P001009/1 and the Advancing Impact Award scheme of the Impact Acceleration Account at King's College London) and the Wellcome EPSRC Centre for Medical Engineering at the School of Biomedical Engineering and Imaging Sciences, King's College London (WT 203148/Z/16/Z).

References

1. Bernard, O., et al.: Deep learning techniques for automatic MRI cardiac multi-structures segmentation and diagnosis: is the problem solved? IEEE Trans. Med. Imaging **37**(11), 2514–2525 (2018)
2. Campello, V.M., et al.: Multi-centre, multi-vendor and multi-disease cardiac segmentation: the M&Ms challenge. IEEE Trans. Med. Imaging **40**(12), 3543–3554 (2021). https://doi.org/10.1109/tmi.2021.3090082
3. Caudron, J., Fares, J., Lefebvre, V., Vivier, P.H., Petitjean, C., Dacher, J.N.: Cardiac MRI assessment of right ventricular function in acquired heart disease. Acad. Radiol. **19**(8), 991–1002 (2012). https://doi.org/10.1016/j.acra.2012.03.022
4. He, K., Zhang, X., Ren, S., Sun, J.: Deep residual learning for image recognition. In: 2016 IEEE Conference on Computer Vision and Pattern Recognition (CVPR). IEEE, June 2016. https://doi.org/10.1109/cvpr.2016.90
5. Huang, G., Liu, Z., Maaten, L.V.D., Weinberger, K.Q.: Densely connected convolutional networks. In: 2017 IEEE Conference on Computer Vision and Pattern Recognition (CVPR). IEEE, July 2017. https://doi.org/10.1109/cvpr.2017.243
6. Iandola, F.N., Han, S., Moskewicz, M.W., Ashraf, K., Dally, W.J., Keutzer, K.: Squeezenet: Alexnet-level accuracy with 50x fewer parameters and <0.5 MB model size (2016)
7. Isensee, F., Jaeger, P.F., Kohl, S.A.A., Petersen, J., Maier-Hein, K.H.: nnU-net: a self-configuring method for deep learning-based biomedical image segmentation. Nat. Methods **18**(2), 203–211 (2020). https://doi.org/10.1038/s41592-020-01008-z
8. Krizhevsky, A., Sutskever, I., Hinton, G.E.: ImageNet classification with deep convolutional neural networks. In: Pereira, F., Burges, C.J.C., Bottou, L., Weinberger, K.Q. (eds.) Advances in Neural Information Processing Systems, vol. 25. Curran Associates, Inc. (2012)

9. Ma, N., Zhang, X., Zheng, H.-T., Sun, J.: ShuffleNet V2: practical guidelines for efficient CNN architecture design. In: Ferrari, V., Hebert, M., Sminchisescu, C., Weiss, Y. (eds.) Computer Vision – ECCV 2018. LNCS, vol. 11218, pp. 122–138. Springer, Cham (2018). https://doi.org/10.1007/978-3-030-01264-9_8

10. Peng, P., Lekadir, K., Gooya, A., Shao, L., Petersen, S.E., Frangi, A.F.: A review of heart chamber segmentation for structural and functional analysis using cardiac magnetic resonance imaging. Magn. Reson. Mater. Phys. Biol. Med. **29**(2), 155–195 (2016). https://doi.org/10.1007/s10334-015-0521-4

11. Ruijsink, B., Puyol-Antón, E., et al.: Fully automated, quality-controlled cardiac analysis from CMR: validation and large-scale application to characterize cardiac function. Cardiovasc. Imaging **13**(3), 684–695 (2020)

12. Sandler, M., Howard, A., Zhu, M., Zhmoginov, A., Chen, L.C.: MobileNetV2: Inverted residuals and linear bottlenecks. In: 2018 IEEE/CVF Conference on Computer Vision and Pattern Recognition. IEEE, June 2018. https://doi.org/10.1109/cvpr.2018.00474

13. Simonyan, K., Zisserman, A.: Very deep convolutional networks for large-scale image recognition. In: International Conference on Learning Representations (2015)

Mesh Convolutional Neural Networks for Wall Shear Stress Estimation in 3D Artery Models

Julian Suk[1]([⊠]), Pim de Haan[2,3], Phillip Lippe[2], Christoph Brune[1], and Jelmer M. Wolterink[1]

[1] Department of Applied Analysis and Technical Medical Centre,
University of Twente, Enschede, The Netherlands
`j.m.suk@utwente.nl`
[2] QUVA Lab, University of Amsterdam, Amsterdam, The Netherlands
[3] Qualcomm AI Research, Qualcomm Technologies Netherlands B.V.,
Nijmegen, The Netherlands

Abstract. Computational fluid dynamics (CFD) is a valuable tool for personalised, non-invasive evaluation of hemodynamics in arteries, but its complexity and time-consuming nature prohibit large-scale use in practice. Recently, the use of deep learning for rapid estimation of CFD parameters like wall shear stress (WSS) on surface meshes has been investigated. However, existing approaches typically depend on a hand-crafted re-parametrisation of the surface mesh to match convolutional neural network architectures. In this work, we propose to instead use mesh convolutional neural networks that directly operate on the same finite-element surface mesh as used in CFD. We train and evaluate our method on two datasets of synthetic coronary artery models with and without bifurcation, using a ground truth obtained from CFD simulation. We show that our flexible deep learning model can accurately predict 3D WSS vectors on this surface mesh. Our method processes new meshes in less than 5 [s], consistently achieves a normalised mean absolute error of ≤ 1.6 [%], and peaks at 90.5 [%] median approximation accuracy over the held-out test set, comparing favourably to previously published work. This demonstrates the feasibility of CFD surrogate modelling using mesh convolutional neural networks for hemodynamic parameter estimation in artery models.

Keywords: Coronary blood flow · Geometric deep learning · Computational fluid dynamics · Surrogate modelling

1 Introduction

In patients suffering from cardiovascular disease, parameters that quantify arterial blood flow could complement anatomical measurements such as vessel diameter and degree of stenosis. For example, the magnitude and direction of wall

Qualcomm AI Research is an initiative of Qualcomm Technologies, Inc.

E. Puyol Antón et al. (Eds.): STACOM 2021, LNCS 13131, pp. 93–102, 2022.
https://doi.org/10.1007/978-3-030-93722-5_11

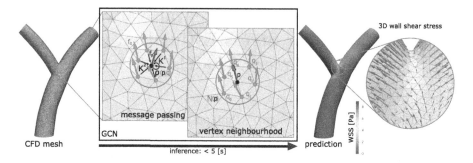

Fig. 1. Our graph convolutional network (GCN) predicts vector-valued quantities on a computational fluid dynamics (CFD) surface mesh in a single forward pass. It consists of convolutional layers that apply filters with kernels K^1, K^2 in a vertex neighbourhood $q \in N(p)$ to vertex features f_p, f_q using parallel transport $\rho(p, q)$ ("message passing").

shear stress (WSS) was found to correlate with atherosclerotic plaque development and arterial remodelling [7,15]. Imaging techniques like particle image velocimetry or 4D flow MRI with which WSS could be measured non-invasively are not widely available and may be less accurate in smaller arteries. Therefore, WSS is often estimated using computational fluid dynamics (CFD). This requires the extraction of a geometric artery model from e.g. CT images, spatial discretisation of this model in a finite-element mesh, and iterative solution of the Navier-Stokes equations within the mesh [17]. However, high-quality CFD solutions require fine meshes, leading to high computational complexity. Therefore, there is a practical demand for surrogate models that trade accuracy for speed.

Deep neural networks are an attractive class of surrogate models. They require generating training data (e.g. by CFD) but once they are trained, inference in new geometries only requires a single forward pass through the network, which can be extremely fast. Previous applications of deep learning to coronary CFD surrogate modelling have studied the use of multilayer perceptrons to estimate fractional flow reserve based on hand-crafted features [8], as well as the use of convolutional neural networks to estimate WSS based on 1D parametrisations of single arteries [16] and 2D views of bifurcated arteries [3]. Furthermore, convolutional neural networks have been applied in stress and hemodynamics prediction in the aorta [10,11].

All of these approaches have in common that they rely on a global or local *parametrisation* of the artery model, which is limited to idealised vessels or does not generalise to more complex vessel trees. This limits their value as a plug-in replacement for simulation in clinical CFD workflows. In contrast, we propose to let deep neural networks learn a (latent) parametrisation of the vessel geometry directly from a surface mesh representing the vessel lumen wall. We describe input meshes as graphs and use graph convolutional networks (GCNs) and their extension, mesh convolutional networks, to predict WSS vectors on the mesh vertices (Fig. 1). This offers a plug-in replacement for CFD simulation operating on a mesh that can be acquired through well-established meshing procedures.

GCNs have been previously studied for general mesh-based simulations [14]. Moreover, several recent works have investigated the use of GCNs in the cardiovascular domain, e.g. for coronary artery segmentation [19], coronary tree labeling [6], or as a surrogate for cardiac electrophysiology models [13]. In a very recent study, Morales et al. [1] showed that GCNs outperform parametrisation-based methods for the prediction of WSS-derived scalar potentials in the left atrial appendage. In this work, we compare isotropic and anisotropic mesh convolution for the prediction of 3D vector quantities on mesh vertices. Additionally, we show how the use of gauge-equivariant mesh convolution [4] results in neural networks that are invariant under translation and equivariant under rotation of the input mesh, two desirable properties for improved data efficiency. The latter is especially relevant in medical applications where access to large datasets is uncommon. Our method is flexible and avoids the loss of information that may occur due to ad-hoc parametrisation of the artery model. We demonstrate the effectiveness of this approach on two sets of synthetic coronary artery geometries.

2   Method

We model the artery lumen wall as a 2D manifold in 3D Euclidean space. Its surface mesh is fully described by a tuple of vertices $\mathcal{V}$ and polygons $\mathcal{F}$, whose undirected edges $\mathcal{E}$ induce the graph $\mathcal{G} = (\mathcal{V}, \mathcal{E})$. We propose to use GCNs to predict WSS *vectors* on the vertices of the graph. Those are defined as the tangential force exerted on the lumen wall by the blood flow. GCNs are trainable operators that map an input signal on a graph $f^{\text{in}} : \mathcal{G} \to \mathbb{R}^{c_{\text{in}}}$ to an output signal $f^{\text{out}} : \mathcal{G} \to \mathbb{R}^{c_{\text{out}}}$ on the same graph. Like many convolutional neural network architectures, they can be composed of convolutional and pooling layers, which are here defined as follows.

Convolution. We implement convolutional layers by spatial message passing in a neighbourhood $N(p)$ around vertex p, expressed in a general form as

$$((K^1, K^2(\cdot, \cdot)) * f)_p := f_p \cdot K^1 + \sum_{q \in N(p)} \rho(p, q) f_q \cdot K^2(p, q), \quad p \in \mathcal{V} \quad (1)$$

where $K^1, K^2 \in \mathbb{R}^{c \times \bar{c}}$ are trainable kernel matrices. These matrices weigh the features of p itself and those of its neighbours, respectively. We let $f_q = f(q) \in \mathbb{R}^c$ denote the feature vector corresponding to $q \in \mathcal{V}$. Equation (1) describes three graph convolutional layers, depending on the choice of $K^2(p, q)$ and $\rho(p, q) \in R^{c \times c}$ as follows. First, we consider GraphSAGE [5] convolution with full neighborhood sampling, which corresponds to choosing an **isotropic** kernel $K^2(p, q) = \frac{1}{|N(p)|} \bar{K}^2$ and $\rho(p, q) = I$, the $c \times c$ identity matrix. Isotropic kernels can accommodate neighbourhoods $N(p)$ of a varying number of vertices without canonical ordering, but sacrifice expressiveness.

Second, we consider FeaSt [18] convolution, which implements an **anisotropic** filter through an attention mechanism $\rho(p, q) = \sigma(w \cdot (f_q - f_p)) I$ with trainable weights $w \in \mathbb{R}^c$ where $\sigma(\cdot)$ is the "softmax" function. The kernel is again chosen $K^2(p,q) = \frac{1}{|N(p)|}\bar{K}^2$.

Third, we employ gauge-equivariant mesh convolution [4]. Here, anisotropic kernels are learned as $K^2(p,q) = K^2(\theta_q^p)$, where the orientation angles $\theta^p \in [0, 2\pi)^{|N(p)|}$ of all neighbours $q \in N(p)$ are measured from one *arbitrary* reference neighbour projected onto the tangent plane at p. This arbitrary choice defines a local coordinate system ("gauge"). Linearly combining vertex features f_q requires expressing them in the same gauge, which is done by the parallel transporter $\rho(p,q) = \rho(\gamma)$ where $\gamma(p,q)$ is the angle between the previously chosen reference neighbours. The representation $\rho(\cdot)$ is determined by the irreducible representations ("irreps") of the planar rotation group SO(2) which compose into the features $f_q, q \in \mathcal{V}$. In order to ensure that the convolution does not depend on the arbitrary choice of local coordinate systems, a linear constraint is placed on the kernel K^2 such that the linearly independent solutions to this *kernel constraint* span the space of possible trainable parameters.

Pooling. Graph pooling consists of clustering followed by a reduction operation. We define clusters in the mesh based on a hierarchy of $n+1$ vertex subsets $\mathcal{V}^0 \supset \mathcal{V}^1 \supset \cdots \supset \mathcal{V}^n$ and assign each vertex in $\mathcal{V}^{i-1}$ to a vertex in $\mathcal{V}^i$ based on geodesic distances. In the gauge-equivariant framework, we average features after parallel transporting to the same gauge. For unpooling, features are distributed over the clusters using parallel transport. There is no notion of parallel transport for general c-dimensional features in GraphSAGE or FeaSt convolution, hence we simply average over the c-dimensional feature space during pooling and distribute statically oriented features over the clusters during unpooling.

Network Architecture. Since WSS depends on the global shape of the blood vessel, we use encoder-decoder networks, akin to the well-known U-Net. These networks are formed by three pooling scales where each scale consists of residual blocks with two convolutional layers and a skip connection. We copy and concatenate signals between corresponding layers in the contracting and expanding pathway. ReLU activation is used throughout the architecture.

We construct input features $f^{\text{in}} : \mathcal{G} \to \mathbb{R}^{c_{\text{in}}}$ that are invariant under translation and equivariant under rotation of the surface mesh. This is done by, for each vertex p, considering a ball $\mathcal{B}_r(p)$ and averaging weighted differences to the vertices $q \in \mathcal{B}_r(p)$ as well as the vertex normals. Subsequently, we take the outer product of the averaged differences with themselves, the averaged normals with themselves, and the differences with the normals. For the GraphSAGE and FeaSt networks, we flatten and concatenate the resulting matrices and use the resulting feature vectors as inputs for each vertex. For the gauge-equivariant network, we instead decompose the matrices into factors of SO(2) irreps. A schematic representation of the employed architecture is shown in Fig. 2.

Gauge-equivariant convolutional layers are by construction equivariant under SO(2) transformation of the local coordinate systems. Representing our network's input features f^{in} in these gauges leads to end-to-end equivariance under

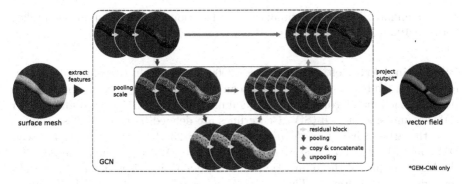

Fig. 2. Schematic representation of the employed network architecture. Coloured vertices signify the ones used for message passing on each of the three pooling scales. Each residual block consists of two convolutional layers and one skip connection. For the gauge-equivariant network, the signal is projected back from the local coordinate systems into Euclidean space at the output.

SO(3) transformation of the global coordinate system. This means that the gauge-equivariant network is a globally rotation-equivariant operator.

Each network is additionally provided with a scalar input feature that contains, for each vertex, the (invariant) shortest geodesic distance to the vessel inlet, giving a rotation-equivariant indication of the flow direction.

3   Datasets

We synthesise two large datasets of coronary artery geometries and corresponding CFD simulations to train and validate our neural networks. The entire shape synthesis and CFD simulation is implemented and automated with the SimVascular [9] Python shell.

Single Arteries. We emulate the dataset by Su et al. [16] and synthesise 2000 idealised 3D coronary artery models. Each model has a single inlet and single outlet, with a centerline defined by control points in a fixed increment along the x-axis and a uniformly random increment along the y-axis. The cross-sectional lumen contour has a random radius $r \sim \mathcal{U}(1.25, 2.0)$ [mm]. Each artery contains up to two randomly-placed, asymmetric stenoses of up to 50%. Each model is lofted and meshed using a global edge length of 0.4 [mm], local edge length refinement proportional to the vessel radius, and five tetrahedral boundary layers. Figure 3 shows an example synthetic geometry. We simulate blood flow under the three-dimensional, incompressible Navier-Stokes equations. We assume dynamic viscosity $\mu = 0.035$ $[\frac{g}{cm \cdot s}]$, blood density $\rho = 1.05$ $[\frac{g}{cm^3}]$, rigid walls, and no-slip boundary condition. The inlet velocity profile is constant and uniform: $u_{in} = 20[\frac{cm}{s}]$. A pressure $p_{out} = 100$ [mmHg] ≈ 13.332 [kPa] is weakly applied at the outlet. A Reynolds number of Re = 180 < 2100 implies the fluid flow is laminar. Blood flow is simulated until stationary and the simulation takes between 10 and 15 [min] on 16 CPU cores (Intel Xeon Gold 5218).

Bifurcating Arteries. We synthesise 2000 3D left main coronary bifurcation models, randomly constructed based on an atlas of coronary measurement statistics [12]. Each model consists of proximal main vessel (PMV), distal main vessel (DMV), and side branch (SB). Centerlines are defined by three angles sampled from the atlas. First, $\beta \sim \mathcal{N}(\mu_\beta = 78.9, \sigma_\beta = 23.1)$ [°] is the angle between DMV and SB. Second, $\beta' \sim \mathcal{N}(\mu_{\beta'} = 61.5, \sigma_{\beta'} = 21.5)$ [°] is the angle between the bisecting line of the bifurcation and SB, describing skew of the bifurcation towards the child branch. Third, $\gamma \sim \mathcal{N}(\mu_\gamma = 9.5, \sigma_\gamma = 21.5)$ [°] is the angle under which the PMV centerline enters the bifurcation plane. We model the vessel lumen with ellipses that are arbitrarily oriented in the plane normal to the centerline curve tangent. The lumen radii are drawn from the coronary atlas:

- $r_{\text{PMV}} \sim \mathcal{N}(\mu_{r_{\text{PMV}}} = 1.75, \sigma_{r_{\text{PMV}}} = 0.4)$ [mm],
- $r_{\text{DMV}} \sim \mathcal{N}(\mu_{r_{\text{DMV}}} = 1.6, \sigma_{r_{\text{DMV}}} = 0.35)$ [mm], and
- $r_{\text{SB}} \sim \mathcal{N}(\mu_{r_{\text{SB}}} = 1.5, \sigma_{r_{\text{SB}}} = 0.35)$ [mm].

We sample lumen radii such that diameters follow a bifurcation law of the form $(d_{\text{PMV}})^a = (d_{\text{DMV}})^a + (d_{\text{SB}})^a + \delta$ with small δ and decrease approximately linear with vessel length. Each model is lofted and meshed using a global edge length of 0.2 [mm] and five tetrahedral boundary layers. Blood flow simulation follows that of the single artery models, with differences as follows. We assume $\mu = 0.04$ $[\frac{\text{g}}{\text{cm}\cdot\text{s}}]$, $\rho = 1.06$, and prescribe a parabolic influx profile corresponding to a uniform inflow velocity $u_{\text{in}} = 11.8[\frac{\text{cm}}{\text{s}}]$ to get WSS values which agree with a healthy range [15] of 1.0 to 2.5 [Pa]. We weakly apply a pressure p_{out} at the union of the outflow surfaces. The Reynolds number for the fluid flow is Re = 100 < 2100 so the flow is laminar. The simulation runs take between 8 and 12 [min] on 32 CPU cores (Intel Xeon Gold 5218).

4   Experiments

We implement three versions of the encoder-decoder network based on Graph-SAGE, FeaSt and gauge-equivariant mesh convolution and refer to these as SAGE-CNN, FeaSt-CNN, and GEM-CNN. Hidden layers are set up so that each network has just under 800,000 trainable parameters. Both datasets are split 80:10:10 into training, validation, and test set. The GCNs are trained to predict a 3D WSS vector at each vertex on the surface mesh. Since GEM-CNN naturally expresses its signal in the tangential planes w.r.t. each vertex, its output vector is here, for WSS, restricted to those tangential planes. However, outputting an additional normal component is also naturally supported. In our experiments, however, this only influenced the performance marginally. SAGE-CNN and FeaSt-CNN operate without the notion of tangential planes, thus they predict an unconstrained 3D vector to avoid overloading the methods and to

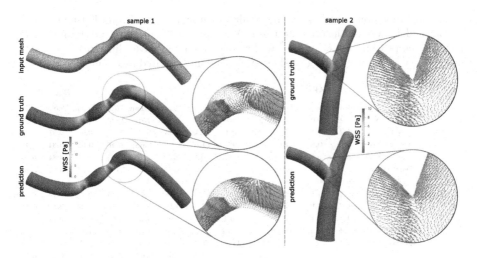

Fig. 3. WSS prediction by our method (GEM-CNN) on two previously unseen artery samples. CFD simulation acts as ground truth. Our method predicts 3D WSS vectors for every mesh vertex. In sample 1, multidirectional WSS can be observed.

save memory. The input meshes are the same surface meshes we use to obtain the ground truth via CFD, so that the CFD solution is directly used for network optimisation. All models are trained with a mean-squared error loss using the Adam optimiser. SAGE-CNN and FeaSt-CNN are trained on an NVIDIA GeForce RTX 3080 (10 GB) for 400 epochs taking 3:15 [h] and 12:30 [h] for the single artery models and twice as long for the bifurcating artery models. GEM-CNN is trained for 160 epochs (ca. 29:00 [h]) on two NVIDIA Quadro RTX 6000 (24 GB) for the single artery models, and 100 epochs (ca. 55:00 [h]) for the bifurcating artery models. Inference in a new and unseen geometry takes less than 5 [s] including pre-processing. We open-source our PyTorch Geometric [2] implementation for the GraphSAGE and FeaSt networks.[1]

We report mean absolute error normalised by the maximum WSS magnitude across the test set (NMAE) and define the approximation error $\varepsilon := \|\triangle\|_2/\|L\|_2$ where elements of $\triangle$ are vertex-wise normed differences between the network output $f^{\mathrm{out}} : \mathcal{G} \to \mathbb{R}^3$ and ground truth $l : \mathcal{G} \to \mathbb{R}^3$ so that $\triangle_p = \|f^{\mathrm{out}}(p) - l(p)\|_2$ for $p \in \mathcal{V}$ and $L_p = \|l_p\|_2$. Additionally, we report the maximum and mean vertex-wise normed difference, i.e. $\triangle_{\max} = \max_p\{\triangle_p\}$ and $\triangle_{\mathrm{mean}} = (\sum_p \triangle_p)/|\mathcal{V}|$.

Prediction Accuracy. Figure 3 shows an example of the reference and predicted WSS in a single and a bifurcating artery model, illustrating that our method does not only predict the WSS magnitude but also the directional 3D vector. Table 1 lists quantitative results of the prediction accuracy over both test sets. Direct comparison of SAGE-CNN, FeaSt-CNN and GEM-CNN on the single artery models suggests strictly better performance when using anisotropic

---

[1] https://github.com/sukjulian/coronary-mesh-convolution.

Table 1. Prediction accuracy on the idealised coronary arteries and left-main bifurcation models over the held-out test sets. We report normalised mean absolute error (NMAE), approximation error ε, as well as maximum and mean difference. CFD simulations act as ground truth. ([†]evaluated on randomly rotated test samples)

		NMAE [%]			ε [%]			$\triangle_{max}$ [Pa]			$\triangle_{mean}$ [Pa]		
		Mean	Median	75th	Mean	Median	75th	Mean	Median	75th	Mean	Median	75th
Single arteries	SAGE-CNN	2.2	2.0	2.6	32.4	30.0	37.0	10.41	7.80	14.65	1.11	1.01	1.32
	FeaSt-CNN	1.2	1.1	1.5	19.0	18.6	22.4	5.83	5.13	8.17	0.60	0.57	0.77
	GEM-CNN	**0.6**	**0.6**	**0.8**	**9.9**	**9.5**	**11.6**	**3.94**	**3.68**	**5.46**	**0.32**	**0.31**	**0.41**
	SAGE-CNN[†]	10.5	9.6	12.8	149.2	128.1	181.2	26.73	23.96	36.17	5.31	4.84	6.50
	FeaSt-CNN[†]	8.3	7.5	10.1	123.7	111.1	152.9	25.63	22.93	34.52	4.22	3.82	5.13
	GEM-CNN[†]	0.6	0.6	0.8	9.8	9.4	11.4	3.80	3.39	5.53	0.32	0.31	0.42
Bifurcating arteries	SAGE-CNN	1.3	1.1	1.5	24.4	21.1	27.3	4.38	4.14	5.50	0.27	0.22	0.29
	FeaSt-CNN	**1.2**	**0.9**	**1.3**	**20.7**	**18.1**	**22.3**	4.10	3.72	4.77	**0.23**	**0.19**	**0.25**
	GEM-CNN	1.3	1.1	1.6	23.3	20.4	28.6	**3.99**	**3.71**	**4.64**	0.27	0.22	0.32
	SAGE-CNN[†]	7.3	6.9	9.6	117.4	115.2	151.3	8.60	8.29	10.10	1.45	1.37	1.91
	FeaSt-CNN[†]	7.4	7.4	10.1	119.6	117.1	161.1	8.39	8.33	9.78	1.48	1.47	2.01
	GEM-CNN[†]	1.3	1.1	1.6	23.3	20.9	28.5	4.00	3.67	4.65	0.26	0.22	0.32

convolution kernels. GEM-CNN's good performance can be attributed to its anisotropic convolution kernels as well as the availability of parallel transport pooling. The results in Table 1 are comparable to previously published results on a similar dataset [16]. To demonstrate our method's flexibility, we train SAGE-CNN, FeaSt-CNN, and GEM-CNN on the bifurcating artery dataset without further adaption of the architecture. We find that all networks perform well, even without task-specific hyperparameter tuning. The performance differences between networks are less pronounced on the bifurcating arteries than on the single arteries. This hints at the input feature (parameters) in conjunction with the network's signal representation lacking expressiveness for the task at hand, since a convolutional neural network should always benefit from directional filters.

Equivariance. Due to its signal representation, GEM-CNN is a globally SO(3)-equivariant operator, i.e. the network output rotates with the input mesh orientation. SAGE-CNN and FeaSt-CNN are not equivariant under SO(3) transformation. This is because their signals are expressed in Euclidean space which makes the networks implicitly learn a reference orientation. To illustrate this, we compare the prediction accuracy on randomly rotated samples of the test sets (see Table 1). GEM-CNN's prediction accuracy is equivalent, with only small numerical deviation due to discretisation of the activation functions. In contrast, the performance of SAGE-CNN and FeaSt-CNN drops drastically, suggesting that they fail to predict useful WSS vectors on rotated samples.

5   Discussion and Conclusion

We have demonstrated the feasibility of GCNs and mesh convolutional networks as surrogates for CFD blood flow simulation in arterial models. We found that

GCNs effortlessly operate on the same meshes used in CFD simulations. Thus, shape parametrisation steps as proposed in other works [3,8,16] may be omitted and GCNs can be seamlessly integrated into existing CFD workflows. We noticed a significant speedup when using our method: inference in a new and unseen 3D geometry took less than 5 [s] including preprocessing, while CFD simulation took up to 15 [min]. The use of the same GCN architecture on two distinctly different datasets shows that our method does not require extensive task-specific fine-tuning and highlights its flexibility.

Our prediction results are based on two datasets which were synthesised with carefully chosen parameters to achieve a high level of realism. In future work, we aim to enrich our training and validation datasets with patient-specific artery geometries. This raises challenges regarding generalisation and uncertainty quantification, also with respect to the sensitivity of the proposed method to different meshing of the same lumen wall. One limitation of our experiments is that we consider steady blood flow, whereas in real applications doctors might be interested in time-averaged WSS over a cardiac cycle of pulsatile blood flow. This may require the incorporation of patient-specific boundary conditions. In the current experiments, boundary conditions were fixed across samples of both the single artery and bifurcating artery models and thus inherently encoded in the datasets. In future work, we will investigate to what extent the networks can be conditioned on user-supplied boundary conditions.

In conclusion, GCNs and mesh convolutional networks are a feasible approach to CFD surrogate modelling with potential applications in personalised hemodynamics estimation.

Acknowledgment. This work is funded in part by the 4TU Precision Medicine programme supported by High Tech for a Sustainable Future, a framework commissioned by the four Universities of Technology of the Netherlands. Jelmer M. Wolterink was supported by the NWO domain Applied and Engineering Sciences VENI grant (18192).

References

1. Ferez, X.M., et al.: Deep learning framework for real-time estimation of in-silico thrombotic risk indices in the left atrial appendage. Front. Physiol. **12**, 694945 (2021)
2. Fey, M., Lenssen, J.E.: Fast graph representation learning with PyTorch geometric. In: ICLR Workshop on Representation Learning on Graphs and Manifolds (2019)
3. Gharleghi, R., Samarasinghe, G., Sowmya, A., Beier, S.: Deep learning for time averaged wall shear stress prediction in left main coronary bifurcations. In: IEEE: International Symposium on Biomedical Imaging, vol. 17 (2020)
4. de Haan, P., Weiler, M., Cohen, T., Welling, M.: Gauge equivariant mesh CNNs: anisotropic convolutions on geometric graphs. In: ICLR (2021)
5. Hamilton, W., Ying, Z., Leskovec, J.: Inductive representation learning on large graphs. In: Neural Information Processing Systems, vol. 30 (2017)
6. Hampe, N., Wolterink, J.M., van Velzen, S.G.M., Leiner, T., Išgum, I.: Machine learning for assessment of coronary artery disease in cardiac CT: a survey. Front. Cardiovasc. Med. **6**, 172 (2019)

7. Hoogendoorn, A., et al.: Multidirectional wall shear stress promotes advanced coronary plaque development: comparing five shear stress metrics. Cardiovasc. Res. **116**, 1136–1146 (2019)
8. Itu, L.M., et al.: A machine learning approach for computation of fractional flow reserve from coronary computed tomography. J. Appl. Physiol. **121**, 42–52 (2016)
9. Lan, H., Updegrove, A., Wilson, N.M., Maher, G.D., Shadden, S.C., Marsden, A.L.: A re-engineered software interface and workflow for the open-source SimVascular cardiovascular modeling package. J. Biomech. Eng. **140**(2), 0255011–02450111 (2018)
10. Liang, L., Liu, M., Martin, C., Sun, W.: A deep learning approach to estimate stress distribution: a fast and accurate surrogate of finite-element analysis. J. R. Soc. Interface **15**, 20170844 (2018)
11. Liang, L., Mao, W., Sun, W.: A feasibility study of deep learning for predicting hemodynamics of human thoracic aorta. J. Biomech. **99**, 109544 (2020)
12. Medrano-Gracia, P., et al.: A study of coronary bifurcation shape in a normal population. J. Cardiovasc. Transl. Res. **10**, 82–90 (2017)
13. Meister, F., et al.: Graph convolutional regression of cardiac depolarization from sparse endocardial maps. In: Statistical Atlases and Computational Models of the Heart. M&Ms and EMIDEC Challenges (2021)
14. Pfaff, T., Fortunato, M., Sanchez-Gonzalez, A., Battaglia, P.: Learning mesh-based simulation with graph networks. In: International Conference on Learning Representations (2021)
15. Samady, H., et al.: Coronary artery wall shear stress is associated with progression and transformation of atherosclerotic plaque and arterial remodeling in patients with coronary artery disease. Circulation **124**, 779–778 (2011)
16. Su, B., Zhang, J.M., Zou, H., Ghista, D., Le, T.T., Chin, C.: Generating wall shear stress for coronary artery in real-time using neural networks: feasibility and initial results based on idealized models. Comput. Biol. Med. **126**, 104038 (2020)
17. Taylor, C.A., Fonte, T.A., Min, J.K.: Computational fluid dynamics applied to cardiac computed tomography for noninvasive quantification of fractional flow reserve. J. Am. Coll. Cardiol. **61**(22), 2233–2241 (2013)
18. Verma, N., Boyer, E., Verbeek, J.: FeaStNet: feature-steered graph convolutions for 3D shape analysis. In: IEEE Conference on Computer Vision and Pattern Recognition (2018)
19. Wolterink, J.M., Leiner, T., Išgum, I.: Graph convolutional networks for coronary artery segmentation in cardiac CT angiography. In: Graph Learning in Medical Imaging (2019)

Hierarchical Multi-modality Prediction Model to Assess Obesity-Related Remodelling

Gabriel Bernardino[1](✉), Patrick Clarysse[1], Álvaro Sepúlveda-Martínez[2,3], Mérida Rodríguez-López[2,4], Susanna Prat-Gonzàlez[5], Marta Sitges[6,7], Eduard Gratacós[3,6], Fàtima Crispi[3,6], and Nicolas Duchateau[1]

[1] Univ Lyon, Université Claude Bernard Lyon 1, INSA-Lyon, CNRS, Inserm, CREATIS UMR 5220, U1294, 69621 Lyon, France
`gabriel.bernardino@creatis.insa-lyon.fr`
[2] Fetal Medicine Research Center, BCNatal - Barcelona Center for Maternal-Fetal and Neonatal Medicine (Hospital Clínic and Hospital Sant Joan de Déu), Institut Clínic de Ginecologia Obstetricia i Neonatologia, Universitat de Barcelona, Centre for Biomedical Research on Rare Diseases (CIBER-ER), Barcelona, Spain
[3] Fetal Medicine Unit, Department of Obstetrics and Gynecology, Hospital Clínico de la Universidad de Chile, Santiago de Chile, Chile
[4] Pontificia Universidad Javeriana Seccional Cali, Cali, Colombia
[5] Centre de Diagnòstic per la Imatge, Hospital Clínic, Universitat de Barcelona, Barcelona, Spain
[6] Institut d'Investigacions Biomèdiques August Pi i Sunyer, Barcelona, Spain
[7] Institut Clínic Cardiovascular, Hospital Clínic, Universitat de Barcelona, Centre for Biomedical Research on CardioVascular Diseases (CIBERCV), Barcelona, Spain

Abstract. The diagnosis of cardiovascular illnesses uses multiple modalities in order to obtain a complete and as robust as possible assessment of the heart. However, when addressing distinct pathologies, not all information might be needed in order to achieve a confident-enough diagnosis.

We propose a probabilistic machine learning method to identify the patients for which the acquisition of more complex data would be useful. We hypothesise that there exists a hierarchical relationship between modalities: echocardiography is more accessible and has a lower economical cost than other modalities (like magnetic resonance imaging (MRI)). The framework consists of two classifier models, each predicting the illness from the echocardiographic and MRI views, and a sample-weighting model that combines both predictions. This weighting model is used to decide which individuals will not need an MRI acquisition additional to the echocardiographic examination.

We illustrated this on a dataset of asymptomatic individuals with an echocardiographic study ($N = 480$), a subset of those also includes a MRI ($N = 159$). We analyse the effect of being overweight on cardiac geometry. We identified that the type of remodelling depended on blood pressure: overweight combined with high blood pressure resulted in an increase of ventricular mass, while only size changes were preserved for low-pressure individuals. With our method, we established that boundary cases of the

© Springer Nature Switzerland AG 2022
E. Puyol Antón et al. (Eds.): STACOM 2021, LNCS 13131, pp. 103–112, 2022.
https://doi.org/10.1007/978-3-030-93722-5_12

former group could be correctly classified after incorporating MRI, while it was not the case for the latter.

Keywords: Data fusion · Probabilistic model · Machine learning · Cardiac imaging · Cardiac remodelling

1   Introduction

Medical imaging is a central component in the diagnosis of cardiovascular illnesses. Several modalities can be used to assess cardiac function and geometry, each offering a different view on the properties of the heart, and each being more or less appropriate for detecting different pathologies. In order to have the most complete assessment of the heart, it could be desirable to use as many modalities as possible, however multiple acquisitions result in higher economic cost, time, discomfort/danger to the patient. Moreover, the heart is a dynamic organ, and changes from beat to beat, difficulting the integration of data acquired in different beats.

Echocardiography is the most common modality used in cardiology, due to its low cost and accessibility, being able to be operated at bedside or in the clinician's office. Other modalities, such as MRI are also available, but are often used after echocardiographic findings are not conclusive enough to produce a diagnosis. The study of what kind of information can be determined from each modality, and when it is more appropriate to use one or the other, has not been quantitatively discussed beyond some consensus established by clinical experts [5].

In this work, we explore the feasibility of identifying the individuals who would benefit from more complex (and therefore costly) acquisitions. Our rationale is that there are individuals for whom echocardiography is not able to provide an accurate diagnosis, yet acquiring MRI will not improve diagnosis. We aim to identify those patients to avoid unnecessary MRI. This problem shares similarities with active learning [6,15], which is a subfield of machine learning that allows to identify in a semi-annotated dataset the unannotated samples that would be more helpful for training a model, and to Bayesian optimisation [4], a black-box optimisation method for functions that are expensive to evaluate.

Our approach is similar to data fusion, which is still a challenging research topic in machine learning [9]. Grossly, there are two strategies: obtaining a fused representation of the different views and then perform predictions in that space, or perform predictions independently and then combine them. Some methods, specially the ones based on a probabilistic framework such as [13] allow missing data (it is mostly assumed that the data is missing at random) and assume no hierarchical relationship between views. This is the main difference with our method, since our assumption is that we can identify the individuals for which the more complex acquisition is unnecessary, and therefore "missing" data will not be at random.

We illustrate the framework on a dataset including echocardiography and MRI from healthy asymptomaptic individuals, for which we study the cardiac changes related to body mass index (BMI). Overweight and obesity are well known cardiovascular risk factors [14]. The most reported risk is an increased probability of myocardial infarct due to the presence of cholesterol in the vessels, but also obesity is related to pressure (due to an increase of systolic blood pressure (SBP)) and volume loading of the left ventricle (LV) (since it needs to produce higher cardiac output (CO) to sustain the increased body mass). Its effect is most noticeable on the left ventricle: myocardial hypertrophy and also a dilation of the heart [7]. The changes are more easily seen in the cardiac geometry, but also affect the function, specially in the diastolic function [12].

2   Methodology

2.1   Echocardiographic and MRI Dataset

More details on the dataset are found in [2]. Its original purpose was to find the effect on the cardiac system of adults (30–40 years old) who were born low-weight. Its effect was found to be subtle, so the individuals with low weight at birth were kept in the current work in order to have a higher population size. We reused this dataset to study the effect of elevated body mass index, using the WHO classification for overweight (BMI > 25) [1]. We used overweight (BMI > 25) over strictly obese (BMI > 30) because our MRI population contained only 20 obese individuals. Overweight definition also includes individuals who are in the transition zone between obesity and appropriate weight, and they are therefore expected to have minor remodelling. A short description of the demographics of the adequate weight and overweight population can be found in Table 1. We did not consider the interaction between weight at birth and overweight, and its effect is treated as noise in this analysis.

Table 1. Demographic description of the population. Variables are described by their mean and STD.

	BMI < 25		BMI ≥ 25	
	Echo	MRI	Echo	MRI
N	320	83	178	64
Weight [kg]	60.6 ± 10	61.5 ± 9.8	84.4 ± 13.0	85.2 ± 11.1
Height [cm]	167 ± 9	168 ± 9	169 ± 10	171 ± 9
BSA [m^2]	1.67 ± 0.18	1.69 ± 0.17	1.99 ± 0.19	2.01 ± 0.16
Male gender [%]	39	36	63	67
Age [y]	30.0 ± 4.9	33.2 ± 3.7	31.7 ± 4.9	33.6 ± 4.1
Heart rate [bpm]	68.4 ± 11.2	66.9 ± 11.3	69.7 ± 12.3	71.2 ± 12.9
Systolic BP [mmHg]	114.6 ± 11.8	115.1 ± 13.4	121.9 ± 12.2	122.8 ± 13.3
Ejection Fraction [%]	65.4 ± 6.8	66.7 ± 7.9	64.5 ± 6.5	64.3 ± 6.7

In short, the data consist of echocardiographic measurements from 480 individuals, and MRI biventricular shapes from a subset 159 individuals. The echocardiographic measurements were: LV long axis, LV basal dimension, RV long axis and LV Mmss (from M-mode), all measured using the recommendations of the European Society of Cardiology [8]. The shapes are the result of fitting a deformable model to each individual, producing meshes at end-diastole (ED) in point-to-point correspondence. Both the measurements and the meshes were indexed by body size area (BSA) to remove cardiac size variability due to the body dimensions, as is commonly used in the clinics. The data was preprocessed using unsupervised dimensionality reduction in order to reduce the noise: principal component analysis (PCA) for the MRI shapes (keeping 5 dimensions), and Isomap for the echocardiographic measurements (keeping 2 dimensions).

2.2   Hierarchical Framework

We propose a probabilistic model to express the probability of the label (Y) given the simple $(\mathbf{X_0})$ and complex $(\mathbf{X_1})$ data features. The estimated probability is a weighted combination of two single-view classification models, with a sample weight. We note $\mathbf{w}$ as the vector containing the weights for all individuals, and w^i its value for the i-th individual. This weight is close to 1 when the information in X_0^i is enough to make the prediction and close to 0 when considering the next level of the hierarchy X_1^i is compulsory. The novelty in our approach lies in these weights, for which we add a prior on $\mathbf{w}$ forcing it to be a smooth function of $\mathbf{X_0} = [X_0^i]$. These weights serve as an estimator of the need of acquiring $\mathbf{X_1} = [X_1^i]$.

$$P(Y|\mathbf{X_0}, \mathbf{X_1}) = (\prod_i (w^i P(Y^i|X_0^i, \theta_0) + (1 - w^i)P(Y^i|X_1^i, \theta_1)))P(\mathbf{w}|\mathbf{X_0}), \quad (1)$$

where θ_0 and θ_1 are the parameters of the single view prediction models. Any probabilistic classification model for the single view models can be chosen, as long as sample-weights can be incorporated in such model. In our case, we arbitrarily chose a Gaussian process (GP) Classifier, with a Bernoulli likelihood [11] using a publicly available implementation [10]; but the framework allows the use of any classification method that estimates probabilities.

To ensure smoothness of $\mathbf{w}$ from $\mathbf{X_0}$, we use a GP prior:

$$P(\mathbf{w}|\mathbf{X_0}) \sim MVN(\mathbf{0}, \mathbf{K_{X_0}}), \quad (2)$$

where MVN is the density function of the multivariate normal, and $\mathbf{K_{X_0}}$ is the classical kernel matrix with a Gaussian kernel.

2.3   Optimisation

After including the prior on w^i and the parameters of the classification models θ_0 and θ_1, the posterior log-probability model for the full dataset is:

$$\log P = \sum_i \log \left(w^i P(Y^i|X_0^i, \theta_0) + (1 - w^i)P(Y^i|X_1^i, \theta_1) \right) + \log P(\mathbf{w}|\mathbf{X_0}). \quad (3)$$

We optimise a lower bound L of the expression obtained by applying Jensen's inequality on the logarithm:

$$\log P \geq \sum_i w^i \log P(Y^i|X_0^i, \theta_0) + (1 - w^i) \log P(Y^i|X_1^i, \theta_1) + \log P(\mathbf{w}|\mathbf{X_0}) = L.$$

(4)

As direct joint optimisation of all the parameters is troublesome and leads to instability, we optimise iteratively the classification model and the combination weights until convergence, until the weights $\mathbf{w}$ of two consecutive iterations are not significantly different. The optimisation of the θ parameters of the classification model is:

$$\max_{\theta_0, \theta_1} \log(L) = \max_{\theta_0} \sum_i w^i \log P(Y^i|X_0^i, \theta_0)$$
$$+ \max_{\theta_1} \sum_i (1 - w^i) \log P(Y^i|X_1^i, \theta_1)) + \sum_i \log(P(w^i)),$$

(5)

which corresponds to two standard sample-weighted classification problems that can be solved using classical classifiers.

The weights $\mathbf{w}$ will be estimated by maximising the joint probability, together with the prior. Putting everything together, we can see that L has terms that do not include the w^i, and therefore do not affect the optimisation. We can pull them out of the optimisation as a term C:

$$\max_{\mathbf{w}} \log(L) = \max_{\mathbf{w}} \sum_i w^i (\log P(Y^i|X_0^i, \theta_0) - \log(P(Y^i|X_1^i, \theta_1)) - \mathbf{w}^t \mathbf{K_{X_0}^{-1}} \mathbf{w}/2 + C, \quad (6)$$

where $C = -d/2 \log(2\pi) - d/2 \log(\det \mathbf{K_{X_0}}) + \log(P(Y^i|X_1^i, \theta_1))$ is independent of $\mathbf{w}$ and does not need to be considered for optimisation. The previous equation corresponds to a quadratic optimisation problem for which the exact solution can be computed.

2.4   Prediction

The previous formulation needs both $\mathbf{X_0}$ and $\mathbf{X_1}$ for all samples to compute the label probability. However, we would like to obtain some of the predictions without having to use $\mathbf{X_1}$, thus avoiding a costly acquisition. For that, we will fix a threshold k for the weights w^i (which can be computed from $\mathbf{X_0}$ only), and only use $\mathbf{X_1}$ when the predicted value of w^i is below that threshold. A higher value of the threshold will result in lower usage of the second level data. The estimated probability is as follows:

$$P(Y^i|X_0^i, X_1^i) = \begin{cases} P(Y^i|X_0^i) & w^i \geq k, \\ w^i P(Y^i|X_0^i) + (1 - w^i) P(Y^i|X_1^i) & w^i < k. \end{cases} \quad (7)$$

Note that since we have imposed a GP prior over the w^i, these weights can be estimated from X_0^i only.

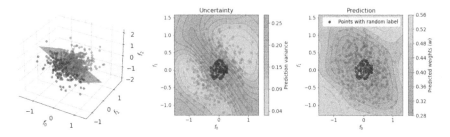

Fig. 1. Scatter plots with the synthetic dataset. We can see that uncertainty (left) is concentrated in the decision boundary $x = -y$, while the weight (right) is also contained near the decision boundary, but excludes the points for which the label is random (black).

3 Results

3.1 Synthetic Experiment

We first verify our framework using a synthetic dataset. This dataset contains two levels of features ($\mathbf{X_0}$ and $\mathbf{X_1}$) and a binary label Y. The label is determined by $\mathbf{X_1}$ and some noise, but to create the situation that not all information can be well predicted from $\mathbf{X_1}$, we set the label of the points of a particular region in the decision boundary of $\mathbf{X_1}$ completely at random. The expected result is that our algorithm gives high weights w^i near the decision boundary, except in the area that the labels are completely random. The data generation is as follows, where $\mathcal{N}$ and Ber denote the normal and Bernoulli distributions respectively:

$$(f_0, f_1, f_2) \sim \mathcal{N}(0,1) \times \mathcal{N}(0,1) \times \mathcal{N}(0,1),$$

$$l \sim \begin{cases} Ber(0.5) & f_0^2 + f_1^2 < 0.3^2, \\ Ber(sigmoid(f_0 + f_1 + f_2)) & \text{otherwise.} \end{cases} \quad (8)$$

The low and high complexity views are defined as the first two coordinates ($\mathbf{X_0} = (f_0, f_1)$) and the three coordinates ($\mathbf{X_1} = (f_0, f_1, f_2)$) respectively. Figure 1 shows the result of applying the hierarchical model to these synthetic data, comparing the predicted weight $\mathbf{w}$ to the model uncertainty, a classical method in active learning/Bayesian optimisation to decide which samples to annotate/acquire. We can observe that the uncertainty is concentrated in the decision boundary, but is not affected by the central area with no information, while the weight submodel predicting the probability of improvement correctly detects that labels in the central area are random and therefore no further acquisition would be beneficial.

3.2 Real Dataset

Using the real dataset, we tested the basic single-view classification models for the echocardiographic measurements and shape embedding only, an average combination of the previous single-view models, and the combination of the

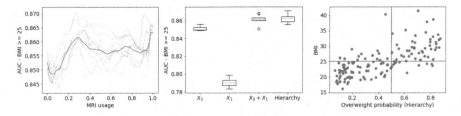

Fig. 2. The *left plot* depicts the AUC for different amounts of MRI data used. Transparent lines are the different bootstrap cross-validations, and the blue line is the mean trace. The maximal AUC is attained at 30%. The *central figure* is a boxplot (orange lines depict the median, the boxes stand for the interquantile range) of the AUC for different models. We can see that there is a small improvement for considering both X_0 and X_1 simultaneously compared to using only X_0 or X_1. This improvement is similar, either when the full population is used, or just a fraction of it. The *right plot* shows the correlation between the probability scores and the raw BMI value. (Color figure online)

two spaces using our weighting approach. 20-fold cross-validation splits were repeated over 10 different seeds to obtain more statistically-stable results, and we compared the area under curve (AUC) of predicting elevated BMI. We also tested the version of our approach which did not use all the X_1^i, with different values for the threshold k in Eq. 7. For each value, we report the AUC and the fraction of weights w^i that fell below the threshold, thus the number of times that X_1 data were used. This AUC as a function of X_1 can be found in Fig. 2, along with the AUC of the different models.

An interesting result is that MRI shapes (X_1) had lower prediction power than simple echocardiographic measurements (X_0). This can be explained as obesity has a global effect on the heart, then just a few measurements are enough to characterise the changes in size and LV hypertrophy related to obesity, so there is little advantage in using shape models instead of simple measurements. However, the combination of both models, either by simple addition or by sample-weights, showed an improvement of prediction power, indicating benefit on the combination of both modalities. For achieving this improvement, it is not necessary to acquire the MRI for the full population, but it could be achieved with approximately 30% of the cases. In the right side of Fig. 2 we can see the correlation between the predicted probability of the hierarchical model, and the BMI. We can see that they share a fair correlation ($\rho = 0.70$), showing that our classification model correctly assigns higher probability to the more obese individuals.

In Fig. 3a, we show the probability contours of the BMI model. The BMI models shows a nice separation between the two classes, with overweights being mostly on the right, and slim individuals on the left. The black line represents the decision boundary of the BMI model. We manually selected synthetic representative points of both thin and overweight individuals, that are depicted in the same figure as stars. In the right part of the figure, a spider plot compares

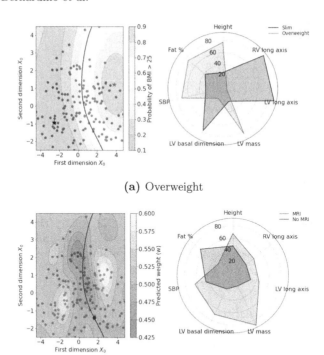

(a) Overweight

(b) Predicted weights

Fig. 3. The subfigures show a scatter plot of the $\mathbf{X}_0$ embedding with contours of the probability of being overweight (top) and of the $\mathbf{w}$ weights indicating the usefulness of the MRI (bottom). The subfigures also contain a spider plot with the characteristics of extreme individuals, selected in the scatter plot as the colored stars. (Color figure online)

the two synthetic representative individuals. The characteristics of the synthetic individuals were interpolated using a GP. We examined the characteristics that our model associated with a high BMI: a higher ventricular mass and smaller ventricle (indexed by BSA), specially in the long axis. While this might seem surprising, the smaller ventricle is mostly due not to a shrinking of the heart but to an increase of the BSA.

We also studied what were the characteristics of the individuals for whom the MRI would add more information. In Fig. 3b, we can see that far from the decision boundary the weight is high, indicating that $\mathbf{X}_0$ is enough for predicting the output. Near the decision boundary, we can observe a differential behaviour: in the upper part the weight is lower than in the lower part. We manually selected two points representative of a low and high w^i, and we studied their characteristics. We found that high weight (i.e. those for which the MRI is most useful) presented a higher LV mass, specially when compared to the ventricular size, and SBP, suggesting hypertrophic remodelling due to pressure loading. Low weight was associated with a higher fat percentage, and a smaller ventricle in both dimensions and mass after indexing by BSA, suggesting a remodelling that is preserving the mass-volume relationship: a change that is mostly size-related. Since the former remodelling involves a change in shape, while the later is mostly in size (which can be already be well approximated by echocardiography), it makes sense that adding MRI shape information helps mostly individuals to belong to the first group [3].

4   Conclusion

We have presented a framework to combine predictions from different models, potentially each trained with different data. The main novelty of this model is its ability to identify the cases that are more likely to improve when adding more data, therefore reducing the number of high-cost data used.

We applied the model to an obesity dataset, and we were able to identify different remodelling patterns in the decision boundary, corresponding to individuals whose diagnosis accuracy would improve after acquiring the MRI data. Our framework reached an accuracy similar to the combination of both models, while only using 30% of the data.

Acknowledgements. This project has been partially funded with support the French ANR (LABEX PRIMES of Univ. Lyon [ANR-11-LABX-0063] and the JCJC project "MIC-MAC" [ANR-19-CE45-0005]), the Erasmus + Programme of the European Union (Framework Agreement number: 2013-0040), "la Caixa" Foundation (LCF/PR/ GN14/10270005, LCF/PR/GN18/10310003), the Instituto de Salud Carlos III (PI14/00226, PI17/00675) integrated in "Plan Nacional de I+D+I" and cofinanciated by ISCIII-Subdirección General de Evaluación and Fondo Europeo de Desarrollo Regional (FEDER) "Una manera de hacer Europa", and AGAUR 2017 SGR grant #1531.

References

1. Obesity: preventing and managing the global epidemic. Report of a WHO consultation. Technical report (2000)
2. Crispi, F., et al.: Exercise capacity in young adults born small for gestational age. JAMA Cardiol. **6**(11), 1308–1316 (2021)
3. Devlin, A., Moore, N., Ostman-Smith, I.: A comparison of MRI and echocardiography in hypertrophic cardiomyopathy. Br. J. Radiol. **72**, 258–264 (1999)
4. Frazier, P.I.: A Tutorial on Bayesian Optimization (2018)
5. Garbi, M., et al.: EACVI appropriateness criteria for the use of cardiovascular imaging in heart failure derived from European National Imaging Societies voting. Eur. Heart J. Cardiovasc. Imaging **17**, 711–721 (2016)
6. Gorriz, M., Carlier, A., Faure, E., Giró i Nieto, X.: Cost-effective active learning for melanoma segmentation. CoRR arXiv:1711.09168 (2017)
7. Grossman, W., Jones, D., McLaurin, L.: Wall stress and patterns of hypertrophy in the human left ventricle. J. Clin. Investig. **56**, 56–64 (1975)
8. Lang, R., Badano, L., Mor-Avi, V., et al.: Recommendations for Cardiac Chamber Quantification by echocardiography in adults. Eur. Heart J. Cardiovasc. Imaging **28**, 1-39.e14 (2015)

9. Li, Y., Yang, M., Zhang, Z.: A survey of multi-view representation learning. IEEE Trans. Knowl. Data Eng. **31**, 1863–83 (2019)

10. Matthews, A., van der Wilk, M., Nickson, T., et al.: GPflow: a Gaussian process library using TensorFlow. J. Mach. Learn. Res. **18**, 1–6 (2017)

11. Nickisch, H., Rasmussen, C.: Approximations for binary Gaussian process classification. J. Mach. Learn. Res. **9**, 2035–78 (2008)

12. Russo, C., Jin, Z., Homma, S., et al.: Effect of obesity and overweight on left ventricular diastolic function. J. Am. Coll. Cardiol. **57**, 1368–74 (2011)

13. Sabuncu, M., Yeo, B., Van Leemput, K., et al.: A generative model for image segmentation based on label fusion. IEEE Trans. Med. Imaging **29**, 1714–29 (2010)

14. Wilson, P., D'Agostino, R., Sullivan, L., et al.: Overweight and obesity as determinants of cardiovascular risk. Arch. Intern. Med. **162**, 1867 (2002)

15. Yang, Y., Loog, M.: Active learning using uncertainty information. In: 2016 23rd International Conference on Pattern Recognition (ICPR), pp. 2646–2651. IEEE (2016)

Neural Angular Plaque Characterization: Automated Quantification of Polar Distribution for Plaque Composition

Hyungjoo Cho[1], Dongmin Choi[2], Hyun-Seok Min[3], Soo-Jin Kang[4], and Hwiyoung Kim[5(✉)]

[1] Seoul National University, Seoul, Republic of Korea
[2] Yonsei University, Seoul, Republic of Korea
[3] Tomocube Inc., Seoul, Republic of Korea
[4] Asan Medical Center, Seoul, Republic of Korea
[5] Yonsei University College of Medicine, Seoul, Republic of Korea
hykim82@yuhs.ac

Abstract. Automated quantification of plaque attenuation and calcification angle within in-vivo coronary arteries using intravascular ultrasound (IVUS) is essential for risk stratification. However, due to the physical limitations of ultrasound, the pixel-level plaque characterization has difficulties in practical application. To overcome these barriers, we propose a novel approach called Neural Plaque Angular Characterizer (NeuPAC), which learns to map an IVUS frame to the polar distribution of plaque composition. This approach inherits the advantages of both labeling and modeling. NeuPAC uses a rough label representing polar information rather than an accurate plaque composition map. NeuPAC directly outputs angular information, which is essential to clinical decision-making. Our empirical results show that NeuPAC performs well in recognizing high-risk coronary lesions for assisting clinicians in real-time.

1 Introduction

Identifying atherosclerotic plaque characteristics using Intravascular Ultrasound (IVUS) is crucial for percutaneous coronary intervention (PCI). Risk stratification for the vessel rupture can be conducted according to the presence and the relative proportion of the attenuated plaque. In calcified plaques that can cause stent underexpansion, treatment strategies are determined based on calcium scoring via calcified plaque characteristics.

Plaque Characterization is the process of identifying plaque characteristics for computer-aided diagnosis. This process aims to reconstruct the plaque composition map for a given IVUS frame by classifying each pixel of that frame to the particular component. To automate it, segmentation methods have been mainly proposed based on machine and deep learning techniques [7,9]. These

© Springer Nature Switzerland AG 2022
E. Puyol Antón et al. (Eds.): STACOM 2021, LNCS 13131, pp. 113–122, 2022.
https://doi.org/10.1007/978-3-030-93722-5_13

models try to learn a mapping from a single IVUS frame to the plaque composition map in a supervised learning fashion. It means that those processes need pixel-level annotations for training a segmentation model.

However, most cases during PCI include atheromatous plaques, which cause the shadow regions on the IVUS frame due to the restriction of acoustic physics [15]. The shadow region is represented as black pixels in the IVUS imaging and does not provide any information about the characteristics of that region (See Fig. 1a). Therefore, it is impossible to label the complete plaque composition map using only IVUS for clinical usage.

To address this problem, cardiologists have developed clinical guidelines based on angle quantification instead of reconstructing plaque composition maps (See Fig. 1b) Following recent studies, a maximal arc of attenuation over 30° on IVUS indicates the presence of thin-cap fibroatheroma, which can contribute to the occurrence of distal embolization and periprocedural myocardial infarction [4]. Calcium scoring also can be based on angle quantification instead of the exact proportion of calcified plaque [3]. In the case of calcified plaque with an arc of more than 90°, this lesion is intermediate calcification requiring high-pressure balloon inflation.

Inspired by these clinician's advances for angle based quantification, we reframe the automated plaque characterization process as the angle-wise classification problem. Our approach, called Neural Plaque Angular Characterizer (NeuPAC), is the unified process for the automated prediction of polar distribution for plaque characterization, including an angle-wise classifier and output calibration. The key idea of NeuPAC is simple; training a deep learning model using an angle-level labeled dataset. NeuPAC allows the user to annotate angle-wise plaque composition per frame efficiently and extracts the corresponding values in the form of a vector for the model training. We call this vector Angular Plaque Composition Vector. Hence NeuPAC directly outputs angular plaque composition vector. For maximize the clinical efficacy reducing the output vibration noise, output calibration is applied in the final stage via Test-Time Augmentation (TTA, [13]).

Because NeuPAC is the first clinically successful method for automated plaque characterization, comparative experiments cannot be performed. Instead, we show the feasibility of the proposed method from the perspective of engineering and clinical performance using a large-scale dataset. In particular, the Pearson correlation scores are greater than 0.9 in most cases of clinical parameters in Fig. 1c for both attenuated and calcified plaques. Our results had been clinically validated and published [1], and this paper describes the methodology in more detail.

2   Neural Plaque Angular Characterizer

In this section, we first formulate the plaque characterization task. Then, we define the angular plaque composition vector for clinical usage. Finally, we introduce the architecture of Neural Plaque Angular Characterizer and the output calibration process, depicted in Fig. 2.

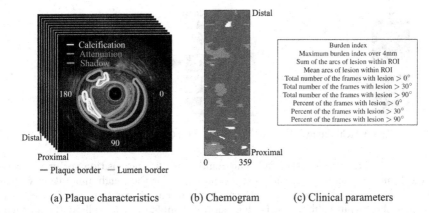

(a) Plaque characteristics   (b) Chemogram   (c) Clinical parameters

Fig. 1. Difference between engineering and clinical goals in Plaque characterization. (a) shows the difficulty of estimating the complete plaque composition map for the severe case. The information needed in PCI is in the form of (b). Cardiologists extract (c) for clinical judgment from (b). The white color and red color indicate calcification and attenuation, respectively. (Color figure online)

2.1 Formulation of Plaque Characterization

Let $\mathbf{x} \in \mathcal{X} \subset \mathbb{R}^{H \times W}$ be the data point of IVUS frame and the corresponding plaque composition map $\mathbf{y} \in \mathcal{Y} \subset \mathbb{R}^{C \times H \times W}$, where (H, W) is the resolution of the IVUS frame and C is the number of plaque types. The plaque characterization model $f^{(m)} : \mathcal{X} \rightarrow \mathcal{Y}$ learns a mapping from $H \times W$-matrix to set of $H \times W$-matrix with $\mathbb{R}^C$ entries. We estimate the performance of $f^{(m)}$ by

$$\mathcal{L}_{\text{map}} := \mathbb{E}_{(\mathbf{x},\mathbf{y}) \sim P(\mathbf{x},\mathbf{y})}[\ell(f^{(m)}(\mathbf{x}), \mathbf{y})] \tag{1}$$

where ℓ is a loss function for segmentation e.g. dice similarity coefficient. Hence one can say $f^{(m)}$ is the pixel-wise classifier.

2.2 Angular Plaque Composition Vector

To extract features from a given IVUS frame $\mathbf{x}^*$ for PCI, one has to train $f^{(m)}$ firstly to predict a plaque composition map $\hat{\mathbf{y}}^*$ which is the output of $f^{(m)}(\mathbf{x}^*)$. Then the transfer function $\mathcal{T} : \mathcal{Y} \rightarrow \mathcal{V}$ is applied to generate the angular plaque composition vector $\mathbf{v}^* \in \mathcal{V} \subset \mathbb{R}^{C \times A}$ according to the $\hat{\mathbf{y}}^*$, where A is the angular resolution. The transfer function $\mathcal{T}$ can be decomposed to the polar coordinate conversion function $\mathcal{P} : \mathbb{R}^{C \times H \times W} \rightarrow \mathbb{R}^{C \times \rho \times A}$ and the pooling function $\mathcal{M} : \mathbb{R}^{C \times \rho \times A} \rightarrow \mathbb{R}^{C \times A}$, such that $\mathcal{T} = \mathcal{P} \circ \mathcal{M}$, where ρ is $\sqrt{(\frac{H}{2})^2 + (\frac{W}{2})^2}$. We estimate the performance of $f^{(m)}$ for the angle-wise classification by

$$\mathcal{L}_{\text{vec}} := \mathbb{E}_{(\mathbf{x},\mathbf{y}) \sim P(\mathbf{x},\mathbf{y})}[\ell(\mathcal{T}(f^{(m)}(\mathbf{x})), \mathcal{T}(\mathbf{y}))] \tag{2}$$

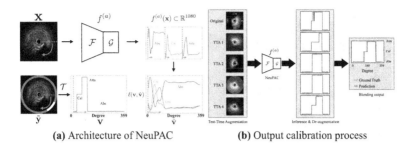

(a) Architecture of NeuPAC   (b) Output calibration process

Fig. 2. (a) Architecture of NeuPAC. The angular plaque composition vector **v** can be derived from the complete plaque composition map **y** or the rough annotation **ỹ**. The white color and red color of **ỹ** indicate calcification and attenuation, respectively; (b) Process of the output calibration. Each generated sample via TTA follows the augmentation process of Sect. 3.2. Cal, Attn, and Abs indicate calcification, attenuation, and absence, respectively. (Color figure online)

2.3 Angle-Wise Classifier

Suppose we have an angular plaque composition vector dataset $\mathcal{D}_a$ denoted by $\mathcal{D}_a := \{(\mathbf{x}, \mathbf{v}) \in \mathcal{X} \times \mathcal{V}\}$. Then we can define the angular plaque characterization model $f^{(a)} : \mathbb{R}^{H \times W} \to \mathbb{R}^{C \times A}$ to directly infer an angular plaque composition vector **v** from input IVUS frame **x**. In this case, we estimate (2) by

$$\mathcal{L}_{\text{vec}} \approx \mathbb{E}_{(\mathbf{x},\mathbf{v}) \sim P(\mathbf{x},\mathbf{v})}[\ell(f^{(a)}(\mathbf{x}), \mathbf{v})] \tag{3}$$

Precisely, cardiologists roughly make annotation images **ỹ** corresponding to the IVUS frames by circumferentially drawing arc lines rather than complete plaque composition map. These annotation images are converted to **v** via transfer function $\mathcal{T}$.

$$\mathbf{v} := \mathcal{T}(\tilde{\mathbf{y}}) \tag{4}$$

We emphasize that labeling **ŷ** is easier and more efficient than labeling **y** to derive **v**.

Angle-wise classifier $f^{(a)}$ can be represented as a composition of two function (See Fig. 2a) Let $\mathcal{F} : \mathbb{R}^{H \times W} \to \mathbb{R}^d$ be the feature extractor of $f^{(a)}$ and $\mathcal{G} : \mathbb{R}^d \to \mathbb{R}^{C \times A}$ be the linear layer, where d is the channel of last feature map. So the predicted plaque probability for each angle **v̂** is defined as follows:

$$\hat{\mathbf{v}} := f^{(a)}(\mathbf{x}) = \mathcal{G}(\mathcal{F}(\mathbf{x})) \tag{5}$$

Under this condition, we can interpret $f^{(a)}$ as the angle-wise classifier for a given IVUS frame.

2.4 Output Calibration

Although plaque components at adjacent angles or frames are strongly correlated in actual vessels, each element of the NeuPAC's output is independent. Due to the nature of this, the output vibration, which is a class of a prediction error that the prediction is repeatedly changed along with the multiple adjacent angles or frames, sometimes occurs. This phenomenon is very harmful for predicting clinical parameters during PCI. To calibrate the output vibration, we apply TTA, as shown in Fig. 2b. During the inference phase, we randomly augment for the input IVUS frame. After inference for all augmented images including the original input, de-augmentation is conducted to remove the effect of transformation operations. Then, outputs are blended through the mean operation to minimize the output vibration and maximize the prediction power.

3   Experimental Results

In this section, we empirically validate our proposed NeuPAC framework for the plaque characterization power and clinical efficacy. We first describe the experimental setup, including dataset, implementation details, and evaluation metrics. Next, we present the prediction power of the NeuPAC and the visualization of how it works. Finally, We demonstrate the effectiveness of our framework via several ablation studies.

3.1   Dataset

IVUS images were obtained from a consecutive series of 598 angina patients at Asan Medical Center, Seoul, Korea, using 3.2 French catheters (Boston Scientific/SCIMED, Minneapolis, MN) consisting of a rotating 40 MHz transducer.

All patients enrolled in this study had at least one coronary lesion and did not undergo stent treatment for that lesion. A region of interest (ROI) was defined as the segment from the ostium to a point 10 mm distal to the lesion (the mean lengths of the ROIs is about 60 mm). A total of 87632 frames were selectively labeled at every 24th frame from these ROIs. The ratio between the frame with plaque and the frame without plaque is about 3:1.

3.2   Implementation Details

NeuPAC's output is a size of 1080 where each element accounts for a prediction whether plaques exist at a particular angle. For example, the output from the 360th to the 720th is for the angle-wise presence of calcification plaque. All frames are resized to 512×512 and EfficinetNet-B3 [12] is used as the backbone. We adopt the Adam [5] optimizer with a initial learning rate of 0.0001 and a cosine annealing strategy [8]. The batch size is set to 24 and the model is trained for 100 epochs. The dataset are split into train, validation, test and unlabeled set (57,588, 14,371, 15,673 and 1,046,124)

Several tricks are conducted to improve the model performance. There is a large amount of variance during the image acquisition process of IVUS. To model this variance, we employ the data augmentation strategy optimized to the IVUS frame characteristics. The other most severe problem of this study is the class imbalance problem, which is that the number of frames with any plaque is significantly lower than without plaque. At the frame level, the number of angles with any plaque is much less than without plaque. To overcome the imbalance problem, we conduct angle-level CutMix, Labeling Smoothing and Pseudo Labeling.

Augmentation. We modify RandAug [2] for our data augmentation strategy. We divide the data augmentation method into two types; transformation and noise. Note that transformation methods requiring label change consist of paired operations for cartesian and polar coordinates. For each mini-batch, one of the transformation methods and one of the noise methods are applied in order.

Angle-Level CutMix. We change the original version of CutMix [14] slightly for our angle-wise prediction framework. Due to the characteristics of intravascular imaging, we convert randomly selected two images into polar coordinates. Notably, only an IVUS image with attenuation or calcification is sampled for the class imbalance problem. The size of mask is $\rho \times \alpha$, where α is randomly generated from $[0, 360]$. After mixing, the newly generated image is converted back to the cartesian coordinates.

Labeling Smoothing. Despite the performance improvement, CutMix creates the artifact at the joint region. This artifact cannot occur in the real dataset and can induce performance bound for angle-wise classification. One can harmonize two images to look naturally for this artifact. Instead, we simply apply Labeling Smoothing technique [11] targeting labels rather than inputs. Labeling smoothing relieves performance limitations by the artifact on joint region.

Pseudo Labeling. Pseudo Labeling [6] is revised for the imbalance issue as well, by changing the sampling condition for the unlabeled dataset. Unlike the original method of sampling the training set based on the maximum class score for unlabeled data, we set the sampling criteria based on the presence of attenuation or calcification for the pseudo label.

Table 1. Performances of NeuPAC at frame levels.

Backbone	Plaque	Sensitivity (%)	Specificity (%)	PPV (%)	NPV (%)	ACC (%)
EfficientNet-B3	Calcified	86.13	97.21	82.05	98.33	96.11
	Attenuated	80.28	95.94	80.41	96.10	93.14

3.3   Evaluation Metrics

To evaluate from the engineering performance to the clinical efficacy, the number of metrics are measured. For the frame level validation, sensitivity, specificity,

PPV, and NPV for presence of each IVUS calcification and attenuation are performed. To evaluate the clinical efficacy, we calculate the clinical parameters from the predicted plaque composition vectors following [1]. These clinical parameters are evaluated by pearson correlation coefficient (PCC).

3.4 Plaque Characterization Performance

Table 1 and 2 validate that the NeuPAC has efficacy in helping clinicians make decisions as well as micro level accuracy. These results are based on the NeuPAC with augmentation, cutmix, label smoothing, pseudo labeling, and TTA. Note that IVUS attenuation is more challenging than IVUS calcification for both clinician and model.

Table 2. Clinical efficacy of NeuPAC. The results over 90% based on PCC are indicated in bold.

Clinical parameter	Calcification (%)	Attenuation (%)
Burden index	**95.24**	89.41
Maximum burden index over 4 mm	**90.59**	82.14
Sum of the arcs of lesion within ROI	**96.53**	**93.26**
Mean arcs of lesion within ROI	**95.22**	89.42
Total number of the frames with lesion >0°	**95.44**	**95.27**
Total number of the frames with lesion >30°	**96.11**	**93.32**
Total number of the frames with lesion >90°	**90.03**	85.92
Percent of the frames with lesion >0°	**94.89**	**93.92**
Percent of the frames with lesion >30°	**96.06**	**91.31**
Percent of the frames with lesion >90°	**90.00**	76.92

In addition, we demonstrate its internal representation to understand what information NeuPAC uses via gradCAM [10]. Figure 3 shows that the NeuPAC tends to pay attention on the relatively high intensity region for IVUS calcification, and the shadow region or the area where the signal gradually disappears for IVUS attenuation. The activation map around zero degrees seems to be disconnected because predictions are made on polar maps.

3.5 Ablation Study

We conducted an ablation study to verify each algorithm of Sect. 3.2 works well. Each score of Table 3 is angle-level DCS. Following the result, every algorithm except label smoothing contribute the performance improvement. The label smoothing works well only with cutmix. This shows that our hypothesis the label smoothing can reduce the paste artifact effect is true.

IVUS frame   GT & Prediction   Calcification   Attenuation

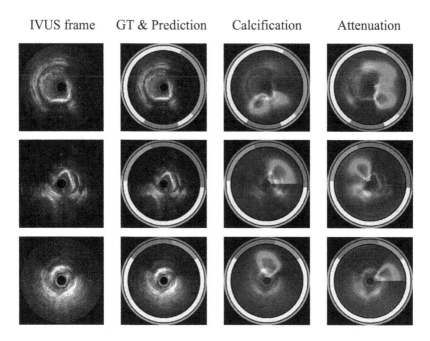

Fig. 3. Visualization of class activation map. In columns 2 to 4, the outer color line is the GT and the inner is the predicted result. The colors purple, red, and green indicate calcification, attenuation, and absence, respectively. (Color figure online)

Table 3. Result of ablation study. Cal, Attn, Abs, and Avg indicate calcification, attenuation, absence and average, respectively. The bold results are the best scores for each metric.

Augmentation	CutMix	Labeling smoothing	Pseudo labeling	TTA	Cal (%)	Attn (%)	Abs (%)	Avg (%)
X	X	X	X	X	72.57	66.61	99.10	79.43
X	O	X	X	X	74.89	68.67	99.17	80.91
O	X	X	X	X	75.18	69.71	99.18	81.36
O	O	X	X	X	75.67	69.75	99.18	81.53
O	O	O	X	X	76.60	70.98	99.20	82.26
O	X	O	O	X	73.76	68.83	99.20	80.60
O	O	O	O	X	78.01	72.62	99.30	83.31
O	**O**	**O**	**O**	**O**	**79.07**	**74.47**	**99.30**	**84.28**

3.6 The Effectiveness of Output Calibration

We evaluate the output calibration qualitatively as well as the performance improvement. We have visualized the results of NeuPAC with TTA comparing

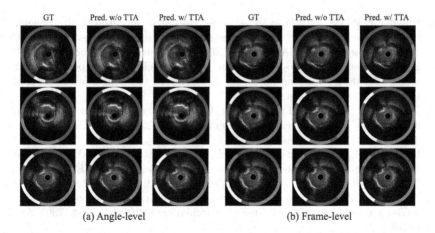

| GT | Pred. w/o TTA | Pred. w/ TTA | GT | Pred. w/o TTA | Pred. w/ TTA |

(a) Angle-level   (b) Frame-level

Fig. 4. Visualization of the effectiveness of output calibration. Pred. w/o TTA indicates the results of NeuPAC without TTA and Pred. w/ TTA indicates the results of NeuPAC with TTA.

with NeuPAC without TTA. Figure 4 shows that the output calibration significantly reduces output vibration in terms of angle and frame level.

4   Conclusion

We introduced a new paradigm of automated plaque characterization for clinical usage. Each component of it improves the model performance from the perspective of both engineering and clinical assistance. We also try to inspect NeuPAC for better understanding via gradCAM visualization. As future research, one can adapt vision transformer, one of the most promising architecture for vision problems, to NeuPAC as a new backbone.

References

1. Cho, H., et al.: Intravascular ultrasound-based deep learning for plaque characterization in coronary artery disease. Atherosclerosis **324**, 69–75 (2021)
2. Cubuk, E.D., Zoph, B., Shlens, J., Le, Q.V.: Randaugment: practical automated data augmentation with a reduced search space. In: Proceedings of the IEEE/CVF Conference on Computer Vision and Pattern Recognition (CVPR) Workshops, June 2020
3. Ehara, S., et al.: Spotty calcification typifies the culprit plaque in patients with acute myocardial infarction: an intravascular ultrasound study. Circulation **110**(22), 3424–3429 (2004)
4. Kang, S.-J., et al.: Multimodality imaging of attenuated plaque using grayscale and virtual histology intravascular ultrasound and optical coherent tomography. Catheter. Cardiovasc. Interv. **88**(1), E1–E11 (2016)

5. Kingma, D.P., Ba, J.: Adam: a method for stochastic optimization. In: Bengio, Y., LeCun, Y. (eds.) 3rd International Conference on Learning Representations, ICLR 2015, San Diego, CA, USA, 7–9 May 2015, Conference Track Proceedings (2015)
6. Lee, D.-H., et al.: Pseudo-label: the simple and efficient semi-supervised learning method for deep neural networks. In: Workshop on Challenges in Representation Learning, ICML, vol. 3, p. 896 (2013)
7. Lee, J., et al.: Automated plaque characterization using deep learning on coronary intravascular optical coherence tomographic images. Biomed. Opt. Express 10, 6497 (2019)
8. Loshchilov, I., Hutter, F.: SGDR: stochastic gradient descent with warm restarts. In: 5th International Conference on Learning Representations, ICLR 2017, Toulon, France, 24–26 April 2017, Conference Track Proceedings. OpenReview.net (2017)
9. Nishi, T., et al.: Deep-learning-based intravascular ultrasound segmentation for the assessment of coronary artery disease. Circulation 142(Suppl_3), A17365–A17365 (2020)
10. Selvaraju, R.R., Cogswell, M., Das, A., Vedantam, R., Parikh, D., Batra, D.: Gradcam: visual explanations from deep networks via gradient-based localization. In: Proceedings of the IEEE International Conference on Computer Vision, pp. 618–626 (2017)
11. Szegedy, C., Vanhoucke, V., Ioffe, S., Shlens, J., Wojna, Z.: Rethinking the inception architecture for computer vision. In: Proceedings of the IEEE Conference on Computer Vision and Pattern Recognition (CVPR), June 2016
12. Tan, M., Le, Q.: EfficientNet: rethinking model scaling for convolutional neural networks. In: Chaudhuri, K., Salakhutdinov, R. (eds.) Proceedings of the 36th International Conference on Machine Learning. Proceedings of Machine Learning Research, vol. 97, pp. 6105–6114. PMLR (2019)
13. Wang, G., Li, W., Aertsen, M., Deprest, J., Ourselin, S., Vercauteren, T.: Aleatoric uncertainty estimation with test-time augmentation for medical image segmentation with convolutional neural networks. Neurocomputing 338, 34–45 (2019)
14. Yun, S., Han, D., Oh, S.J., Chun, S., Choe, J., Yoo, Y.: Cutmix: regularization strategy to train strong classifiers with localizable features. In: International Conference on Computer Vision (ICCV) (2019)
15. Zhang, X., McKay, C.R., Sonka, M.: Tissue characterization in intravascular ultrasound images. IEEE Trans. Med. Imaging 17(6), 889–899 (1998)

Simultaneous Segmentation and Motion Estimation of Left Ventricular Myocardium in 3D Echocardiography Using Multi-task Learning

Kevinminh Ta[1]([✉]), Shawn S. Ahn[1], John C. Stendahl[2], Jonathan Langdon[4], Albert J. Sinusas[2,4], and James S. Duncan[1,3,4]

[1] Department of Biomedical Engineering, Yale University, New Haven, CT, USA
kevinminh.ta@yale.edu
[2] Department of Internal Medicine, Yale University, New Haven, CT, USA
[3] Department of Electrical Engineering, Yale University, New Haven, CT, USA
[4] Department of Radiology and Biomedical Imaging, Yale University, New Haven, CT, USA

Abstract. Motion estimation and segmentation are both critical steps in identifying and assessing myocardial dysfunction, but are traditionally treated as unique tasks and solved as separate steps. However, many motion estimation techniques rely on accurate segmentations. It has been demonstrated in the computer vision and medical image analysis literature that both these tasks may be mutually beneficial when solved simultaneously. In this work, we propose a multi-task learning network that can concurrently predict volumetric segmentations of the left ventricle and estimate motion between 3D echocardiographic image pairs. The model exploits complementary latent features between the two tasks using a shared feature encoder with task-specific decoding branches. Anatomically inspired constraints are incorporated to enforce realistic motion patterns. We evaluate our proposed model on an *in vivo* 3D echocardiographic canine dataset. Results suggest that coupling these two tasks in a learning framework performs favorably when compared against single task learning and other alternative methods.

Keywords: Echocardiography · Motion estimation · Segmentation · Multi-task learning

1 Introduction

Reductions in myocardial blood flow due to coronary artery disease (CAD) can result in myocardial ischemia or infarction and subsequent regional myocardial dysfunction. Echocardiography provides a non-invasive and cost-efficient tool for clinicians to visually analyze left ventricle (LV) wall motion and assess regional dysfunction in the myocardium. However, such a qualitative method of analysis is subjective by nature and, as a result, is prone to high intra- and interobserver variability.

© Springer Nature Switzerland AG 2022
E. Puyol Antón et al. (Eds.): STACOM 2021, LNCS 13131, pp. 123–131, 2022.
https://doi.org/10.1007/978-3-030-93722-5_14

Motion estimation algorithms provide an objective method for characterizing myocardial contractile function, while segmentation algorithms assist in localizing regions of ischemic tissue. Traditional approaches treat these tasks as unique steps and solve them separately, but recent efforts in the medical image analysis and computer vision fields suggest that these two tasks may be mutually beneficial when optimized simultaneously [1–3].

In this paper, we propose a multi-task deep learning network to simultaneously segment the LV and estimate its motion between time frames in a 3D echocardiographic (3DE) sequence. The main contributions of this work are as follows: 1) We introduce a novel multi-task learning architecture with residual blocks to solve both 3D motion estimation and volumetric segmentation using a weight sharing feature encoder with task-specific decoding branches; 2) We incorporate anatomically inspired constraints to encourage realistic cardiac motion estimation; 3) We apply our proposed model to 3DE sequences which typically provides additional challenges over Magnetic Resonance (MR) and Computed Tomography (CT) due to lower signal-to-noise ratio as well as over 2DE due to higher dimensionality and smaller spatial and temporal resolution.

2   Related Works

Classic model-based segmentation methods such as active shape and appearance models usually require *a priori* knowledge or large amounts of feature engineering to achieve adequate results [4,5]. In recent years, data-driven deep learning based approaches have shown promising results, but still suffer challenges due to inherent ultrasound image properties such as low signal-to-noise ratio and low image contrast [4,6]. This becomes increasingly problematic as cardiac motion estimation approaches often rely on accurate segmentations to act as anatomical guides for surface and shape tracking or for the placement of deformable grid points [7,8]. Errors in segmentation predictions can be propagated to motion estimations. While it is conceivable that an expert clinician can manually segment or adjust algorithm predictions prior to the motion estimation step, this is a tedious and infeasible workaround. Several deep learning based approaches have been shown to be successful in estimating motion in the computer vision field, but the difficulty in obtaining true LV motion in clinical data makes supervised approaches challenging. Unsupervised approaches which seek to maximize intensity similarity between image pairs have been successful in MR and CT, however there are still limited applications in 3DE [9–12].

In recent years, efforts have been made to combine the tasks of motion estimation and segmentation. Qin *et al.* [2] proposed a Siamese-style joint encoder network using a VGG-16 based architecture that demonstrates promising results when applied to 2D MR cardiac images. The work done in [13] adapts this idea to 2D echocardiography by adopting a feature bridging framework [1] with anatomically inspired constraints. This is further expanded to 3DE in [3] through the use of an iterative training approach where results from one task influence the training of the other [14]. In this work, we propose an alternative novel framework for combining motion estimation and segmentation in 3DE that uses a

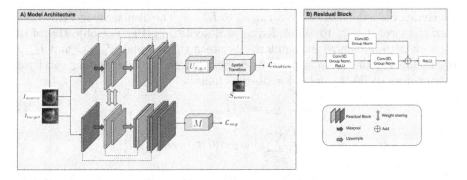

Fig. 1. The proposed network and its components. (A) Motion estimation and segmentation tasks are coupled in a multi-task learning framework. (B) An overview of the residual block.

shared feature encoder to exploit complementary latent representations in the data, which the iterative style of [3] is not capable of doing. In addition to being able to estimate 3D motion and predict volumetric LV segmentations, our model further differs from [2] through the implementation of a 3D U-Net-style architecture [15] with residual blocks [21] for the encoding and decoding branches as opposed to a VGG-16 and FCN architecture (Fig. 1).

3   Methods

3.1   Motion Estimation Branch

Motion estimation algorithms aim to determine the voxel-wise displacement between two sequential images. Given a source image I_{source} and a target image I_{target}, motion estimation algorithms can be described by their formulation of the mapping function F such that $F(I_{source}, I_{target}) \rightarrow U_{x,y,z}$ where $U_{x,y,z}$ is the displacement along the x-y-z directions. Supervised deep learning formulations of F seek to directly learn the regression between the image pairs and ground truth displacement fields. However due to the scarcity of ground truth in cardiac motion, the motion branch of our network is designed and trained in an unsupervised manner similar to the framework presented in [10], which utilizes a spatial transformation to maximize a similarity metric between a warped source image and a target image.

Our proposed motion branch is comprised of a 3D U-Net inspired architecture with a downsampling analysis path followed by an upsampling synthesis path [15,16]. Skip connections are used to branch features learned in the analysis path with features learned in the synthesis path. Our model also uses residual blocks in order to improve model performance and training efficiency [21]. The downsampling analysis path serves as a feature encoder which shares its weights with the segmentation branch. The input to the motion branch is a pair of 3D images, I_{source} and I_{target}. The branch outputs a displacement field $U_{x,y,z}$

which describes the motion from I_{source} to I_{target}. The displacement field is then used to morph I_{source} to match I_{target} as described in [10]. The objective of the network is to maximize the similarity between the morphed I_{source} and I_{target} by minimizing the mean squared error between each corresponding voxel p in the two frames. This can be described as follows:

$$I_{morphed} = \mathcal{T}(I_{source}, U_{x,y,z}) \tag{1}$$

$$\mathcal{L}_{sim} = \frac{1}{\Omega} \sum_{p \in \Omega} (I_{target}(p) - I_{morphed}(p))^2 \tag{2}$$

where $\Omega \subset \mathbb{R}^3$ and $\mathcal{T}$ describes the spatial transforming operator that morphs I_{source} using $U_{x,y,z}$.

Anatomical Constraints. In order to enforce realistic cardiac motion patterns, we incorporate some anatomical constraints. Cardiac motion fields should be generally smooth. That is, there should be no discontinuities or jumps within the motion field. To discourage this behavior, we penalize the L^2-norm of the spatial derivatives in a manner similar to [8,10] as follows:

$$\mathcal{L}_{smooth} = \frac{1}{\Omega} \sum_{p \in \Omega} \|\nabla U_{x,y,z}(p)\|_2^2 \tag{3}$$

Additionally, it is expected that the LV myocardium preserves its general shape through time. In order to enforce this notion, a shape constraint is added which morphs the manual segmentation of I_{source} using $U_{x,y,z}$ and compares it against the manual segmentation of I_{target} in a manner similar to [11] as follows:

$$\mathcal{L}_{shape} = (1 - \frac{2 \mid S_{target} \cap \mathcal{T}(S_{source}, U_{x,y,z}) \mid}{\mid S_{target} \mid + \mid \mathcal{T}(S_{source}, U_{x,y,z}) \mid}) \tag{4}$$

where S_{source} and S_{target} are the manual segmentations of I_{source} and I_{target}, respectively. We can then define the full motion loss as:

$$\mathcal{L}_{motion} = \lambda_{sim}\mathcal{L}_{sim} + \lambda_{smooth}\mathcal{L}_{smooth} + \lambda_{shape}\mathcal{L}_{shape} \tag{5}$$

3.2   Segmentation Branch

The objective of segmentation is to assign labels to voxels in order to delineate objects of interest from background. The segmentation branch of our proposed model follows the same 3D U-Net architectural style of the motion branch [15]. A downsampling analysis path shares weights with the motion branch and separates to a segmentation-specific upsampling synthesis path. The goal of this branch is to minimize a combined dice and binary cross entropy loss between predicted segmentations and manual segmentations as follows:

$$\mathcal{L}_{dice} = (1 - \frac{2 \mid S \cap M \mid}{\mid S \mid + \mid M \mid}) \tag{6}$$

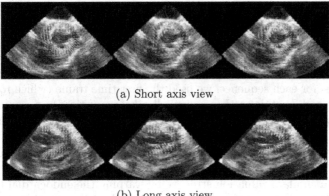

(a) Short axis view

(b) Long axis view

Fig. 2. Motion estimations from end-disatole to end-systole for healthy baseline canines. From left to right for both (a) and (b): *optical flow, motion only, proposed model.*

$$\mathcal{L}_{bce} = -(y_i log(P_i) + (1 - y_i)(log(1 - P_i)))   \tag{7}$$

$$\mathcal{L}_{seg} = \lambda_{dice}\mathcal{L}_{dice} + \lambda_{bce}\mathcal{L}_{bce}   \tag{8}$$

where M is the model prediction of the LV mask, S is the manually traced ground truth mask. y is a binary indicator for if a voxel is correctly labeled, and P is the predicted probability a voxel is part of the LV mask.

3.3   Shared Feature Encoder

Multi-task learning is a popular method for combining closely related tasks in a unified framework. In this work, we adopt a soft parameter sharing approach [18]. Inspired by the success of [2] on 2D MR images, we employ a similar Siamese-style model (using a 3D U-Net style architecture with residual blocks) for feature encoding by sharing the weights between the downsampling analysis path of the motion estimation and segmentation branches. These shared weights are then concatenated to each task-specific upsampling synthesis path (feature decoding), thereby using features learned in both tasks to influence the final output. In this way, the model allows each branch to exploit the complementary latent representations for each task during training. Both branches are trained simultaneously and optimized using a composite loss function, weighted by α and β:

$$\mathcal{L}_{total} = \alpha\mathcal{L}_{motion} + \beta\mathcal{L}_{seg}   \tag{9}$$

4   Experiments and Results

4.1   Datasets and Evaluation

In vivo studies were conducted on 8 anesthetized open-chested canines, which were each imaged under 5 conditions: healthy baseline, mild LAD stenosis, mild

LAD stenosis with low-dose dobutamine (5μ g/kg/min), moderate LAD stenosis, and moderate LAD stenosis with low-dose dobutamine [19]. Images were captured using a Philips iE33 scanner with an X7-2 probe. In total, we had 40 3D echocardiographic sequences which we then sampled into image pairs to be inputted to the network. Image pairs were sampled in a 1-to-Frame manner. In other words, for each sequence, we used the first time frame (which roughly corresponds to end-diastole) as I_{source} and all subsequent time frames as I_{target}. All experiments conducted in support of this work were approved under Institutional Animal Care and Use Committee policies.

Due to the scarcity of true cardiac motion, quantitative evaluation of motion performance is often done by comparing propagated labels [25]. In this work, we employ a similar evaluation strategy by warping the endocardial (endo) and epicardial (epi) contours of the source mask and evaluating the mean contour distance (mcd) from the expected (manually traced) target mask contours. We compare our model against a conventional motion estimation approach (denoted as *Optical flow* - as formulated in [20]) as well as a state-of-the-art deep learning based model (denoted as *Motion only* - which resembles the VoxelMorph framework described in [10]). Results displayed on Fig. 2 and Table 1a suggest that the proposed model performs favorably against the alternative methods. Wilcoxon rank sum test indicates significant increase in performance ($p < 0.05$) of the proposed model [24].

To evaluate the segmentation predictions, we compare the Jaccard index and Hausdorff distance (HD) between model predictions and manually traced segmentations [22]. We evaluate the performance of the proposed model against a segmentation specific branch without feature sharing (denoted as *Segmentation only* - which resembled the 3D U-Net architecture described in [15]). Additionally, since both the segmentation and motion branches of the proposed model produce segmentation predictions as either part of their main task or in support of the shape constraint, we report the average values from these predictions. Results displayed on Fig. 3 and Table 1b suggest that the proposed model performs favorably in predicting LV myocardium segmentation. Wilcoxon rank sum test indicates significant improvement in performance ($p < 0.05$) of the proposed for both Jaccard and HD over the segmentation only model [24].

4.2   Implementation Details

Of the 8 canine studies, we set aside 1 entire study, which consisted of all 5 conditions, for testing. Of the 7 remaining studies, we randomly divided the image pairs such that 90% were used for training and 10% were used for validation and parameter searching. The acquired images were resampled from their native ultrasound resolutions so that each voxel corresponded to 1 mm^3. During training and testing, the images were further resized to 64^3 due to computational limitations. Images were resized back to 1 mm^3 resolution prior to evaluation. An Adam optimizer with a learning rate of $1e^{-4}$ was used. The model was trained with a batch size of 1 for 50 epochs. Due to the small batch size, group normalization was used in place of standard batch normalization [23]. Hyperparameters and loss weights

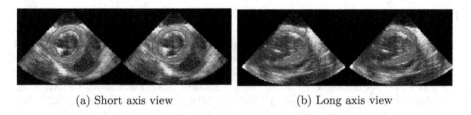

| (a) Short axis view | (b) Long axis view |

Fig. 3. Predicted left ventricular masks. From left to right for both (a) and (b): *segmentation only, proposed model.*

Table 1. Quantitative evaluation. (a) Lower mcd indicates better performance. (b) Higher Jaccard (up to 1) and lower HD indicates better performance

Methods	Endo (mm)	Epi (mm)
Optical flow	0.78 ± 0.41	1.89 ± 1.37
Motion only	0.66 ± 0.31	1.62 ± 1.20
Proposed model	0.49 ± 0.19	0.61 ± 0.32

(a) Motion estimation: mean contour distances (mcd)

Methods	Jaccard	HD (mm)
Seg only	0.61 ± 0.06	4.00 ± 0.52
Proposed model	0.71 ± 0.04	3.65 ± 0.29

(b) Segmentation: Jaccard index and Hausdorff distance (HD)

were empirically selected. The model was developed using PyTorch and trained on an NVIDIA GeForce RTX 2080 Ti. Pre and post processing were done using MATLAB 2019b.

5   Conclusion

In this paper, we proposed a novel multi-task learning architecture that can simultaneously estimate 3D motion and predict volumetric LV myocardial segmentations in 3D echocardiography. This is accomplished through a weight-sharing feature encoder that is capable of learning latent representations in the data that is mutually beneficial to both tasks. Anatomical constraints are incorporated during training in order to encourage realistic cardiac motion patterns. Evaluations using an *in vivo* canine dataset suggest that our model performs favorably when compared to single task learning and other alternative methods. Future work includes further evaluation such as cross-validation using our existing dataset or validation against a larger or different dataset. Furthermore, we will explore the potential clinical applications of our model in estimating cardiac strain and detecting and localizing myocardial ischemia.

Acknowledgements. This work was funded by the following grants: R01 HL121226, R01 HL137365, and HL 098069. Additionally, we are grateful toward Drs. Nripesh Parajuli and Allen Lu and the Fellows and staff of the Yale Translational Research Imaging Center, especially Drs. Nabil Boutagy and Imran Al Khalil, for their technical support and assistance with the *in vivo* canine imaging studies.

References

1. Cheng, J., et al.: SegFlow: joint learning for video object segmentation and optical flow. In: IEEE International Conference on Computer Vision (2017)
2. Qin, C., et al.: Joint learning of motion estimation and segmentation for cardiac MR image sequences. In: Frangi, A.F., Schnabel, J.A., Davatzikos, C., Alberola-López, C., Fichtinger, G. (eds.) MICCAI 2018. LNCS, vol. 11071, pp. 472–480. Springer, Cham (2018). https://doi.org/10.1007/978-3-030-00934-2_53
3. Ta, K., Ahn, S.S., Stendahl, J.C., Sinusas, A.J., Duncan, J.S.: A semi-supervised joint network for simultaneous left ventricular motion tracking and segmentation in 4D echocardiography. In: Martel, A.L., et al. (eds.) MICCAI 2020. LNCS, vol. 12266, pp. 468–477. Springer, Cham (2020). https://doi.org/10.1007/978-3-030-59725-2_45
4. Chen, C., et al.: Deep learning for cardiac image segmentation: a review. Front. Cardiovasc. Med. **7**, 25 (2020)
5. Huang, X., et al.: Contour tracking in echocardiographic sequences via sparse representation and dictionary learning. Med. Image Anal. **18**, 253–271 (2014)
6. Dong, S., et al.: A combined fully convolutional networks and deformable model for automatic left ventricle segmentation based on 3D echocardiography. BioMed Res. Int. **2018**, 1–16 (2018)
7. Papademetris, X., et al.: Estimation of 3-D left ventricular deformation from medical images using biomechanical models. IEEE Trans. Med. Imaging **21**, 786–800 (2002)
8. Parajuli, N., et al.: Flow network tracking for spatiotemporal and periodic point matching: applied to cardiac motion analysis. Med. Image Anal. **55**, 116–135 (2019)
9. Qiu, H., Qin, C., Le Folgoc, L., Hou, B., Schlemper, J., Rueckert, D.: Deep learning for cardiac motion estimation: supervised vs. unsupervised training. In: Pop, M., et al. (eds.) STACOM 2019. LNCS, vol. 12009, pp. 186–194. Springer, Cham (2020). https://doi.org/10.1007/978-3-030-39074-7_20
10. Balakrishnan, G., et al.: An unsupervised learning model for deformable medical image registration. In: The IEEE Conference on Computer Vision and Pattern Recognition (2018)
11. Zhu, W., et al.: NeurReg: neural registration and its application to image segmentation. In: Winter Conference on Applications of Computer Vision (2020)
12. Ahn, S.S., et al.: Unsupervised motion tracking of left ventricle in echocardiography. In: Medical Imaging 2020: Ultrasonic Imaging and Tomography, International Society for Optics and Photonics (2020)
13. Ta, K., et al.: A semi-supervised joint learning approach to left ventricular segmentation and motion tracking in echocardiography. In: IEEE International Symposium on Biomedical Imaging (2020)
14. Tsai, Y.-H., et al.: Video segmentation via object flow. In: IEEE Conference on Computer Vision and Pattern Recognition (2016)
15. Çiçek, Ö., Abdulkadir, A., Lienkamp, S.S., Brox, T., Ronneberger, O.: 3D U-Net: learning dense volumetric segmentation from sparse annotation. In: Ourselin, S., Joskowicz, L., Sabuncu, M.R., Unal, G., Wells, W. (eds.) MICCAI 2016. LNCS, vol. 9901, pp. 424–432. Springer, Cham (2016). https://doi.org/10.1007/978-3-319-46723-8_49
16. Ronneberger, O., Fischer, P., Brox, T.: U-net: convolutional networks for biomedical image segmentation. In: Navab, N., Hornegger, J., Wells, W.M., Frangi, A.F. (eds.) MICCAI 2015. LNCS, vol. 9351, pp. 234–241. Springer, Cham (2015). https://doi.org/10.1007/978-3-319-24574-4_28

17. Lu, A., et al.: Learning-based regularization for cardiac strain analysis with ability for domain adaptation. arXiv preprint arXiv:1807.04807 (2018)
18. Ruder, S.: An overview of multi-task learning in deep neural networks. arXiv:1706.05098 (2017)
19. Stendahl, J.C., et al.: Regional myocardial strain analysis via 2D speckle tracking echocardiography: validation with sonomicrometry and correlation with regional blood flow in the presence of graded coronary stenoses and dobutamine stress. Cardiovasc. Ultrasound. **18**(1), 2 (2020). PMID: 31941514; PMCID: PMC6964036. https://doi.org/10.1186/s12947-019-0183-x
20. Besnerais, G.L., et al.: Dense optical flow by iterative local window registration. In: IEEE International Conference on Image Processing (2005)
21. He, K., et al.: Deep residual learning for image recognition. arXiv:1512.03385 (2015)
22. Yushkevich, P.A., et al.: User-guided 3D active contour segmentation of anatomical structures: significantly improved efficiency and reliability. Neuroimage **31**, 1116–1128 (2006)
23. Wu, Y., He, K.: Group normalization. In: Ferrari, V., Hebert, M., Sminchisescu, C., Weiss, Y. (eds.) ECCV 2018. LNCS, vol. 11217, pp. 3–19. Springer, Cham (2018). https://doi.org/10.1007/978-3-030-01261-8_1
24. Gibbons, J.D., et al.: Nonparametric Statistical Inference, 5th edn. Chapman & Hall/CRC Press, Taylor & Francis Group, Boca Raton (2011)
25. Yu, H., et al.: FOAL: fast online adaptive learning for cardiac motion estimation. In: Proceedings of the IEEE/CVF Conference on Computer Vision and Pattern Recognition, June 2020

Statistical Shape Analysis
of the Tricuspid Valve in Hypoplastic Left
Heart Syndrome

Jared Vicory[1], Christian Herz[2], David Allemang[2], Hannah H. Nam[2],
Alana Cianciulli[2], Chad Vigil[2], Ye Han[1], Andras Lasso[3],
Matthew A. Jolley[2], and Beatriz Paniagua[1(✉)]

[1] Kitware Inc, Carrboro, NC, USA
beatriz.paniagua@kitware.com
[2] Children's Hospital of Philadelphia, Philadelphia, PA 19104, USA
[3] Queen's University, Kingston, ON, Canada

Abstract. Hypoplastic left heart syndrome (HLHS) is a congenital heart disease characterized by incomplete development of the left heart. Children with HLHS undergo a series of operations which result in the tricuspid valve (TV) becoming the only functional atrioventricular valve. Some of those patients develop tricuspid regurgitation which is associated with heart failure and death and necessitates further surgical intervention. Repair of the regurgitant TV, and understanding the connections between structure and function of this valve remains extremely challenging. Adult cardiac populations have used 3D echocardiography (3DE) combined with computational modeling to better understand cardiac conditions affecting the TV. However, these structure-function analyses rely on simplistic point-based techniques that do not capture the leaflet surface in detail, nor do they allow robust comparison of shapes across groups. We propose using statistical shape modeling and analysis of the TV using Spherical Harmonic Representation Point Distribution Models (SPHARM-PDM) in order to generate a reproducible representation, which in turn enables high dimensional low sample size statistical analysis techniques such as principal component analysis and distance weighted discrimination. Our initial results suggest that visualization of the differences in regurgitant vs. non-regurgitant valves can precisely locate populational structural differences as well as how an individual regurgitant valve differs from the mean shape of functional valves. We believe that these results will support the creation of modern image-based modeling tools, and ultimately increase the understanding of the relationship between valve structure and function needed to inform and improve surgical planning in HLHS.

Keywords: Statistical shape modeling · Statistical shape analysis · Hypoplastic left heart syndrome · Pediatric cardiac imaging · 3D echocardiography

E. Puyol Antón et al. (Eds.): STACOM 2021, LNCS 13131, pp. 132–140, 2022.
https://doi.org/10.1007/978-3-030-93722-5_15

1   Introduction

Hypoplastic left heart syndrome (HLHS) is a form of congenital heart disease characterized by incomplete development of the left heart, yielding a left ventricle that is incapable of supporting the systemic circulation. HLHS affects over 1,000 live born infants in the US per year and would be uniformly fatal without intervention [5]. Staged surgical treatment allows children with HLHS to survive and flourish, but the right ventricle (RV) remains the sole functioning ventricle and the tricuspid valve (TV) the sole functioning atrioventricular valve. Tricuspid regurgitation (TR) is highly associated with heart failure and death and necessitates surgical intervention in almost 30% of HLHS patients [7]. Defining the mechanisms of TR is difficult, with both 2D echocardiography (2DE) and surgical inspection having significant limitations [13]. We instead analyze the structure of the TV in 3D echocardiography (3DE) images with the goal of detecting and quantifying TR.

Gross metrics that attempt to model the geometry of a structure, such as diameter, area, or volume, have been used as intuitive measurements in medical imaging studies. Previous work has investigated the use of such metrics to quantitatively detect the presence and severity of TR. For the TV, these metrics include annular area [3,11], septolateral diameter [3,4,11], bending angle [8], anterior leaflet prolapse [4], and anterior papillary muscle location [3,11]. While some of these measures are restricted to a certain region or leaflet of the TV, they are still summarizing these regions rather than modeling them directly and so can not always reflect localized structural changes.

In contrast, shape is a proven biomarker that is more robust and, in many cases, more clinically relevant than traditional 2D or 3D-based metrics. Statistical Shape Modeling (SSM) can be used to quantitatively characterize shape and produce models that represent the average shape of a population as well as the principal modes of variation. Further, these modeling techniques can precisely quantify the location and magnitude of differences between two populations (e.g. regurgitant vs. non-regurgitant valves) and can do so with fewer assumptions and less bias than traditional approaches.

In this paper we describe a semi-automatic pipeline for creating 3D models of TV leaflet surface geometry suitable for statistical analysis and demonstrate its usefulness in several experiments. The ultimate goal of this modeling is to compare the shape of the TV in patients with moderate or greater TR to those with mild or lower to correlate structure to TV dysfunction.

2   Materials

2.1   Subjects

Acquisition of transthoracic 3DE images of the TV is part of the standard clinical echo lab protocol for HLHS at the Children's Hospital of Philadelphia (CHOP). An institutional database was used to retrospectively identify patients with HLHS with a Fontan circulation in whom 3DE of the TV had been previously performed. Exclusion criteria included presence of significant stitch artifact and inability to delineate the TV. This study was performed according to a protocol approved by the CHOP Institutional Review Board (IRB). We have 100 3DE scans with age range 2.14 years to 30.64 years, with an average of 10.36 years. Images were acquired using sector narrowed Full Volume or 3D Zoom mode with a wide FOV. EKG gated acquisitions were obtained when patient cooperation allowed. Transthoracic X7 or X5 probes were used with the Philips IE33 and EPIQ 7 ultrasound systems.

3   Methods

3.1   Image Segmentation and Model Creation

Images were exported to Philips DICOM, converted to cartesian DICOM in QLAB, and imported into 3D Slicer [1] using the SlicerHeart [2,3] Philips DICOM converter. A single mid-systolic frame was chosen for static 3D modeling of the TV. TV segmentation was performed using the 3D Slicer Segments module (Fig. 1).

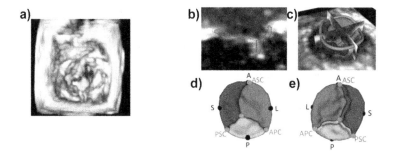

Fig. 1. A. Volume rendering of 3DE (left ventricular view) of TV in patient with HLHS; B and C. Segmentation of TV; D and E. Atrial and ventricular views of valve model with landmark annotations (A = anterior, P = posterior, S = septal, L = lateral, ASC, APC, PSC = commisures)

3.2   Modeling the TV

We propose to represent TV geometry via point distribution models (PDMs) of each leaflet. We generate these models by first parametrizing the segmentations using spherical harmonics [6] and then use icosahedron subdivision to obtain densely-sampled boundary PDMs using the SPHARM-PDM software package available in SlicerSALT [14]. When applied to modeling the TV, this modeling technique does not require user input other than providing the number of structures in the model (three leaflets for a tricuspid valve).

While SPHARM-PDM can often produce high-quality correspondence with no post-processing, this is not the case for the TV. To bring the SPHARM-PDM models into correspondence, we adapt the approach of Lyu [9] and use specific anatomical landmarks to guide correspondence improvement. For each leaflet, we map three manually identified landmarks (the two commisures and the valve center) on the SPHARM mesh back to their location on the underlying spherical parameterization. Then, choosing one case as a template, we rotate the spherical parameterization of the other cases to minimize the distances between the sets of landmark locations in parameter space. We then remesh each leaflet using its rotated parameterization to obtain PDMs with the same geometry but with vertices reindexed to have better correspondence. Quality control of the correspondences for each leaflet type is performed using the color-coded spherical parameterization. Equally colored areas represent equal corresponding areas. Figure 2 shows the result of this procedure for a subset of 40 septal leaflets.

Fig. 2. Visualization of the SPHARM-PDM correspondence using the ϕ-attribute shown on forty randomly selected septal leaflets. Corresponding anatomical locations have similar colors, indicating good correspondence.

Finally, models of the TV were built by merging the SPHARM-PDM models of the septal, anterior and posterior leaflets into a multi-object PDM of the entire TV which can be analyzed using the methods described in the next sections.

Before statistical shape analysis, we pre-process the population of SPHARM-PDM correspondent models by aligning and scaling it using standard Procrustes analysis. Second, we replaced the Procrustes computed scaling with body surface area (BSA) for normalization. We normalize shape geometry of the TV using BSA instead of by scale because that metric has a been used in cardiology before to predict physiologic outcomes such as flow rates.

3.3   Principal Component Analysis

We use Principal Component Analysis, a method for computing an efficient parameterization of the variability of linear data (in this case the 3-dimensional correspondent points contained in our spharm models), to build low dimensional statistical shape spaces for each one of the TR/severity subgroups as well as the healthy group. Thanks to our interactive tools, we can explore the generated PCA space and to evaluate the quality of the generated models.

3.4 Distance Weighted Discrimination

DWD is a binary classification method designed to address shortcomings with support vector machine (SVM) performance when applied to high-dimension, low-sample-size (HDLSS) data [10, 15] by considering the effects of all data on the separating hyper-plane rather than just a limited set of support vectors. Here, we apply DWD to classify the PDMs of each TV by regurgitation severity into trivial/mild or moderate/severe severity. Like SVM, DWD performs classification by computing the distance of each sample to a separating hyperplane, with samples laying on the same side of the hyperplane being classified together.

4 Results

4.1 Shape Modeling

During our initial experiments with SPHARM-PDM we realized that the heuristic methods applied to ensure the spherical topology of the segmentation were significantly (as large as 0.8 mm) changing the thin structures of the individual TV leaflets. Due to this fact, we decided to bypass this step and directly compute surface meshes corresponding and the spherical parametrizations on the original segmentation. Following the correspondence optimization process described in Sect. 3.2, high-quality models were successfully generated for all three leaflets in all 100 TV images.

4.2 Principal Component Analysis

We use principal component analysis (PCA) to compute the mean and major modes of variation of the population of TV models and explore the existing phenotypes in our population. Figure 3 shows the mean and four most significant modes of variation for our sample. The first mode is essentially pure scaling, that is present after rescaling all geometry by BSA, while the next three show scaling of each of the individual leaflets. While these modes intuitively make sense they do not turn out to be particularly relevant for correlating structural changes with functional ones. Instead, we investigate the use of DWD for this purpose.

4.3 Distance Weighted Discrimination

Thanks to DWD we were able to examine the effect of different regurgitation levels in TV geometry. In order to do this, we group the existing four categories in two groups, i.e. trivial/mild cases and moderate/severe, because the boundaries between the four categories can be somewhat arbitrary.

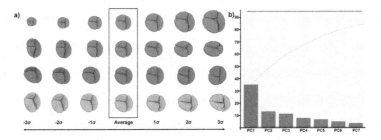

Fig. 3. a) Axes of geometric variation in the TV population. (red) First principal component (PC) showing TV scale, even after normalization. (blue) Second PC capturing sizing on the septal leaflet. (green) Third PC capturing size of the anterior leaflet. (yellow) Fourth PC capturing size of the posterior leaflet. b) Scree plot showing the variability contained in the first seven principal components does not reach 85% of explained variability, indicating a complex problem (Color figure online)

Figure 4 shows the effects of different scale normalization schemes in classification results. For each subject we compute the distance to the separating hyperplane and use kernel density estimation to fit probability distributions to each severity group. The discrimination power between different regurgitation levels is greatly reduced by not normalizing the correspondent point based models or by just normalizing using the gross scale computed by Procrustes (see Fig. 4a and 4b). Surprisingly normalizing with both gross scale and BSA tend to incorrectly classify valves as Trivial/Mild, indicating that the distance to the hyperplane is not independent of scale or BSA, while the unscaled models show relatively good separation between the classes. Additionally, we see a strong discrimination between trivial and severe groups when scaling using BSA is performed, indicating a correlation between regurgitation degree and TV geometry.

Projecting samples orthogonally to the separating hyperplane, also called the separation direction, is also an excellent visualization tool that relates our functional marker, regurgitation, to valve geometry. In Fig. 5 we show shapes from both sides of the separating hyperplane as you move from far from the plane on one side through the plane and far to the other side. This shows a clear progression from tenting on the trivial/mild side to billowing on the severe side, consistent with the clinical observation that regurgitant valves tend to have more billow.

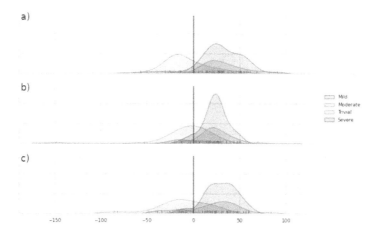

Fig. 4. Kernel density estimate of the signed distance to DWD separating hyperplane for a) No scaling, b) Scaling using BSA, and c) Scaling using gross geometric scale.

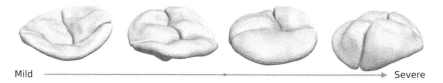

Fig. 5. Example shapes sorted by distance along the DWD separation direction from the trivial/mild side (left) to the moderate/severe side (right). The red dot shows the location of the separating hyperplane. Shapes show more billow as regurgitation gets more severe. (Color figure online)

5   Discussion

In this paper we introduce a pipeline for semi-automatically creating representations of tricuspid valve leaflet boundaries suitable for statistical analysis using SPHARM-PDM followed by a novel correspondence-optimizing post-processing. We demonstrate the effectiveness of this modeling approach by using statistical shape analysis approaches such as PCA and DWD to examine the difference between valves with trivial or mild regurgitation and those with moderate or severe regurgitation. PCA results are of limited usefulness due to the challenging nature of the problem, natural normal variability within this population, and our normalization strategy, but DWD analysis is able to show clear linkages between the geometry captured between the best separating axis between these two clinical populations and functional variables such as the tenting volume on the whole TV or the individual tenting volume of each individual leaflet, indicating that our representations are able to capture clinically relevant information.

We have identified several paths forward for future work on this problem. Regarding modeling, our current semi-automatic pipeline could be fully automated by developing methods for automatically placing landmarks at the commisures and center of the valve. We are also exploring methods for automatically segmenting the TV from ultrasound, further reducing the manual effort required to get from data acquisition to analysis. We will also continue to refine our analysis pipeline by investigating other clinical variables as well as investigating additional statistical shape analysis approaches. Finally, we will investigate the relationship between the various traditional metrics used to detect TR to changes in shape to determine if these effects can be further localized or if shape analysis can provide additional power than these metrics alone.

We will continue to investigate different normalization approaches because, as our PCA results show, our current approach using BSA is not sufficient for fully separating the relationship between the size of a TV and its shape and may in fact be introducing additional dependence between the two. We will investigate normalizing by other metrics of the TV, including various transformations of BSA along with more direct measures such as annular diameter. We will also investigate using metrics based on the full heart, such as ventricular length or volume, though these are also problematic due to potential ventricular dilation correlated with TR. We will also investigate using age-specific normalization strategies to attempt to separate normal growth from abnormal annulus dilation which can be associated with TR.

In addition to the SPHARM-PDM boundary models described here, we are also investigating the use of skeletal models [12] for representing and analyzing TVs. Skeletal models have been shown to be powerful representations for statistical shape analysis and due to the thin nature of the TV leaflets we hypothesize that they will be particularly well-suited to this problem.

Acknowledgements. Research reported in this publication was supported by the National Institute of Biomedical Imaging and Bioengineering from the National Institutes of Health under Award Number R01HL153166 as well as the National Institute of Biomedical Imaging and Bioengineering under award number R01EB021391. The content is solely the responsibility of the authors and does not necessarily represent the official views of the National Institutes of Health. The methods presented in this work are distributed as part of SlicerSALT [14], an open-source, free, comprehensive software that will allow biomedical scientists to precisely locate shape changes in their imaging studies, and SlicerHeart [2,3], a 3DSlicer extension containing tools for cardiac image import (3D/4D ultrasound, CT, MRI), quantification, and implant placement planning and assessment.

References

1. Fedorov, A., et al.: 3D Slicer as an image computing platform for the quantitative imaging network. Magn. Reson. Imaging **30**(9), 1323–1341 (2012). https://doi.org/10.1016/J.MRI.2012.05.001. ISSN 1873-5894

2. Scanlan, A.B., et al.: Comparison of 3D echocardiogram-derived 3D printed valve models to molded models for simulated repair of pediatric atrioventricular valves. Pediatr. Cardiol. **39**(3), 538–547 (2018). https://doi.org/10.1007/S00246-017-1785-4. ISSN 1432-1971

3. Nguyen, A.V., et al.: Dynamic three-dimensional geometry of the tricuspid valve annulus in hypoplastic left heart syndrome with a fontan circulation. J. Am. Soc. Echocardiogr. **32**(5), 655–666.e13 (2019). https://doi.org/10.1016/J.ECHO.2019. 01.002. ISSN 1097-6795

4. Bautista-Hernandez, V., et al.: Mechanisms of tricuspid regurgitation in patients with hypoplastic left heart syndrome undergoing tricuspid valvuloplasty. J. Thorac. Cardiovasc. Surg. **148**(3), 832–840 (2014)

5. Gordon, B.M., et al.: Decreasing number of deaths of infants with hypoplastic left heart syndrome. J. Pediatr. **153**(3), 354–358 (2008). https://doi.org/10.1016/J. JPEDS.2008.03.009. ISSN 1097-6833

6. Brechbühler, C., Gerig, G., Kübler, O.: Parametrization of closed surfaces for 3-D shape description. Comput. Vis. Image Underst. **61**(2), 154–170 (1995)

7. Barber, G., et al.: The significance of tricuspid regurgitation in hypoplastic left-heart syndrome. Am. Heart J. **116**(6), 1563–1567 (1988). https://doi.org/10.1016/ 0002-8703(88)90744-2. ISSN 0002-8703

8. Kutty, S., et al.: Tricuspid regurgitation in hypoplastic left heart syndrome: mechanistic insights from 3-dimensional echocardiography and relationship with outcomes. Circ. Cardiovasc. Imaging **7**(5), 765–772 (2014)

9. Lyu, I., et al.: Robust estimation of group-wise cortical correspondence with an application to macaque and human neuroimaging studies. Front. Neurosci. **9**, 210 (2015)

10. Marron, J.S., Todd, M.J., Ahn, J.: Distance-weighted discrimination. J. Am. Stat. Assoc. **102**(480), 1267–1271 (2007). https://doi.org/10.1198/016214507000001120

11. Nii, M., et al.: Three-dimensional tricuspid annular function provides insight into the mechanisms of tricuspid valve regurgitation in classic hypoplastic left heart syndrome. J. Am. Soc. Echocardiogr. **19**(4), 391–402 (2006)

12. Pizer, S.M., et al.: Object shape representation via skeletal models (s-reps) and statistical analysis. In: Riemannian Geometric Statistics in Medical Image Analysis, pp. 233–271. Elsevier (2020)

13. Bharucha, T., et al.: Right ventricular mechanical dyssynchrony and asymmetric contraction in hypoplastic heart syndrome are associated with tricuspid regurgitation. J. Am. Soc. Echocardiogr. **26**(10), 1214–1220 (2013). https://doi.org/10. 1016/J.ECHO.2013.06.015. ISSN 1097-6795

14. Vicory, J., et al.: SlicerSALT: shape AnaLysis toolbox. In: Reuter, M., Wachinger, C., Lombaert, H., Paniagua, B., Lüthi, M., Egger, B. (eds.) ShapeMI 2018. LNCS, vol. 11167, pp. 65–72. Springer, Cham (2018). https://doi.org/10.1007/978-3-030-04747-4_6

15. Wang, B., Zou, H.: Another look at distance-weighted discrimination. J. R. Stat. Soc. Series B Stat. Methodol. **80**(1), 177–198 (2018). https://doi.org/10.1111/rssb. 12244. https://rss.onlinelibrary.wiley.com/doi/pdf/10.1111/rssb.12244

An Unsupervised 3D Recurrent Neural Network for Slice Misalignment Correction in Cardiac MR Imaging

Qi Chang[1(✉)], Zhennan Yan[2], Meng Ye[1], Kanski Mikael[3], Subhi Al'Aref[4], Leon Axel[3], and Dimitris N. Metaxas[1]

[1] Rutgers University, Piscataway, NJ, USA
{qc58,dnm}@cs.rutgers.edu
[2] SenseBrain Technology Ltd. LLC, Princeton, NJ, USA
[3] Department of Radiology, New York University, New York, NY, USA
[4] Department of Medicine, University of Arkansas for Medical Sciences, Little Rock, AR, USA

Abstract. Cardiac magnetic resonance (CMR) imaging is the most accurate imaging modality for cardiac function analysis. However respiration misalignment can negatively impact the accuracy of the cardiac wall 3D segmentation and the assessment of cardiac function. A learning based misalignment correction method is needed, in order to build an end-to-end accurate cardiac function analysis pipeline. To this end, we proposed an unsupervised misalignment correction network to solve this challenge problem. We validated the proposed framework on synthetic and real CMR segmented images, and the result prove the efficiency of misalignment correction and the improvement with the corrected CMR image. Experimental results using our approach show that it: 1) could more efficiently correct the misalignment of CMR images compared with the traditional optimization process. 2) incorporated an unsupervised loss named "intersection distance" loss to guide the network output to the accurate correction prediction. 3) is the first to use the unsupervised learning based method for Cardiac MR slices' misalignment problem and achieved more accurate results.

Keywords: Unsupervised learning · RNN · Cardiac MR image · Respiration correction

1 Introduction

Cardiovascular diseases are the world's most deadly killer, globally accounting for over 17.9 million deaths in the global every year [9,16]. Cardiac magnetic resonance (CMR) imaging is one of the best imaging modality for cardiac function quantification and disease diagnosis [10], as well as for planning and guidance of cardiac interventions [2] due to its good soft tissue contrast, high temporal and spatial resolution, and absence of radiation exposure. High-resolution 3D

© Springer Nature Switzerland AG 2022
E. Puyol Antón et al. (Eds.): STACOM 2021, LNCS 13131, pp. 141–150, 2022.
https://doi.org/10.1007/978-3-030-93722-5_16

MR sequences enable whole-heart structural imaging but are time-consuming, expensive to acquire and they often require long breath holds that are not suitable for patients [1]. In clinical practice, 2D cine CMR is routinely used. The multi-planar short-axis (SAX) view which shows cross-section of the left and right ventricle across the cardiac cycle can provide high temporal and in-plane spatial resolution. The long-axis (LAX) view, orthogonal to SAX, which shows perpendicular views of the ventricles, provides complementary information of the shape and function of the heart, especially on the basal and apical borders. In order to quantitatively analyze the global function and the regional myocardial motion, the 3D cardiac volume needs to be precisely reconstructed from a set of 2D CMR images combining both LAX and SAX views.

However, the precision of 3D reconstruction from multi-planar CMR images strongly relies on the alignment of the cardiac geometry between image planes. The acquisition of cine CMR images covering the heart using the balanced steady state free precession (bSSFP) technique typically requires several breath-holds [12]. Variation in breath-hold locations during the acquisition of different image plane will change the position of the heart in the images, resulting in misalignment of the cardiac geometry across all planes. These respiration-related position offsets between different MRI sequences can cause errors in the 3D reconstruction of cardiac structures and adversely affect associated cardiac function analysis.

Using machine learning-based methods to address the misalignment between 2D slices in CMR imaging has not been well studied and can be the missing building block for an end-to-end accurate cardiac function analysis pipeline. Many recent researches [11,17] have introduced the machine learning-based pipeline for cardiac function analysis. The analysis starts from machine learning-based full cycle segmentation, and then progresses to 3D volume and motion reconstruction while using an iterative alignment process for misalignment correction in between. By using a machine learning based methods, rather than the optimization method, we could bridge the gap between 2/3D cardiac segmentation and 3D aligned volume reconstruction and improve the overall accuracy.

Related Work. Image-based registration has been used in some approaches to correct the misalignment [3,7,15,19]. However, these methods are vulnerable to non-rigid deformation of tissues around the heart under different breath-holds. Some of them also require an additional 3D scan as the reference, which is time-consuming.

On the other hand, some other works have shown that segmentation of the heart could be used to correct the respiratory motion without original image information [8,14,18]. Sinclair et al. [14] treated the segmentations as binary images to conduct image-based registration. Yang et al. [18] treated the segmentations as contours and aligned the slices by the differences between corresponding intersection points on different contours. Since the correspondence of intersection points is not always valid, [8] used an iterative closest point (ICP) algorithm instead to compute the parameters of translation.

Contributions: In this paper, we firstly propose a recurrent neural network for the misalignment correction of joint multi-view weighted fusion attention segmentation in Cardiac MR images, named **MC-Net**. Inspired by the traditional optimization-based algorithms [8,14,18], we proposed an unsupervised loss named "**intersection distance loss**" to guide the network output for accurate misalignment correction prediction. To the best of our knowledge, we are the first to use this unsupervised learning-based method for CMR misalignment problem, and have achieved more accurate results.

2   Method

In this section, we introduce our unsupervised 3D recurrent neural network and the proposed intersection distance loss. An overview of the proposed architecture is shown in Fig. 2. Next, we will introduce the formation of the input data of the model, and then we will describe the details of the architecture design of the MC-Net and the unsupervised object function.

2.1   Formation of the Input

The 3D recurrent neural network takes the multi-channel segmentation masks of the myocardium wall from short-axis and long-axis 2D views of CMR images x_{t-1} and the offset prediction y_{t-1} from the last iteration $t-1$ as input. Unlike the traditional optimization processes, we adopt a neural network which directly outputs the offset of each 2D view.

Reformat Multi-channel 3D CMR. We reformat both the segmentations of the SAX and the LAX CMR images to adapt to the network and the object function calculation. We focus here on the left ventricle (LV), although the methods can be readily extended to other chambers. Instead of directly using the binary segmentation mask of the short axis (SAX) images, we convert the segmentation map into the Euclidean Distance Map (EDM) which is defined as:

$$\mathcal{E}(x^{sa}) = \begin{cases} \min_{z \in \mathcal{B}(\mathcal{O})} d(x^{sa}, z) & if \, x^{sa} \notin \mathcal{O} \\ 0 & if \, x^{sa} \in \mathcal{O} \end{cases} \quad (1)$$

Where $\mathcal{O}$ denotes the target organ, $\mathcal{B}(\mathcal{O})$ is the points set that includes the boundary voxels of the myocardium, $d(x^{sa}, z)$ is the Euclidean distance from x^{sa} to z. Instead of directly using a binary mask, the value of the Euclidean distance map at the cross-section location of various planes indicates the distance of the misalignment. The Fig. 1 (b) demonstrates the multi-channel input of the MC-Net. The color in each pixel indicates the distance from the position of this pixel to the surface of the myocardium wall. The darker of the blue background, the nearer to the contour of the mask. The lines across the contour indicates the intersection of the long axis (LAX) view masks with the SAX masks.

We then intersect the binary mask of the LAX image with each SAX plane as $\mathcal{P}_{x^{la}}$, which indicates the boundary of the myocardium wall.

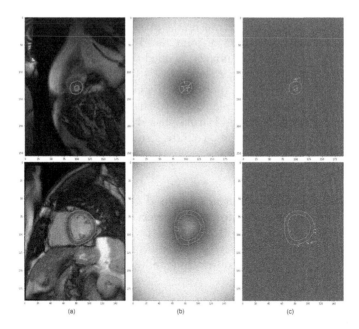

Fig. 1. The sample images show the multi-channel input format and the calculation of intersection distance error. (a) shows slices of the original MR image and associated segmentation of the LV myocardium. (b) shows conversion of the short axis segmentation (green) into Euclidean Distance Map (EDM) and the intersection lines (red) of the long axis with the short axis masks. (c) shows after the Intersection Distance Loss calculation, the misalignment errors accumulated from the intersections. (Color figure online)

Offset Prediction. The offset prediction at step $t-1$ comprises the SAX offset y_{t-1}^{sa} and the LAX offset y_{t-1}^{la}.

2.2   Architecture

An overview of MC-Net is shown in Fig. 2. It contains a 3D VGG-like network [13] which predicts a single offset vector y_t that indicates the shifted coordinates of each 2D slices from the original 3D volume. Initially at time $t-1$, there are multi-channel short axis SAX and intersected LAX volumes x_{t-1} and the offset vector y_{t-1} which by default is a zero vector. The network takes all the input and outputs the offset vector y_t. At time t, the original volume x_{t-1} will be applied spatial transform by the output offset vector y_t, and then the intersection distance loss will be calculated. The misalignment corrected volume x_t and the output offset vector y_t will then be sent into the network and used to fine-tune the offset prediction. Based on the offset prediction of the last iteration, the model can then align the 2D slices more accurately.

Specifically, the output y_t can then be split into two offset vectors: y_t^{sa}, and y_t^{la} for SAX and LAX correction separately. The SAX branch outputs $B \times 13 \times 2$

vector which will adjust 13 2D slices of the SAX, and the LAX branch outputs $B \times 3 \times 2$ vector which will adjust 3 LAX slices, and B is the batch size.

Our network consists of 6 convolutional blocks, and one fully connected layers. Every convolution block contains 2 or 4 convolutional layers, and each convolutional layer contains a 3D convolutional layer followed by batch normalization [6] and the ReLU [5] activation. Due to the relatively low resolution along z-index of the input tensor, the last three 3D max-pooling layers use $2 \times 2 \times 1$ kernel while the rest of 3D max-pooling layers adopt $2 \times 2 \times 2$ kernel instead.

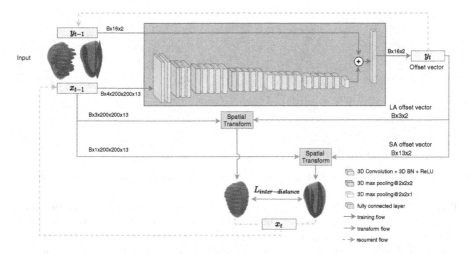

Fig. 2. The overview of MC-Net framework. We use modified 3D VGG architecture as our backbone of the network. The blue lines are training workflow, the green flows are spatial transform and loss calculation process, the orange dot lines indicate the recurrent process. (Color figure online)

2.3 Objective of MC-Net

The Intersection Distance Loss ($\mathcal{L}_{ID}$) measures the average distance between LV myocardium wall in the SAX view and LAX view; it is defined as:

$$\mathcal{L}_{ID} = \sum_{t}^{n} \lambda_t \mathcal{T}(\mathcal{E}(x_{t-1}^{sa}), y_t^{sa}) \circ \mathcal{T}(\mathcal{I}(x_{t-1}^{la}), y_t^{la}) \qquad (2)$$

Where λ_t is the weight for each iteration, x_{t-1}^{sa} and x_{t-1}^{la} denote segmentation volume of SAX and LAX from the last iteration, y_t^{sa} and y_t^{la} denote the offset prediction of the current iteration. Function $\mathcal{T}(\cdot, \cdot)$ denotes spatial transform operation, function $\mathcal{E}(\cdot)$ is the euclidean distance map function and function $\mathcal{I}$ is the intersection function indicate the segmentation image computed from the LAX view intersecting SAX view described in Sect. 2.1, ∘ is the element-wised product. The first transform operation takes the SAX volume from the last iteration: $t-1$ and the offset prediction of the current iteration. The second transform operation takes the binary intersected LAX segmentation from the last iteration and the offset prediction of the current iteration. The intersection distance loss will accumulate the loss for n iterations with different weights λ_t. The Fig. 1 (c) demonstrates the calculation results of intersection distance error. The darker of the blue background, the nearer to the contour of the mask. The lines across the contour indicates the intersection of the LAX view masks.

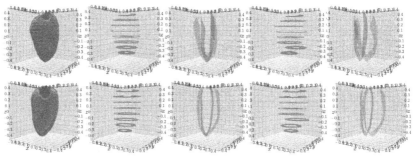

(a)CT Segmentation Volume (b)Short Axis Subsample (c)Long Axis Subsample (d)SAX Randomly Shifted (e)LAX Randomly Shifted

Fig. 3. The overview of synthetic dataset. (a) shows the original CT segmentation volume. (b) and (c) generated by sub-sampling the volume. (b) select one short axis slice every other 10 ± 3 slices, and (c) select two perpendicular slices along sagittal and coronal plane as long axis. (d) and (e) applied the randomly shift to simulate the misalignment of each Cardiac MRI series.

3   Experiments and Results

In this section, we evaluate the MC-Net on a synthetic dataset to illustrate how MC-Net learn the offset of the misalignment between the 2D CMR slices, and then apply the MC-NET to the Cardiac Atlas Project (CAP) [4] dataset.

3.1   Datasets

Synthetic Dataset. The synthetic dataset is generated from the 3D cardiac CT volumes of the MMWHS dataset [20,21] and contains 20 3D segmentation volumes acquired in real clinical environment. The CT images provide higher 3D spatial resolution compared to the CMR images, and overall accurate alignment without respiration misalignment. The synthetic dataset shown in Fig. 3 simulated the CMR imaging by selecting one short axis slice every other 10 ± 3 slices, and selecting two perpendicular slices along sagittal and coronal plane as LAX views. We then shifted the slices to simulate the misalignment in CMR images. Since the initial synthetic MRI volumes don't have misalignment error, by manually shifting the slices known amounts, we could able to quantitatively evaluate the performance of MC-Net.

CAP Dataset. This dataset is provided by the Cardiac Atlas Project (CAP) [4] Left Ventricular Segmentation Challenge with 100 cases. There are 1–7 LAX and 8–24 SAX views for each case. CMR protocol parameters varied between cases giving a heterogeneous mix of scanner types and imaging parameters. MR scanner systems were GE Medical Systems (Signa 1.5T), Philips Medical Systems (Achieva 1.5T, 3.0T, and Intera 1.5T), and Siemens Medical Systems (Avanto 1.5T, Espree 1.5T and Symphony 1.5T). Typical SAX slice parameters were either a 6 mm slice thickness with 4 mm gap or 8 mm slice thickness with 2 mm gap. Image size ranged from 138×192 to 512×512 pixels. The temporal resolution was between 19 and 30 frames. LAX images in the four- and two-chamber orientations were also available.

3.2 Evaluation Metrics

We adopt metrics similar to those described in Sinclair, et al. [14] to evaluate the misalignment correction performance of our MC-Net. The mean contour distance (MCD) in pixels between the contours from the LAX versus SAX planes was used to assess the misalignment correction results of the different methods. We also use one-side Hausdorff distance to evaluate the distance between boundaries of ground-truth and segmented masks:

$$HD_{oneside}(G, S) = \sup_{x \in \partial G} \inf_{y \in \partial S} d(x, y) \tag{3}$$

3.3 Experiment on Synthetic Dataset

In this section, we show that the proposed MC-Net can learn the offsets of the given 3D volume and restore shifted 2D slices to the original positions given different misalignment conditions.

Settings. We compare the performance of learning to correct offset in 3 settings: (1) **None**: The original dataset, (2) **MC-Net (n = 1)**: iterate 1 time based on the 2 loss function. (3) **MC-Net (n = 3)**: iterate 3 times and accumulate losses with weights $[0.2, 0.5, 1]$. For the synthetic dataset, we adopt different shift variance: a. MMWHS ± 5 px, shift all slices range from $-5px$ to $5px$ b. MMWHS $\pm 10\,px$, shift all slices range from $-10px$ to $10px$.

Results. The quantitative misalignment correction results are shown in Table 1.

3.4 Experiment on CAP Dataset

In this section, we show that the proposed MC-Net can learn to correct the offset of the given volume from the real CAP dataset, and compare to the other image based optimization method in accuracy and efficiency.

Table 1. The Mean Absolute Distance (MAD), the DICE results for the different experiments setting

Experiments	MMWHS±5px		MMWHS±10px	
	MCD (pixels) ↓	DICE ↑	MCD (pixels) ↓	DICE ↑
(a) None	3.02px	0.963	5.48px	0.928
(b) MC-Net (n = 1)	0.550px	0.986	0.793px	0.978
(c) MC-Net (n = 3)	0.499px	0.994	0.515px	0.992

Settings. There are 80 training and 20 testing cases in the CAP dataset, each case contains 19–29 frames. In total, we separately extracted 1781 training and 435 testing segmentation volumes separately. We conducted the following experiments: (1) None: the mean contour distance was calculated for the original segmentation volumes, as the baseline. (2) Optim: the optimization method described in [14,18]. (3) MC-Net: our method with recurrent iteration $n = 1$, $n = 3$, $n = 5$ respectively.

Results. The quantitative misalignment correction results are shown in Table 2.

Table 2. The average intersection distance (AvgID), the mean contour distance (MCD) for the different experiments. The optimization method (Optim) directly use mask rather than the EDM, so the AvgID is unavailable.

Method	AvgID (pixels)	MCD (pixels)
(a) None	4.53 ± 1.04	4.59 ± 0.75
(b) Optim	–	0.61 ± 0.29
(c) MC-Net (n = 1)	0.53 ± 0.30	0.75 ± 0.31
(d) MC-Net (n = 3)	0.46 ± 0.25	0.58 ± 0.28
(e) MC-Net (n = 5)	0.54 ± 0.30	0.77 ± 0.31

4   Discussion and Conclusion

In this work, we proposed an unsupervised misalignment correction network called MC-Net. Our method could efficiently predict the misaligned shift for SAX and LAX images and is the first to use the unsupervised learning based method for Cardiac MR slices' misalignment problem and achieved more accurate results. The future work would focus on integrate the misalignment network with the segmentation or 3D myocardium volume reconstruction network to build an end-to-end pipeline for more accurate and robust cardiac analysis method.

References

1. Biffi, C., et al.: 3D high-resolution cardiac segmentation reconstruction from 2D views using conditional variational autoencoders. In: 2019 IEEE 16th International Symposium on Biomedical Imaging (ISBI 2019), pp. 1643–1646. IEEE (2019)
2. Cavalcante, J.L., Lalude, O.O., Schoenhagen, P., Lerakis, S.: Cardiovascular magnetic resonance imaging for structural and valvular heart disease interventions. JACC: Cardiovasc. Interv. **9**(5), 399–425 (2016)
3. Chandler, A.G., et al.: Correction of misaligned slices in multi-slice cardiovascular magnetic resonance using slice-to-volume registration. J. Cardiovasc. Magn. Reson. **10**(1), 1–9 (2008)
4. Fonseca, C.G., et al.: The cardiac atlas project-an imaging database for computational modeling and statistical atlases of the heart. Bioinformatics **27**(16), 2288–2295 (2011)
5. Glorot, X., Bordes, A., Bengio, Y.: Deep sparse rectifier neural networks. In: Proceedings of the Fourteenth International Conference on Artificial Intelligence and Statistics, pp. 315–323 (2011)
6. Ioffe, S., Szegedy, C.: Batch normalization: accelerating deep network training by reducing internal covariate shift. arXiv preprint arXiv:1502.03167 (2015)
7. Lötjönen, J., Pollari, M., Kivistö, S., Lauerma, K.: Correction of movement artifacts from 4-D cardiac short- and long-axis MR data. In: Barillot, C., Haynor, D.R., Hellier, P. (eds.) MICCAI 2004. LNCS, vol. 3217, pp. 405–412. Springer, Heidelberg (2004). https://doi.org/10.1007/978-3-540-30136-3_50
8. Metaxas, D.N., Yan, Z.: Chapter 12 - deformable models, sparsity and learning-based segmentation for cardiac MRI based analytics. In: Zhou, S.K., Rueckert, D., Fichtinger, G. (eds.) Handbook of Medical Image Computing and Computer Assisted Intervention, pp. 273–292. Academic Press (2020)
9. Moriyama, I.M., Krueger, D.E., Stamler, J.: Cardiovascular diseases in the United States, vol. 10. Harvard University Press (1971)
10. Pazos-López, P., et al.: Value of CMR for the differential diagnosis of cardiac masses. JACC: Cardiovasc. Imaging **7**(9), 896–905 (2014)
11. Ruijsink, B., et al.: Fully automated, quality-controlled cardiac analysis from CMR: validation and large-scale application to characterize cardiac function. Cardiovasc. Imaging **13**(3), 684–695 (2020)
12. Scheffler, K., Lehnhardt, S.: Principles and applications of balanced SSFP techniques. Eur. Radiol. **13**(11), 2409–2418 (2003)
13. Simonyan, K., Zisserman, A.: Very deep convolutional networks for large-scale image recognition. arXiv preprint arXiv:1409.1556 (2014)
14. Sinclair, M., Bai, W., Puyol-Antón, E., Oktay, O., Rueckert, D., King, A.P.: Fully automated segmentation-based respiratory motion correction of multiplanar cardiac magnetic resonance images for large-scale datasets. In: Descoteaux, M., Maier-Hein, L., Franz, A., Jannin, P., Collins, D.L., Duchesne, S. (eds.) MICCAI 2017. LNCS, vol. 10434, pp. 332–340. Springer, Cham (2017). https://doi.org/10.1007/978-3-319-66185-8_38
15. Villard, B., Zacur, E., Dall'Armellina, E., Grau, V.: Correction of slice misalignment in multi-breath-hold cardiac MRI scans. In: Mansi, T., McLeod, K., Pop, M., Rhode, K., Sermesant, M., Young, A. (eds.) STACOM 2016. LNCS, vol. 10124, pp. 30–38. Springer, Cham (2017). https://doi.org/10.1007/978-3-319-52718-5_4
16. Wilkins, E., et al.: European cardiovascular disease statistics 2017 (2017)

17. Yang, D., Huang, Q., Mikael, K., Al'Aref, S., Axel, L., Metaxas, D.: MRI-based characterization of left ventricle dyssynchrony with correlation to CRT outcomes. In: 2020 IEEE 17th International Symposium on Biomedical Imaging (ISBI), pp. 1–4. IEEE (2020)
18. Yang, D., Wu, P., Tan, C., Pohl, K.M., Axel, L., Metaxas, D.: 3D motion modeling and reconstruction of left ventricle wall in cardiac MRI. In: Pop, M., Wright, G.A. (eds.) FIMH 2017. LNCS, vol. 10263, pp. 481–492. Springer, Cham (2017). https://doi.org/10.1007/978-3-319-59448-4_46
19. Zakkaroff, C., Radjenovic, A., Greenwood, J., Magee, D.: Stack alignment transform for misalignment correction in cardiac MR cine series. Technical report. Citeseer (2012)
20. Zhuang, X.: Challenges and methodologies of fully automatic whole heart segmentation: a review. J. Healthc. Eng. **4**, 371–407 (2013)
21. Zhuang, X., Shen, J.: Multi-scale patch and multi-modality atlases for whole heart segmentation of MRI. Med. Image Anal. **31**, 77–87 (2016)

Unsupervised Multi-modality Registration Network Based on Spatially Encoded Gradient Information

Wangbin Ding[1], Lei Li[2,3], Liqin Huang[1($\boxtimes$)], and Xiahai Zhuang[2($\boxtimes$)]

[1] College of Physics and Information Engineering, Fuzhou University, Fuzhou, China
hlq@fzu.edu.cn
[2] School of Data Science, Fudan University, Shanghai, China
zxh@fudan.edu.cn
[3] School of Biomedical Engineering, Shanghai Jiao Tong University, Shanghai, China

Abstract. Multi-modality medical images can provide relevant or complementary information for a target (organ, tumor or tissue). Registering multi-modality images to a common space can fuse these comprehensive information, and bring convenience for clinical application. Recently, neural networks have been widely investigated to boost registration methods. However, it is still challenging to develop a multi-modality registration network due to the lack of robust criteria for network training. In this work, we propose a multi-modality registration network (MMReg-Net), which can perform registration between multi-modality images. Meanwhile, we present spatially encoded gradient information to train MMRegNet in an unsupervised manner. The proposed network was evaluated on the public dataset from MM-WHS 2017. Results show that MMRegNet can achieve promising performance for left ventricle registration tasks. Meanwhile, to demonstrate the versatility of MMRegNet, we further evaluate the method using a liver dataset from CHAOS 2019. Our source code is publicly available (https://github.com/NanYoMy/mmregnet).

Keywords: Multi-modality registration · Left ventricle registration · Unsupervised registration network · Gradient information

1 Introduction

Registration is a critical technology to establish correspondences between medical images [9]. The study of registration algorithm enables tumor monitoring [17], image-guided intervention [1], and treatment planning [7]. Multi-modality images, such as CT, MR and US, capture different anatomical information. Alignment of multi-modality images can help a clinician to improve the disease

X. Zhuang and L. Huang—Co-senior. This work was funded by the National Natural Science Foundation of China (Grant No. 61971142), and Shanghai Municipal Science and Technology Major Project (Grant No. 2017SHZDZX01).

© Springer Nature Switzerland AG 2022
E. Puyol Antón et al. (Eds.): STACOM 2021, LNCS 13131, pp. 151–159, 2022.
https://doi.org/10.1007/978-3-030-93722-5_17

diagnosis and treatment [6]. For instance, Zhuang [21] registered multi-modality myocardium MR images to fuse complementary information for myocardial segmentation and scar quantification. Heinrich et al. [10] performed registration of intra-operative US to pre-operative MR, which could aid image-guided neurosurgery.

Over the last decades, various methods have been proposed to perform multi-modality image registration. The most common methods are based on statistical similarity metric, such as mutual information (MI) [15], normalized MI (NMI) [18] and spatial-encoded MI (SEMI) [22]. Registrations are performed by maximizing these similarity metrics between the moved and fixed images. However, these metrics usually suffer from the loss of spatial information [16]. Other common methods are based on invariant representation. Wachinger et al. [19] presented the entropy and Laplacian image which are invariant structural representations across the multi-modality image, and registrations were achieved by minimizing the difference between the invariant representations. Zhuang et al. [23] proposed the normal vector information of intensity image for registration, which obtained comparable performance to the MI and NMI. Furthermore, Heinrich et al. [9] designed a handcraft modality independent neighborhood descriptor to extract structure information for registrations. Nevertheless, these conventional methods solved the registration problem by iterative optimizing, which is not applicable for time-sensitive scenarios.

Recently, several registration networks, which could efficiently achieved registrations in an one-step fashion, have been widely investigated. Hu et al. [11] proposed a weakly supervised registration neural network for multi-modality images by utilizing anatomical labels as the criteria for network training. Similarly, Balakrishnan et al. [4] proposed a learning-based framework for image registration. The framework could extend to multi-modality images when the anatomical label is provided during the training. Furthermore, Luo et al. [14] proposed a group-wise registration network, which could jointly register multiple atlases to the target image. Nevertheless, these methods required extensive anatomical labels for network training, which prevents them from unlabeled datasets.

At present, based on image-to-image translation generative adversarial network (GAN) [12], several unsupervised registration networks had been proposed. Qin et al. [16] disentangled a shape representation from multi-modality images via GAN, then a convenient mono-modality similarity metric could be applied on the shape representation for registration network training. Arar et al. [2] connected registration network with a style translator. The network could jointly perform spatial and style transformation on a moving image, and was trained by minimizing the difference between the transformed moving image and fixed image. The basic idea of these GAN-based methods is converting the multi-modality registration problem into a mono-modality one. Unfortunately, GAN methods easily cause geometrically distortions and intensity artifacts during image translation [20], which may lead to unrealistic registration results.

In this work, we propose an end-to-end multi-modality 3D registration network (MMRegNet). The main contributions are: (1) We present a spatially encoded gradient information (SEGI), which can provide a similarity criteria to train the registration network in an unsupervised manner. (2) We evaluated our method on multi-modality cardiac left ventricle and liver registration tasks and obtained promising performance on both applications.

2   Method

Registration Network: Let I_m and I_f be a moving and fix image, respectively. Here, I_m and I_f are acquired via different imaging protocols, and are defined in a 3-D spatial domain Ω. We construct the MMRegNet based on a U-shape convolution neural network [5], which takes a pair of I_m and I_f images as input, and predicts forward U and backward V dense displacement fields (DDFs) between them simultaneously. Therefore, MMRegNet is formulated as follows,

$$(U, V) = f_\theta(I_m, I_f) \tag{1}$$

where θ is the parameter of MMRegNet. Each voxel $x \in \Omega$ in I_m and I_f can be transformed by U and V as follows,

$$(I_m \circ U)(x) = I_m(x + U(x)), \tag{2}$$

$$(I_f \circ V)(x) = I_f(x + V(x)), \tag{3}$$

where $\circ$ is spatial transformation operation, and $I_m \circ U$, $I_f \circ V$ denote the moved image of I_m, I_f, respectively.

Spatially Encoded Gradient Information: Generally, the parameter of MMRegNet could be optimized by minimizing the intensity-based criteria, such as mean square error of intensity between the moved image and fixed image. However, such metrics are ill-posed when applied in multi-modality scenarios. This is because the intensity distribution of an anatomy usually varies across different modalities of images. Normalized gradient information (NGI) [8] is widely explored for conventional multi-modality registration methods. The basic idea of NGI is based on the assumption that image structures can be defined by intensity changes. Let G_I be the NGI of an intensity image I, each element of G_I is calculated as follows,

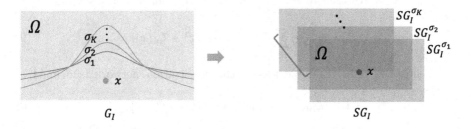

Fig. 1. A visual demonstration of the SEGI.

$$G_I(x) = \frac{\nabla I(x)}{\|\nabla I(x)\|_2}, \tag{4}$$

where $x \in \Omega$, and ∇I refers to the gradient of image I. Ideally, MMRegNet can be trained by minimizing the difference between $G_{I_m \circ U}$ and G_{I_f}. However, such a criteria is sensitive to the noises or artifacts of intensity images. It is still error-prone to train a registration network via NGI criteria in our practical experiments. To overcome this, we extend NGI to SEGI. Figure 1 illustrates the idea of SEGI, it is achieved by introducing a set of spatial variables $\Sigma = \{\sigma_1, \sigma_2 \cdots, \sigma_K\}$ to the standard NGI (G_I). For each spatial variable σ_k, we compute its associated SEGI ($SG_I^{\sigma_k}$) as follows,

$$SG_I^{\sigma_k}(x) = \sum_{p \in \Omega} \mathcal{N}(p|x, \sigma_k^2) \frac{\nabla I(p)}{\|\nabla I(p)\|_2}, \tag{5}$$

where $x \in \Omega$, and $\mathcal{N}(p|x, \sigma_k^2)$ denotes Gaussian distribution. Notably, we accumulate gradient information around x for a more robust representation of the intensity change. Finally, given a set of spatial variables Σ, the SEGI of an intensity image I is defined as,

$$SG_I = \{SG_I^{\sigma_1}, SG_I^{\sigma_2}, \cdots, SG_I^{\sigma_K}\}. \tag{6}$$

Loss Function: We train the network by minimizing the cosine distance between the SEGI of moved ($SG_{I_m \circ U}$) and fixed (SG_{I_f}) images,

$$\mathcal{L}_{SG} = \frac{1}{K} \sum_{k=1}^{K} \mathcal{D}(SG_{I_m \circ U}^{\sigma_k}, SG_{I_f}^{\sigma_k}), \tag{7}$$

$$\mathcal{D}(SG_{I_m \circ U}^{\sigma_k}, SG_{I_f}^{\sigma_k}) = \frac{-1}{|\Omega|} \sum_{x \in \Omega} cos(SG_{I_m \circ U}^{\sigma_k}(x), SG_{I_f}^{\sigma_k}(x)), \tag{8}$$

where $|\Omega|$ counts the number of voxels in an image, and $cos(\boldsymbol{A}, \boldsymbol{B})$ calculates the cosine distance between vector $\boldsymbol{A}$ and $\boldsymbol{B}$.

Meanwhile, MMRegNet is designed to simultaneously predict U and V for each pair of I_m and I_f. Normally, U and V should be inverse of each other. Hence, we employ the cycle consistent constraint [5] for the DDFs such that each I_m can be restored to its original one after transforming by U and V in succession,

$$\mathcal{L}_{CC} = \frac{1}{|\Omega|} \sum_{x \in \Omega} \|I_m \circ U \circ V(x) - I_m(x)\|_1. \tag{9}$$

Finally, the total trainable loss of the registration network is defined as follows,

$$\mathcal{L} = \mathcal{L}_{SG} + \lambda_1 \mathcal{L}_{CC} + \lambda_2 \{\Psi(U) + \Psi(V)\}, \tag{10}$$

where $\Psi(U)$ and $\Psi(V)$ are smoothness regularization terms for DDFs, and λ_1, λ_2 are the hyper-parameters.

3   Experiments and Results

Experimental Setups: MMRegNet was implemented by the TensorFlow on an NVIDIA P100. We tested it on two public datasets, i.e., the MM-WHS[1] [24] and CHAOS[2] [13].

- MM-WHS: MM-WHS contains multi-modality (CT, MR) cardiac medical images. We utilized 20 MR and 20 CT images for left ventricle registration task. MMRegNet was trained to perform the registration of MR to CT images.
- CHAOS: CHAOS contains multi-modality abdominal images from healthy volunteers. For each volunteer, the dataset includes their T1, T2 and CT images. We adopted 20 T1 MR, 20 T2 MR and 20 CT images for liver registration.

During the training phase, we employed ADAM optimizer to optimize the network parameters for 5000 epochs. The spatial variables Σ were given to $\{1, 1.5, 3\}$ practically, aiming to capture multi-scale of robust gradient information for registration. Meanwhile, we tested λ_1 and λ_2 with four different weighting values, i.e., 0.01, 0.1, 1, 10. According to the corresponding results of different setups, we set $\{\lambda_1 = 0.1, \lambda_2 = 10\}$ and $\{\lambda_1 = 0.1, \lambda_2 = 1\}$ for MM-WHS and CHAOS dataset, respectively. To evaluate the performance of MMRegNet, we computed the Dice (DS) and average symmetric surface distance (ASD) between the corresponding label of moved and fix images. All experimental results were reported by 4-fold cross-validation.

Results: We compared our registration method with three state-of-the-art multi-modality registration methods.

- Sy-NCC: The conventional affine + deformable registration, which is based on the symmetric image normalization method with normalized cross-correlation (NCC) as optimization metric [3]. We implemented it based on the popular ANTs software package.[3]
- Sy-MI: The Sy-NCC method which uses the MI instead of the NCC as optimization metric.
- VM-NCC: The state-of-the-art registration network [4], which was trained by using the NCC as training criteria. We adopted their official online implementation.[4]

Table 1 shows the results on MM-WHS dataset. Compared with the conventional methods (Sy-NCC and Sy-MI), MMRegNet could achieve better performance on both left ventricle cavity (LVC) and left ventricle myocardium (Myo). Notably, compared to the state-of-the-art registration network, i.e., VM-NCC, MMRegNet obtained comparable results in terms of DS and ASD. This reveals

---

[1] www.sdspeople.fudan.edu.cn/zhuangxiahai/0/mmwhs/.

[2] https://chaos.grand-challenge.org/.

[3] https://github.com/ANTsX/ANTsPy.

[4] https://github.com/voxelmorph/voxelmorph.

Table 1. The performance of different multi-modality registration methods on MM-WHS dataset.

Method	LVC (MR → CT)		Myo (MR → CT)	
	DS (%)↑	ASD (mm)↓	DS (%)↑	ASD (mm)↓
Sy-NCC [3]	70.07 ± 16.57	4.51 ± 2.67	50.66 ± 16.02	4.10 ± 1.77
Sy-MI [3]	69.16 ± 15.25	4.66 ± 2.54	49.00 ± 16.21	4.34 ± 2.04
VM-NCC [4]	79.46 ± 8.73	**2.81 ± 1.05**	62.77 ± 9.51	**2.49 ± 0.61**
MMRegNet	**80.28 ± 7.22**	3.46 ± 1.30	**62.92 ± 8.62**	3.01 ± 0.74

Table 2. The performance of different multi-modality registration methods on CHAOS dataset.

Method	Liver (T1 → CT)		Liver (T2→CT)	
	DS (%)↑	ASD (mm)↓	DS (%)↑	ASD (mm)↓
Sy-NCC [3]	74.94 ± 11.05	8.46 ± 4.10	75.46 ± 9.42	8.41 ± 3.86
Sy-MI [3]	73.88 ± 10.08	8.84 ± 3.70	75.82 ± 7.23	8.32 ± 2.73
VM-NCC [4]	74.63 ± 6.54	8.25 ± 2.17	71.10 ± 6.09	9.30 ± 2.01
MMRegNet	**79.00 ± 8.06**	**7.03 ± 2.55**	**76.71 ± 8.80**	**7.87 ± 1.75**

that MMRegNet is applicable for multi-modality registration tasks, and the proposed SEGI could serve as another efficient metric, such as MI and NCC, for multi-modality registration.

Table 2 shows the results on CHAOS dataset. We independently reported the registration result of T1 or T2 to CT images. MMRegNet achieved comparable accuracy to the state-of-the-art conventional methods, i.e., Sy-MI and Sy-NCC. Meanwhile, compared to VM-NCC, MMRegNet obtained average 4.99% (T1 → CT: 4.37%, T2 → CT: 5.61%) and 1.33 mm (T1 → CT: 1.22 mm, T2 → CT: 1.43 mm) improvements in terms of DS and ASD, respectively. This indicates that MMRegNet could achieve promising performance for multi-modality registration tasks.

Additionally, Fig. 2 visualizes four representative cases from the two datasets. On MM-WHS dataset, one can observe that both VM-NCC and MMRegNet achieved better visual results than Sy-MI and Sy-NCC, which is consistent with the quantitative results in Table 1. On CHAOS dataset, the yellow arrows highlight where MMRegNet could obtain relative reasonable results than other methods.

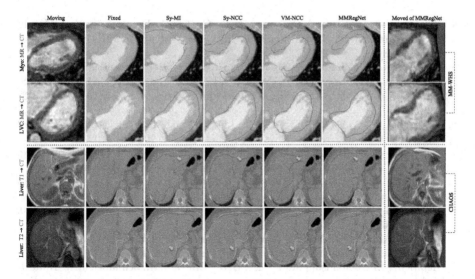

Fig. 2. Visualization of different methods on MM-WHS and CHAOS datasets. The showed images are the representative cases in terms of DS by MMRegNet. The blue contours are the gold standard label of the fixed images, while the red contours delineate the label of moving or moved images. We indicate the advantage of MMRegNet via yellow arrows. Moreover, the last column presents the moved images of MMRegNet. (The reader is referred to the online version of this article)

4 Conclusion

In this paper, we present an end-to-end network for multi-modality registration. The network is both applicable for heart and liver registration tasks. Meanwhile, we propose SEGI to obtain a robust structural representation for multi-modality images, and then applied it as the loss function for unsupervised registration network training. The results showed that MMRegNet could achieve promising performance when comparing with the state-of-the-art registration methods. Further work will extend MMRegNet to other multi-modality datasets.

References

1. Alam, F., Rahman, S.U., Ullah, S., Gulati, K.: Medical image registration in image guided surgery: issues, challenges and research opportunities. Biocybern. Biomed. Eng. **38**(1), 71–89 (2018)
2. Arar, M., Ginger, Y., Danon, D., Bermano, A.H., Cohen-Or, D.: Unsupervised multi-modal image registration via geometry preserving image-to-image translation. In: Proceedings of the IEEE/CVF Conference on Computer Vision and Pattern Recognition, pp. 13410–13419 (2020)
3. Avants, B.B., Tustison, N., Song, G.: Advanced normalization tools (ANTS). Insight J. **2**(365), 1–35 (2009)

4. Balakrishnan, G., Zhao, A., Sabuncu, M.R., Guttag, J., Dalca, A.V.: VoxelMorph: a learning framework for deformable medical image registration. IEEE Trans. Med. Imaging **38**(8), 1788–1800 (2019)
5. Ding, W., Li, L., Zhuang, X., Huang, L.: Cross-modality multi-atlas segmentation using deep neural networks. In: Martel, A.L., et al. (eds.) MICCAI 2020. LNCS, vol. 12263, pp. 233–242. Springer, Cham (2020). https://doi.org/10.1007/978-3-030-59716-0_23
6. Fu, Y., Lei, Y., Wang, T., Curran, W.J., Liu, T., Yang, X.: Deep learning in medical image registration: a review. Phys. Med. Biol. **65**(20), 20TR01 (2020)
7. Giesel, F., et al.: Image fusion using CT, MRI and pet for treatment planning, navigation and follow up in percutaneous RFA. Exp. Oncol. **31**(2), 106 (2009)
8. Haber, E., Modersitzki, J.: Intensity gradient based registration and fusion of multi-modal images. In: Larsen, R., Nielsen, M., Sporring, J. (eds.) MICCAI 2006. LNCS, vol. 4191, pp. 726–733. Springer, Heidelberg (2006). https://doi.org/10.1007/11866763_89
9. Heinrich, M.P., et al.: MIND: modality independent neighbourhood descriptor for multi-modal deformable registration. Med. Image Anal. **16**(7), 1423–1435 (2012)
10. Heinrich, M.P., Jenkinson, M., Papież, B.W., Brady, S.M., Schnabel, J.A.: Towards realtime multimodal fusion for image-guided interventions using self-similarities. In: Mori, K., Sakuma, I., Sato, Y., Barillot, C., Navab, N. (eds.) MICCAI 2013. LNCS, vol. 8149, pp. 187–194. Springer, Heidelberg (2013). https://doi.org/10.1007/978-3-642-40811-3_24
11. Hu, Y., et al.: Weakly-supervised convolutional neural networks for multimodal image registration. Med. Image Anal. **49**, 1–13 (2018)
12. Huang, X., Liu, M.Y., Belongie, S., Kautz, J.: Multimodal unsupervised image-to-image translation. In: Proceedings of the European Conference on Computer Vision (ECCV), pp. 172–189 (2018)
13. Kavur, A.E., et al.: CHAOS challenge-combined (CT-MR) healthy abdominal organ segmentation. Med. Image Anal. **69**, 101950 (2021)
14. Luo, X., Zhuang, X.: MvMM-RegNet: a new image registration framework based on multivariate mixture model and neural network estimation. In: Martel, A.L., et al. (eds.) MICCAI 2020. LNCS, vol. 12263, pp. 149–159. Springer, Cham (2020). https://doi.org/10.1007/978-3-030-59716-0_15
15. Maes, F., Collignon, A., Vandermeulen, D., Marchal, G., Suetens, P.: Multimodality image registration by maximization of mutual information. IEEE Trans. Med. Imaging **16**(2), 187–198 (1997)
16. Qin, C., Shi, B., Liao, R., Mansi, T., Rueckert, D., Kamen, A.: Unsupervised deformable registration for multi-modal images via disentangled representations. In: Chung, A.C.S., Gee, J.C., Yushkevich, P.A., Bao, S. (eds.) IPMI 2019. LNCS, vol. 11492, pp. 249–261. Springer, Cham (2019). https://doi.org/10.1007/978-3-030-20351-1_19
17. Seeley, E.H., et al.: Co-registration of multi-modality imaging allows for comprehensive analysis of tumor-induced bone disease. Bone **61**, 208–216 (2014)
18. Studholme, C., Hill, D.L., Hawkes, D.J.: An overlap invariant entropy measure of 3D medical image alignment. Pattern Recogn. **32**(1), 71–86 (1999)
19. Wachinger, C., Navab, N.: Entropy and Laplacian images: structural representations for multi-modal registration. Med. Image Anal. **16**(1), 1–17 (2012)
20. Zhang, Z., Yang, L., Zheng, Y.: Translating and segmenting multimodal medical volumes with cycle-and shape-consistency generative adversarial network. In: Proceedings of the IEEE Conference on Computer Vision and Pattern Recognition, pp. 9242–9251 (2018)

21. Zhuang, X.: Multivariate mixture model for myocardial segmentation combining multi-source images. IEEE Trans. Pattern Anal. Mach. Intell. **41**(12), 2933–2946 (2018)
22. Zhuang, X., Arridge, S., Hawkes, D.J., Ourselin, S.: A nonrigid registration framework using spatially encoded mutual information and free-form deformations. IEEE Trans. Med. Imaging **30**(10), 1819–1828 (2011)
23. Zhuang, X., Gu, L., Xu, J.: Medical image alignment by normal vector information. In: Hao, Y., et al. (eds.) CIS 2005. LNCS (LNAI), vol. 3801, pp. 890–895. Springer, Heidelberg (2005). https://doi.org/10.1007/11596448_132
24. Zhuang, X., et al.: Evaluation of algorithms for multi-modality whole heart segmentation: an open-access grand challenge. Med. Image Anal. **58**, 101537 (2019)

In-silico Analysis of Device-Related Thrombosis for Different Left Atrial Appendage Occluder Settings

Eric Planas[1], Jordi Mill[1], Andy L. Olivares[1], Xabier Morales[1],
Maria Isabel Pons[1], Xavier Iriart[2], Hubert Cochet[2], and Oscar Camara[1(✉)]

[1] PhySense, Department of Information and Communication Technologies,
Universitat Pompeu Fabra, Barcelona, Spain
oscar.camara@upf.edu
[2] Hôpital de Haut-Lévêque, Bordeaux, France

Abstract. Atrial fibrillation (AF) is one of the most common cardiac arrhythmias and is associated to an increasing risk of stroke. Most AF-related strokes are formed in the left atrial appendage (LAA). To prevent thrombus formation, LAA occlusion (LAAO) is considered a suitable alternative for AF patients with contraindications for anti-coagulation treatment. Nevertheless, LAAO is linked to a non-negligible risk of generating thrombus at the surface near the device (i.e., device-related thrombus, DRT), depending on the implantation settings. For instance, it has been shown that not covering the pulmonary ridge (PR) with the LAAO increases the risk of DRT. In-silico analysis is a useful tool to better understand the blood flow patterns after LAAO and predict the risk of DRT for a given patient and device configuration. In the present work we designed a modelling pipeline based on fluid simulations, including a thrombus model using discrete phase modelling, to analyse the risk of DRT in six patient-specific geometries for different LAAO settings. In particular, we studied the possible incidence of DRT depending on the device positioning (covering/uncovering the PR) and type (Amplatzer Amulet and Watchman FLX). The resulting in-silico indices demonstrated that covering the PR entails less thrombogenic patterns than uncovering it. In our study, disk-based devices had better adaptability to complex LAA morphologies and a slightly minor associated risk of DRT than non-disk devices.

Keywords: Atrial appendage · Occluder device · Thrombus · Particle model · Fluid dynamic

1 Introduction

The risk of thrombus formation in the left atrial appendage (LAA) is increased in patients with atrial fibrillation (AF). When oral anticoagulants (OAC) are contraindicated to AF patients, left atrial appendage occlusion (LAAO) stands as a

© Springer Nature Switzerland AG 2022
E. Puyol Antón et al. (Eds.): STACOM 2021, LNCS 13131, pp. 160–168, 2022.
https://doi.org/10.1007/978-3-030-93722-5_18

suitable alternative. Nevertheless, some complications are associated to LAAO interventions, including peri-device bleeding, device migration and device-related thrombus (DRT). Regarding DRT, a large variability on its prevalence has been reported, ranging from 1% to 7% of the patients [1]. According to these studies, DRT seems to be ruled by multiple factors, such as an incomplete endothelialization of the device and the presence of thrombogenic patterns (i.e., low velocity re-circulations and vortex structures, high wall shear stress). For instance, recent investigations [1,2] have demonstrated that in most patients developing DRT, the thrombus was formed in the PR region, with the LAAO not covering it. Additionally, low velocities ($\leq$0.2 m/s) near the LAA have been associated with an increased risk of DRT [3].

In LAAO planning, it is crucial to accurately explore and understand the LA/LAA geometry of the patient since it is highly variable and fully determine which device, and where, is going to be implanted. To characterize LAA morphology, the most used imaging techniques are transesophageal echocardiography (TEE) and X-ray, but Computed Tomography (CT) is starting to have a relevant role for this purpose as well [4]. Unfortunately, standard imaging techniques provide limited information about LA and LAA haemodynamics, which is critical for DRT analysis. Several in-silico models with fluid simulations in the LA are available in the literature (e.g., [5–7] as some recent ones), including thrombus models with Discrete Phase Modelling (DPM) approaches [8], which contribute to better understand blood flow patterns in the LA under different conditions. Some investigators [9–11] have also incorporated LAAO devices to study the influence of device settings on post-procedure haemodynamic outcomes such as DRT, but only in a limited number of LAAO options and LA geometries, without considering thrombus models. The main objective of this study was the evaluation of DRT after LAAO through the in-silico analysis of simulated blood flow patterns and platelet deposition (with DPM) with different LAAO types (Amplatzer Amulet from Abbott Vascular, Santa Clara, CA, USA; and Watchman FLX from Boston Scientific, Malborough, Massachusetts, USA) and positions (covering or not the PR) on six different patient-specific LA/LAA geometries extracted from CT images.

2   Materials and Methods

2.1   Data Acquisition and Modelling Pipeline

The LA geometries of the six cases that were assessed in this work were obtained from the segmentation of CT images of patients from Hôpital Cardiologique du Haut-Lévêque (Bordeaux, France). Three-dimensional cardiac CT studies were performed on a 64-slice dual Source CT system (Siemens Definition, Siemens Medical Systems, Forchheim, Germany). CT scans were reconstructed into isotropic voxel sizes (ranged between 0.37–0.5 mm; 512 × 512 × [270–403] slices). The Hôpital Cardiologique du Haut-Lévêque ethical committees approved this study and patients had given written informed consent.

Different scenarios were tested using the modelling pipeline shown in Fig. 1. First, the LA was segmented from pre-procedural CT images using region-growing and thresholding tools available in 3D Slicer.[1] The surface meshes including the deployed device (built using Computational Aided Design software) in the LAA were built using Meshmixer.[2] Some simplifications needed to be made to achieve high-quality meshes (e.g., removal of anchors in Amulet device). During implantation, we considered rigid walls for the LA surface and a deformable device. Then, volumetric meshes were generated and, after conducting the CFD simulations, the resulting flow field and associated risk of DRT were assessed using different techniques of flow visualization and quantification.

For each LA geometry, the two studied positions (covering and uncovering the PR) were previously defined and assessed before implanting the device, with the help of interventional cardiologists from the Hôpital de Haut-Lévêque (Bordeaux). The size of the implanted device in each case was assessed using the VIDAA platform [9], based on morphological measurements such as the LAA ostium size and LAA depth. A total of 24 different configurations were successfully completed after placing the deployed devices at the corresponding positions in all studied geometries.

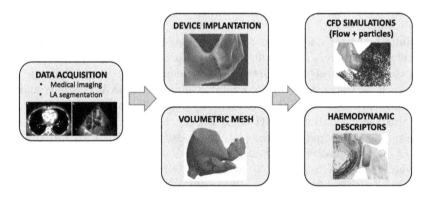

Fig. 1. Modelling pipeline, based on Computational Fluid Dynamics (CFD) simulations, for the in-silico analysis of device-related thrombus after left atrial appendage occlusion (LAAO).

2.2   Haemodynamic Simulations

Blood flow was simulated using Ansys Fluent 19.2 (Ansys, Inc. Pennsylvania, USA[3]). Blood was considered as an incompressible Newtonian fluid in a laminar regime, with a density and viscosity of $\rho = 1060$ kg/m^3 and $\mu = 0.0035$ kg/m·s.

---

[1] https://www.slicer.org/.

[2] meshmixer.com.

[3] https://www.ansys.com/products/fluids/ansys-fluent.

Three cardiac beats were simulated. In all cases the same boundary conditions were assumed: pressure inlet boundary conditions were set in all pulmonary veins (PV) with a profile extracted from an AF patient (in sinus rhythm); velocity outlet boundary conditions were set in the mitral valve (MV) from Doppler data from each studied patient. A spring-based dynamic mesh model was applied to include LA wall movement, considering the MV ring plane as a moving boundary as defined in [12]. Out of the resulting CFD simulations, blood flow velocities and several in-silico indices were estimated to assess the risk of DRT. For instance, as in previous works [11], we estimated the endothelial cell activation potential (ECAP) index, which is the ratio between the time average wall shear stress (TAWSS) and oscillatory shear index (OSI), since it points to regions with low velocities and complex flows, thus prone to DRT. Additionally, the Lambda-2 and Q-criterion were computed to visually identify regions with vortices in blood flow patterns. Both metrics depend of the rotational (i.e., vorticity) and strain rate tensors. For this study, a Lambda-2 with negative value (-50) was used for vortex identification, as in other studies focused in the human heart [13].

2.3 Platelet Adhesion Model

A discrete phase was added to the continuous phase (blood flow) in the simulations to assess the platelet attachment to the LAA and better evaluate DRT risk. During simulations, particles were injected through the PV inlet, which were then dragged by the flow until they got attached to the LAA or escaped though the MV. Particles were assumed to be clusters of platelets, thus the values of density ρ_p, viscosity μ_p, surface tension σ_p and platelet concentration in blood c_p were taken accordingly. Particle trajectories were computed individually by integrating the force balance of each particle within the continuous phase, which reads:

$$m_p \frac{d\mathbf{u}_p}{dt} = m_p \frac{\mathbf{u} - \mathbf{u}_p}{\tau_r} + \mathbf{F}, \tag{1}$$

where m_p is the particle mass, $\mathbf{u}$ the fluid phase velocity, $\mathbf{u}_p$ the particle velocity and ρ the fluid density. The term $m_p \left(\mathbf{u} - \mathbf{u}_p\right)/\tau_r$ corresponds to the drag force, where τ_r is the droplet relaxation time. An extra term, $\mathbf{F}$, with the Saffman lift force was included due to the influence of shear rate in the particle attachment process.

Particles were injected through the PV surfaces during the first 10 time steps at the beginning of each beat. We assumed that, in the fist injection, the number of injected particles, n_{part}, corresponded to a platelet concentration in blood of $c_p = 2 \cdot 10^8$ mL^{-1}, which is within the normal range in physiological conditions.[4] In order to fulfill this requirement, the following parameters were computed: the number of platelets per cluster, n_{ppc}; the particle diameter, d_p; and the total flow rate, Q. For the platelet models, the following values were assumed: number of time steps during the injection $n_{ts} = 10$, diameter of platelets $d_{plat} = 3$ μm, particle density $\rho_p = 1550$ kg m^{-3} and time steps for the injection $\Delta t = 0.01$ s. As

---

[4] https://bionumbers.hms.harvard.edu/search.aspx.

Table 1. Mean velocities during the second cardiac cycle in a region near the device, in systole (Sys) and diastole (Dias). In bold, values below 0.2 m/s. PR: pulmonary ridge.

Average velocities near the device (m/s)							
	PR covered				PR uncovered		
	Amulet		Watchman		Amulet		Watchman	
	Sys	Dias	Sys	Dias	Sys	Dias	Sys	Dias
Case 1	**0.11**	0.29	**0.12**	0.30	**0.10**	**0.17**	**0.13**	**0.19**
Case 2	**0.16**	0.28	**0.18**	0.29	**0.11**	**0.18**	**0.07**	**0.17**
Case 3	**0.16**	0.39	**0.17**	0.37	**0.14**	0.34	**0.16**	0.29
Case 4	**0.19**	0.36	**0.18**	0.36	**0.11**	0.22	**0.11**	0.25
Case 5	**0.09**	**0.13**	**0.07**	**0.12**	**0.05**	**0.08**	**0.02**	**0.05**
Case 6	0.22	0.29	**0.18**	0.25	**0.09**	**0.16**	**0.13**	**0.15**

we fixed concentration, all these parameters were dependent on the LA volume (V_{LA}). Therefore, the number of platelets per injection n_{plat}, was computed as: $n_{plat} = c_p \cdot V_{LA}$. In order to assess DRT, the number of particles attached to the LAA and their location were analyzed at the end of the simulations. To do so, the LAA surface was separated from the rest of the LA and a wall-film condition was assumed. This condition implied that any particle that touched the LAA, independently of its velocity, was stuck to the surface; they could eventually be detached from the LAA if the flow would drag them to the LAA boundary.

3   Results and Discussion

Table 1 displays the mean velocities obtained from fluid simulations for all studied cases and LAAO configurations. Mean velocities were below 0.2 m/s in almost all cases during systole, with the exception of Case 6 in the covered PR position with Amulet device. Low velocities during diastole were less common, due to the proximity of the LAA to the MV, which is open during this phase. Additionally, a higher incidence of low velocities in uncovered PR configurations was encountered, thus a higher risk of DRT. On the other hand, no significant differences between the two devices (i.e., Amulet and Watchman FLX) were encountered in this regard.

Figure 2 shows the four analyzed LAAO configurations (i.e., pulmonary ridge covered/uncovered and Amulet/Watchman FLX devices) for one LA geometry (Case 5), in the form of blood flow streamlines colored by the velocity magnitude and vorticity indices. As a general rule, flow re-circulation patterns with low velocities (i.e., conditions required for DRT) were commonly detected at the edge of Watchman FLX at some point during the cardiac cycle, independently of the position of the occluder. Same conclusions were drawn from the analysis of vorticity indices. Conversely, vortices with low velocities were rarely detected in the covered PR configurations with the Amulet device. However, vortices were

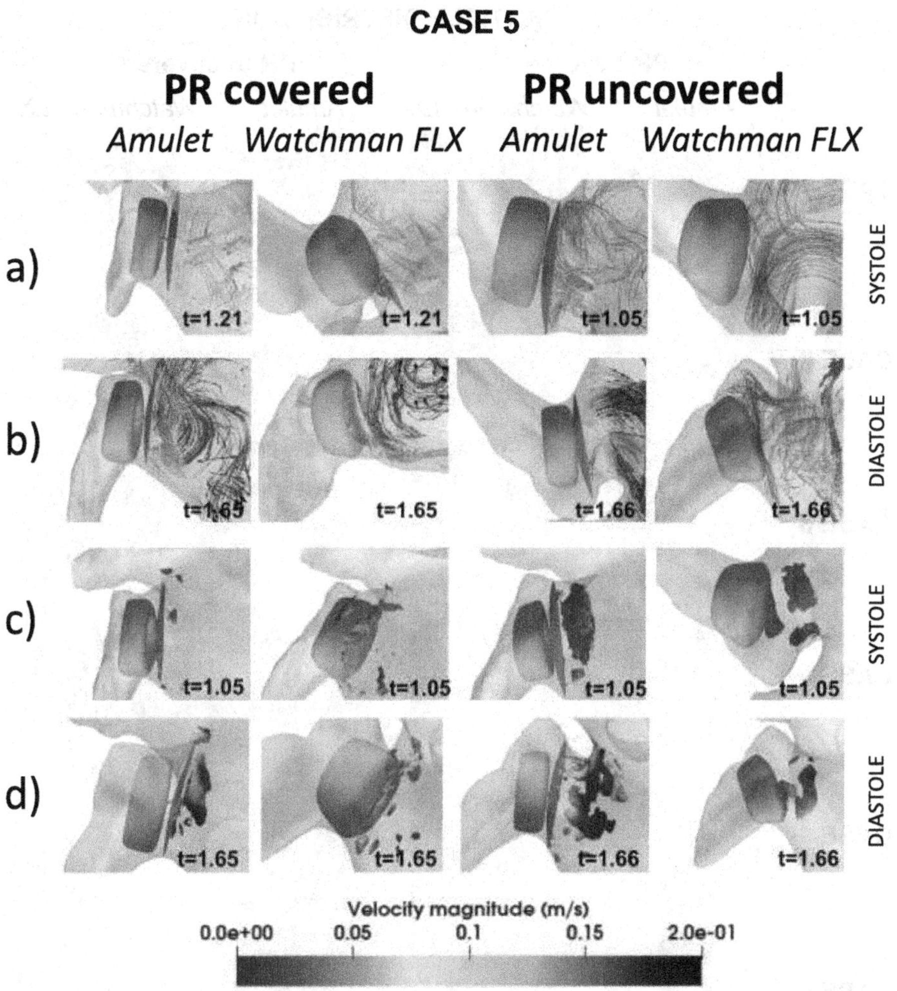

CASE 5

PR covered   PR uncovered
Amulet   Watchman FLX   Amulet   Watchman FLX

Fig. 2. Results for covered/uncovered pulmonary ridge (PR) for one of the studied geometries (Case 5). a)b) Blood flow streamlines colored by velocity magnitude (systole and diastole, respectively). c)d) Vorticity index, with isovolumes defined by $\lambda_2 = -50$ (systole and diastole, respectively). Time is expressed in seconds.

present in the uncovered PR position both for Amulet and Watchman FLX devices, being slightly more common in systole than in diastole. The Watchman FLX device was also related with higher ECAP values (i.e., more risk of DRT) in the neighborhood of the device than the Amulet device. As a general rule no relevant differences were appreciated between the ECAP maps of the covered and uncovered PR positions.

Figure 3 shows the platelet distribution at end diastole of the second cardiac cycle for some of the studied cases. It can be observed that platelets did not only

PARTICLE DISTRIBUTION

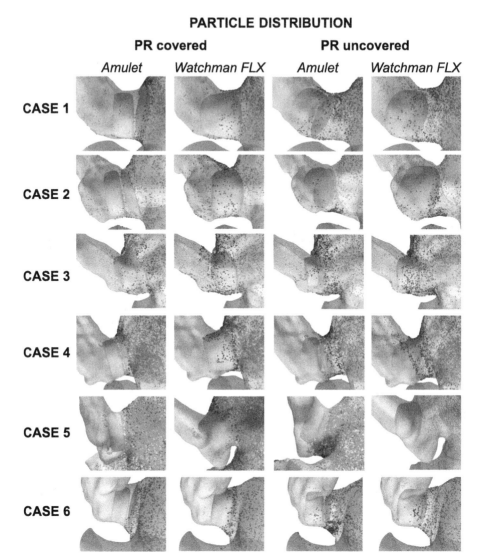

Fig. 3. Particle distribution in the left atrial appendage in diastole for the two analyzed devices (Amulet and Watchman FLX) in the covered and uncovered pulmonary ridge (PR) configuration. A higher density of attached particles (red) would be associated to a higher risk of device-related thrombus. Blue: free particles. (Color figure online)

stack in the PR region, but they got attached to other uncovered regions of the LAA. However, for the uncovered position for geometries 2 and 6, a significant amount of platelets clustered in the PR region rather than in other zones, in agreement with literature [2]. Overall, 75% of the analyzed configurations has more platelets attached to the LAA (i.e., more risk of DRT) with an uncovered

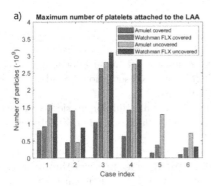
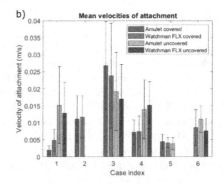

Fig. 4. a) Maximum number of platelets attached to the left atrial appendage (LAA). b) Mean velocity of attachment for all the studied configurations (Cases 1–6): Amulet-uncovered pulmonary ridge (PR) (blue); Watchman FLX-uncovered PR (orange); Amulet - covered PR (yellow); and Watchman FLX - covered PR (purple). (Color figure online)

PR, as can be seen in Fig. 4a (yellow and purple bars higher than blue and red ones in all cases). Also, the Watchman FLX device appeared to have more platelets prone to get attached to the LAA than the Amulet one.

4 Conclusions

A modelling pipeline has been developed to predict the risk of DRT after the implantation of different LAAO devices in several positions within six LA/LAA morphologies. According to the resulting haemodynamic in-silico indices, uncovered PR device configurations generated more thrombogenic patterns, which was confirmed by the DPM analysis. Moreover, the Amplatzer Amulet device showed less cases with low-velocity swirls and vortices than the Watchman FLX, thus being less prone to DRT after implantation, arguably due to the presence of the disk in the device design. Future work will mainly focus on establishing a robust and rigorous validation framework for the fluid simulations, using advanced imaging such as 4D flow magnetic resonance imaging and blood speckle tracking as well as in-vitro set-ups like the one recently proposed by Dueñas-Pamplona et al. [14].

Acknowledgments. This work was supported by the Agency for Management of University and Research Grants of the Generalitat de Catalunya under the Grants for the Contracting of New Research Staff Programme - FI (2020 FI_B 00608) and the Spanish Ministry of Economy and Competitiveness under the Programme for the Formation of Doctors (PRE2018-084062), the Maria de Maeztu Units of Excellence Programme (MDM-2015-0502) and the Retos Investigación project (RTI2018-101193-B-I00). Additionally, this work was supported by the H2020 EU SimCardioTest project (Digital transformation in Health and Care SC1-DTH-06-2020; grant agreement No. 101016496).

References

1. Aminian, A., Schmidt, B., Mazzone, P., et al.: Incidence, characterization, and clinical impact of device-related thrombus following left atrial appendage occlusion in the prospective global AMPLATZER Amulet Observational Study. JACC: Cardiovasc. Interv. **12**(11), 1003–1014 (2019)
2. Freixa, X., Cepas-Guillen, P., Flores-Umanzor, E., et al.: Pulmonary ridge coverage and device-related thrombosis after left atrial appendage occlusion. EuroIntervention **16**(15), E1288–E1294 (2021)
3. Tamura, H., Watanabe, T., Hirono, O., et al.: Low wall velocity of left atrial appendage measured by trans-thoracic echocardiography predicts thrombus formation caused by atrial appendage dysfunction. J. Am. Soc. Echocardiogr. **23**(5), 545–552 (2010)
4. Chow, D.H.F., Bieliauskas, G., Sawaya, F.J., et al.: A comparative study of different imaging modalities for successful percutaneous left atrial appendage closure. Open Heart **4**(2), e000627–e000627 (2017)
5. García-Isla, G., Olivares, A.L., Silva, E., et al.: Sensitivity analysis of geometrical parameters to study haemodynamics and thrombus formation in the left atrial appendage. Int. J. Numer. Meth. Biomed. Eng. **34**(8), 1–14 (2018)
6. Masci, A., Alessandrini, M., Forti, D., et al.: A proof of concept for computational fluid dynamic analysis of the left atrium in atrial fibrillation on a patient-specific basis. J. Biomech. Eng. **142**(1), 011002 (2020)
7. García-Villalba, M., Rossini, L., Gonzalo, A., et al.: Demonstration of patient-specific simulations to assess left atrial appendage thrombogenesis risk. Front. Physiol. **12**, 596596 (2021)
8. Wang, Y., Qiao, Y., Mao, Y., Jiang, C., Fan, J., Luo, K.: Numerical prediction of thrombosis risk in left atrium under atrial fibrillation. Math. Biosci. Eng. **17**(3), 2348–2360 (2020)
9. Aguado, A.M., et al.: In silico optimization of left atrial appendage occluder implantation using interactive and modeling tools. Front. Physiol. **10**(MAR), 237 (2019)
10. Mill, J., et al.: Impact of flow dynamics on device-related thrombosis after left atrial appendage occlusion. Can. J. Cardiol. **36**(6), 13–968 (2020)
11. Mill, J., et al.: Patient-specific flow simulation analysis to predict device-related thrombosis in left atrial appendage occluders. REC: Interv. Cardiol. (Engl. Ed.) (2021)
12. Veronesi, F., Corsi, C., Sugeng, L., et al.: Quantification of mitral apparatus dynamics in functional and ischemic mitral regurgitation using real-time 3-dimensional echocardiography. J. Am. Soc. Echocardiogr. **21**(4), 347–354 (2008)
13. ElBaz, M.S.M., Lelieveldt, B.P.F., Westenberg, J.J.M., van der Geest, R.J.: Automatic extraction of the 3D left ventricular diastolic transmitral vortex ring from 3D whole-heart phase contrast MRI using Laplace-Beltrami signatures. In: Camara, O., Mansi, T., Pop, M., Rhode, K., Sermesant, M., Young, A. (eds.) STACOM 2013. LNCS, vol. 8330, pp. 204–211. Springer, Heidelberg (2014). https://doi.org/10.1007/978-3-642-54268-8_24
14. Dueñas-Pamplona, J., Sierra-Pallares, J., García, J., Castro, F., Munoz-Paniagua, J.: Boundary-condition analysis of an idealized left atrium model. Ann. Biomed. Eng. **49**(6), 1507–1520 (2021). https://doi.org/10.1007/s10439-020-02702-x

Valve Flattening with Functional Biomarkers for the Assessment of Mitral Valve Repair

Paula Casademunt[1]([✉]), Oscar Camara[1], Bart Bijnens[2,3], Èric Lluch[1], and Hernan G. Morales[4]

[1] Physense, BCN MedTech, Department of Information and Communication Technologies, Universitat Pompeu Fabra, Barcelona, Spain
paula.casademunt01@estudiant.upf.edu
[2] ICREA, Barcelona, Spain
[3] IDIBAPS, Barcelona, Spain
[4] Philips Research Paris, Suresnes, France

Abstract. Computational modelling is becoming a crucial aid to better understand valve physiopathology. It allows experts to gain deeper insights on valve biomechanics and deformation, thus helping in the planning of therapies and assessing the efficacy of cardiovascular devices. However, there is a lack of proper visualization techniques to facilitate the interpretation of simulation results. In this work, Smoothed Particle Hydrodynamics (SPH) was used to model mitral valve regurgitation (MVR) and a common minimally-invasive intervention, an edge-to-edge repair. Furthermore, a flattening visualization of the mitral valve (MV) was implemented to ease the analysis of the obtained in-silico indices in different stages. The obtained results show the relevance of proper planning prior to the edge-to-edge repair procedure, improving the safety and efficacy of the devices, while decreasing the risk of re-intervention.

Keywords: Mitral valve regurgitation · SPH · Meshless · Cardiac computational modelling · Edge-to-edge repair · Flattening

1 Introduction

The main role of the mitral valve (MV) is to ensure unidirectional flow from the left atrium to the left ventricle. If the valve does not properly close, it can lead to a flow leakage, known as mitral valve regurgitation (MVR) [1]. Dysfunction or altered anatomy of any of the components of the MV system (i.e. leaflets, chordae tendinae, papillary muscles and mitral annulus) can lead to MVR.

Percutaneous therapies for MVR, such as edge-to-edge repair, increase the likelihood of successfully treating patients with complex presentations [2]. During this procedure, the anterior and posterior mitral leaflets are "clipped", re-establishing leaflet coaptation. It consistently reduces MVR, increasing the quality of life [2]. Nevertheless, there is a risk of treatment failure and re-intervention,

© Springer Nature Switzerland AG 2022
E. Puyol Antón et al. (Eds.): STACOM 2021, LNCS 13131, pp. 169–178, 2022.
https://doi.org/10.1007/978-3-030-93722-5_19

which can be reduced with proper patient selection and planning of the procedure. Clinical decision support tools using the predictive behaviour of modelling can be very useful in this multi-disciplinary and multi-dimensional problem.

Several computational models have been developed to assess the effect of the edge-to-edge repair on the mitral valve, ranging from Fluid-Structure Interaction to Finite Element Method models, using simplified geometries or patient-specific ones [3–5]. These works study the effect the clip has on the valve in terms of stress, structure, and regurgitation. Nevertheless, two difficulties can be mentioned: 1) mesh-related tasks, as surface preparation, interpolation or mesh adaptation; 2) complex result visualization and interpretation due to the convoluted structure of the MV.

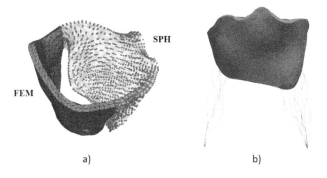

a)   b)

Fig. 1. a) SPH (Smoothed Particle Hydrodynamics) vs. FEM (Finite Element Method): Differences in domain discretization. b) Valve geometry with chordae linking the valve leaflets with the papillary muscles.

Although mesh-based methods are very mature and have successfully been used to model the MV, they have several challenges when dealing with complex structures, related to the need for a mesh (see Fig. 1). Firstly, mesh-based methods require a smooth surface where the volumetric mesh can be generated. This surface mesh may be difficult to obtain from medical images and manual pre-processing may be required. Secondly, for large valve deformations, mesh quality needs to be ensured for accurate calculations. Third, self contact can occur and therefore, the algorithm needs to handle this condition [6]. For the later challenge, the usual strategy is to manually divide the valve in segments that can collide.

Meshless methods, such as Smoothed Particle Hydrodynamics (SPH), do not have these limitations, and can be of great use in applications with large degrees of deformations [7]. Furthermore, irregular segmentations can be used as geometrical inputs, thanks to the implicit smoothness of the algorithm. Although the maturity of these methods has yet to improve, recent developments in SPH have allowed the modelling of a wider range of problems, including cardiac biomechanics [7–9].

In this work, we propose SPH to overcome the inherent mesh-related difficulties in valve modelling, combined with a flattening of the valve to visualize the results. In this way, the functional biomarkers from the simulation of healthy, diseased and treated valves can be mapped on the flattened valve view for later comparison.

2   Materials and Methods

2.1   SPH Continuum Mechanics

A total Lagrangian SPH numerical scheme was used to numerically solve the partial differential equations governing the valve motion (please refer to [7,9] for further information). In SPH, the domain is defined as a set of moving particles without the need of a mesh. A kernel function defines a set of neighbouring particles for each particle and weights how they interact. The use of SPH for MV solid mechanics requires four equations to be solved, namely: the conservation of momentum, the continuity equation, the deformation gradient tensor, and the Neo-Hookean constitutive law for a quasi incompressible material.

Contact between particles is detected when the particle i is near a non-neighbour particle k within a given distance δ_c. The algorithm used in this work, proposed by Wang et al. [10], was initially thought for sliding frictional problems, but it can also be used to avoid particle penetration without the need of a problem-dependent penalization coefficient commonly used with penalty-based contact algorithms. Moreover, the authors performed validation tests to proof the efficiency, accuracy and stability of their algorithm. Once contact between two particles (i, k) is detected, the contact force is computed by their relative velocity, weighted by their mass ratio. To ensure symmetry within the collisions, the average normal vector of the particles is computed following Seo and Min [11]. In that work, SPH was also used and validated. However, the contact algorithm from Seo and Min was not adopted here due to its dependency with penalty value calculation and the need for an accurate particle penetration, which makes relatively more expensive calculations.

2.2   Mitral Valve Model

The geometry used for the computational simulations was obtained from an open access database [12], consisting of patient-specific valves segmented out of Computed Tomography (CT) images from patients with MVR. To obtain a less stenotic valve, Ros et al. [9] manually remodeled case 03 user 01. This mesh was a surface model, and so, two additional layers were added to the original one to generate a valve with a uniform thickness of 1 mm, since SPH needs volumes. The resulting geometry included 2694 regularly distributed particles and a valve surface area of $4.65\,\text{cm}^2$.

MV leaflets were defined to be hyperelastic, according to the description of the Neo-Hookean constitutive model. Young's Modulus was defined as 1.5

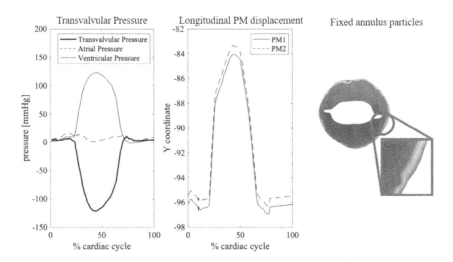

Fig. 2. The boundary conditions to consider: a physiological transvalvular pressure (left), PM (papillary muscle) motion over one cardiac cycle (middle), and fixed annular ring particles (right). PM1 and PM2 stand for the left and right papillary muscles, respectively.

MPa and Poisson's Ratio was set to 0.49, accordingly to [13], to model nearly incompressible behaviour.

Three boundary conditions were applied to deform the mitral valve model. First, the chords were modeled as a set of 130 linear mass-spring systems (that only act in traction) connecting particles with two points in space, which represent the papillary tips and move over time. The movement of the papillary muscle tips followed the dynamics described in Sanfilippo et al. [14]. Secondly, a time-varying transvalvular pressure (extracted from Kaiser et al. [15]) was uniformly applied to the ventricle layer of the valve leaflets, illustrated in Fig. 2. Finally, the mitral annulus ring particles were set to be fixed.

Pathological Valve Models. Two pathological states were studied. First, Primary Mitral Valve Regurgitation (PMVR) was modeled in terms of chordae degeneration. To represent this mechanism, the stiffness parameter of 46 chords of the P1-P2 scallops was decreased from 10^5 to 10^4 N/m. The second pathological case was Traumatic Mitral Valve Regurgitation (TMVR), which is a more severe case of PMVR, posterior to a traumatic injury. In this scenario, there is a total rupture of the degenerated chords, resulting in a valve without chords attached to the P1-P2 scallops.

2.3   Edge-to-Edge Repair

The edge-to-edge repair intervention was modeled as springs linking a group of particles from the free edge of opposing leaflets. The springs were set to have an initial length $L_0 = 0$ mm, and a stiffness parameter of $K = 1.5$ MN/m, to ensure the coaptation of that region throughout the cardiac cycle. The width of the clipped region was approximately 5 mm, mimicking the width of a MitraClip, a device used for this intervention. Two independent clip repairs were performed, one in the P1/A1 area (where coaptation was observed), and another repair in the P3/A3 (positions displayed in Fig. 3).

Fig. 3. Positions of placement of the MitraClip, in the P1/A1 and P3/A3 scallops.

2.4 Flattening Visualisation of the Mitral Valve

As a mean to improve the visualization, comparison and quantification of all the simulations, the valves were flattened based on the work of Lichtenberg *et al.* [18]. The general idea of the algorithm is to cut the MV along its lateral commissure and to unroll it along the diameter. Three steps can be identified:

1. **Annulus parametrization:** The annulus' height was plotted along the y-axis, and its length along the x-axis. The correspondence of the x-axis in 3D was a reference plane that crossed the two commissure points of the annulus and the barycenter of the posterior annulus.
2. **Leaflet initialization:** The leaflet geometry was placed below the annulus, making use of spline-like lines. The points in the spline-like line were mapped to a shared x value, and the distance between points was preserved vertically.
3. **Leaflet relaxation optimization:** Since the obtained projection had some deformations in the area, an energy term that described the distortion of the edge lengths was minimized. This per-vertex energy compares the distance between a point and its neighbours in 3D and 2D.

The original work uses the flattened views to depict structural differences between MVs and in-vivo measurements from patients, such as the prolapsed regions at peak systole. In this work, we made use of this flattened view to easily depict the outcomes of a simulation, identifying differences in a specific in-silico metric between two different scenarios. In other words, combining the predictive capabilities of SPH, with the visualization scheme proposed by Lichtenberg *et al.* [18], we could compare diseased valves with post-interventional ones, using metrics that could not have been obtained from imaging techniques. Hence, it becomes a very powerful tool to identify the changes produced by a certain intervention, allowing a deeper insight on the impacts of the procedure, helping in the decision-making process.

2.5   Simulation Analysis

The original mesh connectivity was used to link the simulation particles and build a surface, and the values of the analysed metrics were interpolated to better visualize the results. The outcomes of the simulations were assessed with the flattened views by means of the contact, curvature and principal stresses of the valve. The latter can give insight regarding the risk of leaflet rupture. The contact can be directly related to the degree of coaptation and the pathological condition of the valve, and the curvature can avoid losing 3D information when doing the flattening.

3   Results

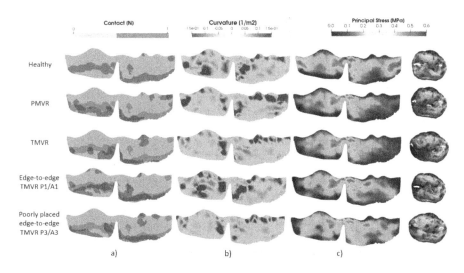

Fig. 4. a) Contact distribution at peak systole in the healthy MV, Primary Mitral Valve Regurgitation (PMVR), Traumatic Mitral Valve Regurgitation (TMVR), TMVR valve after an optimal edge-to-edge intervention, and post-interventional TMVR valve placing the clip in a non-optimal place. b) Curvature at peak systole of the four modeled conditions. c) Peak systolic principal stress distributions of the simulated scenarios in a flattened view and atrial view. (Color figure online)

Figure 4 depicts the outcome of the edge-to-edge therapy when placing the clip between the prolapsing scallop and the opposite leaflet, and the outcome of placing the clip in a suboptimal position. In the healthy valve, a continuous area of contact is observed in both leaflets (green zone in Fig. 4). This contact pattern is lost in both PMVR and TMVR, which is the indication of the MVR. However, this contact pattern is almost restored when the P1/A1 intervention is used but not for the P3/A3 one.

As for the curvature, the distribution changed. The post-interventional valve placing the clip in the P1/A1 scallops closely resembled the curvature of the healthy valve at peak systole, while the poorly placed clip had almost no effect on the curvature, resembling the structural conformation of the diseased valve. Finally, with regard to the stresses, the clip led to an increase in the principal stress of the prolapsing leaflet in both scenarios, as well as a local increase in the region where the clip was placed. The latter column shows the stress distributions at peak systole on the modeled scenarios in a 3D view. The healthy valve showed a mean stress of 110 KPa at peak systole, while the mean stress in the diseased valve increased to 160 KPa. Placing the clip in the prolapsing region led to a mean stress of 113 KPa, and in the poorly placed scenario, this value was of 160 KPa.

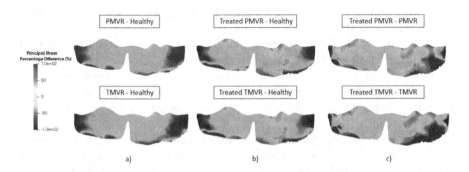

Fig. 5. Percentage differences of the principal stress distribution at peak systole between a healthy valve, pre-interventional regurgitant mitral valves (Primary Mitral Valve Regurgitation or PMVR, and Traumatic Mitral Valve Regurgitation or TMVR), and after properly placing the edge-to-edge clip. Differences between a) diseased and healthy valves; b) treated and healthy valves; c) post-interventional and diseased valves. (Color figure online)

Figure 5 shows the percentage difference in terms of the principal stresses between regurgitant valves, a healthy valve and post-interventional regurgitant valves. When comparing the diseased valves to the healthy one (Fig. 5a), a decrease in the stresses of the prolapsing scallop and the opposed one can easily be identified. The clip increases the stresses of the neighbouring tissue and the prolapsing scallop by over a 100% in both cases (Fig. 5c), but cannot completely restore the lost stresses due to the regurgitation, as depicted in the middle column (5b), comparing the post-interventional valve to the healthy scenario.

4   Discussion

Flattening the MV overcomes several limitations of the 3D visualization, which is not able to depict the whole valve, and requires several angles to have a full understanding of the results. Using the flattened views, one can easily compare between valves, even compute the local differences between scenarios, which aids in the analysis of the outcomes. Furthermore, flattening the valve allows the visualization of metrics such as the contact, that are arduous to see in 3D.

The healthy valve has been considered as the initial simulation, used to build upon it and develop the pathological valves. It is observed that the stress values at peak systole agree with the results of previous mesh-based computational simulations [16,17], resembling the spatial distribution of the stresses in the leaflets, as well as the absolute principal stress values, being around 200 KPa in most regions.

In both diseased cases, PMVR and TMVR, an increase in the anterior leaflet stresses was found, as well as a decrease in the stresses of the prolapsing scallop, as observed in [16,17]. Caballero et al. [17] developed an FSI model and found that the leaflet stress increased during chordae rupture, agreeing with our results: the average principal stress value for PMVR was 125 KPa, while, for a healthy valve, it was 110 KPa.

Placing the edge-to-edge intervention in optimal locations led to a considerable improvement of the pathology of the valve (see Fig. 4). The contact was enhanced, almost achieving total leaflet coaptation; and the curvatures at peak systole resembled the ones of the healthy valve. In other words, the dynamics of the post-interventional valves were partially restored, even in terms of the structural conformation.

With regard to the principal stress distribution, the intervention managed to recover some of the lost stresses in the prolapsing leaflet, agreeing with the results of [19,20]. Specifically, Caballero et al. [20] reported a significant increase (50–210%) in all post-clip models, especially in the vicinity of clip insertion. In addition, the mean peak systolic principal stress was restored, having values similar to the ones of a healthy valve. The clip successfully increased the stresses by over 100% in the prolapsing leaflet, together with a local increase of the stresses in the scallop end where the clip was placed.

On the other hand, the clip placed further from the prolapsing region led to non-desirable results, potentially inducing treatment failure: the mean peak systolic principal stress was increased, agreeing with [21]; and the intervention had almost no effect on the curvature of the valve at peak systole. Kamakoti et al. [21] studied the effect of different placements of the clip, and reported that the stresses were smallest for the ideal location of the MitraClip, as the chordae tendinae do not pull the leaflets back when the MC is placed. However, the stresses increased as the MitraClip was placed further away from the missing chords. Therefore, it is of utmost importance to place the edge-to-edge clip in a proper location, since failing to place it in the prolapsing region may require a reintervention, adding another clip. This could lead to increased stiffness of the valve, not being able to properly open during the diastole, and potentially becoming stenotic [20].

The computational time required to run the simulations varied depending on the specifications, such as the number of particles, the kernel size or the use of external springs. It must be noted that the in-house SPH solver can further be optimized, decreasing the computational requirements. For instance, the contact is computed in a fairly inefficient manner. Namely, without the contact algorithm, a cardiac cycle could be computed in 8 min, being of 29 min if the contact is considered. All the simulations were run using an Asus UX530UX-FY021T Intel Core i7-7500U with 2 cores and 4 logical processors.

5   Conclusion

This project shows the potential use of SPH in cardiac applications related to the MV, potentially contributing to guide planning of interventions and improving clinical-decision making. In addition, the flattened views of the valve offer a direct and straightforward analysis. When coupled with the simulation results, these views allow the examination of measurements that cannot be properly inspected with the 3D valve. Nevertheless, it must be noted that additional patient data, such as papillary tips and pressure condition is required to aim for patient-specificity. In addition, only one valve example was used, so the generalizability of this pipeline needs to be further assessed. Also, a sensitivity analysis of the particle resolution as well as other SPH parameters should have been carried out. Finally, regarding the contact computation algorithm, we rely on the validations that were done in the original papers, although we need to perform dedicated validations in the case of the MV with real data. All in all, the model proposed in this work can become the first steps towards the implementation of patient-specific modelling software to aid in the clinical decision-making process for the treatment of mitral valve regurgitation.

References

1. Apostolidou, E., et al.: Primary mitral valve regurgitation: update and review. Glob. Cardiol. Sci. Pract. **2017**(1), e201703 (2017)
2. Del Forno, B., et al.: Mitral valve regurgitation: a disease with a wide spectrum of therapeutic options. Nat. Rev. Cardiol. **17**, 807–827 (2020)
3. Lau, K.D., Díaz-Zuccarini, V., Scambler, P., Burriesci, G.: Fluid-structure interaction study of the edge-to-edge repair technique on the mitral valve. J. Biomech. **44**(13), 2409–2417 (2011)
4. Avanzini, A.: A computational procedure for prediction of structural effects of edge-to-edge repair on mitral valve. J. Biomech. Eng. **130**(3), 0301015 (2008)
5. Kong, F., Caballero, A., McKay, R., Sun, W.: Finite element analysis of Mitra-Clip procedure on a patient-specific model with functional mitral regurgitation. J. Biomech. **104**, 109730 (2020)
6. Zhang, L., Ademiloye, A., Liew, K.: Meshfree and particle methods in biomechanics: prospects and challenges. Arch. Comput. Meth. Eng. **26**, 1547–1576 (2019)
7. Lluch, E., et al.: Breaking the state of the heart: meshless model for cardiac mechanics. Biomech. Model. Mechanobiol. **18**(6), 1549–1561 (2019). https://doi.org/10.1007/s10237-019-01175-9
8. Hermida, U., Lluch, E., De Craene, M., Bijnens, B., Morales, H.G.: Aortic valve dynamic modelling with a meshless method. Congreso Anual de la Sociedad Española de Ingeniería Biomédica (2019)

9. Ros, J., Camara, O., Hermida, U., Bijnens, B., Morales, H.G.: Towards mesh-free patient-specific mitral valve modeling. In: Puyol Anton, E., et al. (eds.) STACOM 2020. LNCS, vol. 12592, pp. 66–75. Springer, Cham (2021). https://doi.org/10.1007/978-3-030-68107-4_7

10. Wang, J., Chan, D.: Frictional contact algorithms in SPH for the simulation of soil-structure interaction. Int. J. Numer. Anal. Meth. Geomech. **38**(7), 747–770 (2014)

11. Seo, S., Min, O.: Axisymmetric SPH simulation of elasto-plastic contact in the low velocity impact. Comput. Phys. Commun. **175**, 583–603 (2006)

12. Tautz, L., Vellguth, K., Sündermann, S., Degener, F., Wamala, I.: CT segmented mitral valves in open state. Zenodo (2019)

13. Hisham, M., Mohd, H.: Degenerative vs rigidity on mitral valve leaflet using fluid structure interaction (FSI) model. J. Biomim. Biomater. Biomed. Eng. **26**, 60–65 (2016)

14. Sanfilippo, A., et al.: Papillary muscle traction in mitral valve prolapse: quantitation by two-dimensional echocardiography. J. Am. Coll. Cardiol. **19**(3), 564–571 (1992)

15. Kaiser, A., McQueen, D., Peskin, C.: modelling the mitral valve. Int. J. Numer. Meth. Biomed. Eng. **35**(11), e3240 (2019)

16. Rim, Y., Laing, S., Kee, P., McPherson, D., Kim, H.: Evaluation of mitral valve dynamics. JACC: Imaging **6**(2), 263–268 (2013)

17. Caballero, A., et al.: New insights into mitral heart valve prolapse after chordae rupture through fluid-structure interaction computational modelling. Sci. Rep. **8**, 17306 (2018)

18. Lichtenberg, N., et al.: Mitral valve flattening and parameter mapping for patient-specific valve diagnosis. Int. J. Comput. Assist. Radiol. Surg. **15**(4), 617–627 (2020). https://doi.org/10.1007/s11548-019-02114-w

19. Zhang, Y., et al.: Mechanical effects of MitraClip on leaflet stress and myocardial strain in functional mitral regurgitation - a finite element modelling study. PLoS ONE **14**(10), e0223472 (2019)

20. Caballero, A., Mao, W., McKay, R., Hahn, R., Sun, W.: A comprehensive engineering analysis of left heart dynamics after MitraClip in a functional mitral regurgitation patient. Front. Physiol. **11**, 430 (2020)

21. Kamakoti, R., Dabiri, Y., Wang, D., Guccione, J., Kassab, G.: Numerical simulations of MitraClip placement: clinical implications. Sci. Rep. **9**, 15823 (2019)

Multi-modality Cardiac Segmentation via Mixing Domains for Unsupervised Adaptation

Fuping Wu[1], Lei Li[2], and Xiahai Zhuang[1(✉)]

[1] School of Data Science, Fudan University, Shanghai, China
zxh@fudan.edu.cn
[2] School of Biomedical Engineering, Shanghai Jiao Tong University, Shanghai, China

Abstract. Unsupervised domain adaptation is very useful for medical image segmentation. Previous works mainly considered the situation with one source domain and one target domain. However in practice, multi-source and/or multi-target domains are generally available. Instead of implementing adaptation one by one, in this work we study how to achieve multiple domain alignment simultaneously to improve the segmentation performance of domain adaptation. We use the VAE framework to transform all domains into a common feature space, and estimate their corresponding distributions. By mixing domains and minimizing the distribution distance, the proposed framework extracts domain-invariant features. We verified the method on multi-sequence cardiac MR images for unsupervised segmentation. Results experimentally demonstrated that mixing target domains together could improve the segmentation accuracy, when the label distributions of mixed target domains are closer to that of the source domain than each unmixed target domain. Compared to state-of-the-art methods, the proposed framework obtained promising results.

Keywords: Domain adaptation · Cardiac segmentation · Multi-modality

1 Introduction

Cardiac segmentation is important for medical assistant diagnosis, with applications such as 3D modeling and functional analysis of the heart [1,2]. In clinics, multi-modality cardiac images are commonly used by physicians. Hence, automatic segmentation for multi-modality images is of great value. However, supervised learning for a segmentation model can be labour intensive and expensive, as it requires the annotation of heart structure of all modalities. The technology of knowledge transfer [3] is effective to alleviate the labour work by learning the structure knowledge from one modality, and then transferring them to others. Formally, the modality with ground truth for training is denoted as source

This work was funded by the National Natural Science Foundation of China (grant no. 61971142, 62111530195 and 62011540404).

domain, and those unpaired and unlabeled images from other modalities as the target domains. The main obstacle of transfer learning from source domain to target domain is the distribution gap between them, which is also as *dataset bias* or *domain shift* [4]. Domain adaptation is one of the most popular technologies tackling this domain shift problem [5]. It aims to extract modality-invariant features by transforming two/multi- domains into a common space, where the discrepancy of their distributions can be minimized.

All previous researches on segmentation via unsupervised domain adaptation (UDA), to the best of our knowledge, are for one source and one target domains. These methods either aligned the distributions on feature level [6], or translated images to images keeping the same anatomic structure [7]. To guarantee the features to be modality-invariant, or the generated images to be realistic, most of them employed adversarial learning strategy with the generative adversarial network (GAN) [8,9]. For example, Kamnitsas *et al.* [10] proposed a multi-connected architecture with multi-layer features for adaptation, and achieved significant improvement in multi-spectral MR segmentation of brain lesions. One of the most related works to ours is from Dou et al. [6]. They proposed a plug-and-play domain adaptation method, and employed a dilated fully convolution network to extract modality-invariant features. Their method was validated on 2D cardiac CT-MR cross-modality segmentation, and obtained promising results. Another related state-of-the-art work is from Chen et al. [11]. They used GANs to implemented adaptation on both of the image and feature levels. More recently, Ouyang *et al.* [12] proposed a data efficient method for multi-domain medical image segmentation with a prior distribution matching technique. An implicit assumption in these works is that the label distributions between two domains should be similar.

For situation where multi-target domains exist, the aforementioned methods can achieve segmentation by adapting the distributions of the target domains to the source domain one by one. However, this strategy could ignore some complementary information among different target domains. For example, as will be shown in the experiments, mixing target domains could make the distribution of the mixed labels closer to that of source domain than any single target domain. Hence, it is possible to improve the segmentation accuracy by taking those complementary domains into account simultaneously in adaptation process.

In this work, we propose a domain adaptation framework for this multi-target situation. We first transform all domains into a common feature space via variational auto-encoder (VAE) [13]. Then, we mix these complementary target domains, and reduce the distribution discrepancy between the mixed domain and the source domain in the common feature space. We validated the method on multi-sequence cardiac MR images (i.e., BSSFP CMR, LGE CMR and T2 CMR) with BSSFP CMR as the source domain and other two modality images as target domains, and consider two situations: (1) the structure distributions of the three domains are similar to each other, and (2) both the structure distributions of LGE CMR and T2 CMR are quite different from that of BSSFP CMR, while their mixing structure distribution is similar to that of latter. The experiments demonstrated that complementary domains could be helpful for improving the unsupervised domain adaptation and segmentation accuracy.

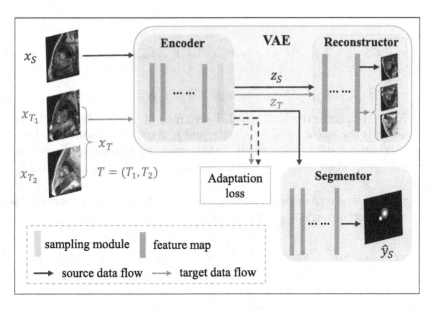

Fig. 1. Framework of the proposed domain adaptation method for multi-task medical image segmentation: the *encoder* module extracts the latent features; the *segmentor* module outputs the segmentation results; the *reconstructor* module and the *sampling* module are the main structure of the VAE; the *Adaptation loss* module computes the domain discrepancy.

2   Methods

Let $X_S = \{x_S^i\}_{i=1}^{N_S}$ be the set of N_S source images, $Y_S = \{y_S^i\}_{i=1}^{N_S}$ the corresponding segmentation, and $X_{T_1} = \{x_{T_1}^i\}_{i=1}^{N_{T_1}}$, $X_{T_2} = \{x_{T_2}^i\}_{i=1}^{N_{T_2}}$ are image samples from two different target modalities T_1 and T_2, whose segmentation, $Y_{T_k} = \{y_{T_k}^i\}_{i=1}^{N_{T_k}}$, $k = 1, 2$, are unknown. The images are randomly sampled from their own distributions $p_S(x)$ and $p_{T_k}(x)$. Our target is to obtain Y_{T_1}.

2.1   Framework

To utilize the complementary information of T_2, we propose a domain adaptation framework, illustrated in Fig. 1, which minimizes the distribution discrepancy between the source domain and the mixed T_1 and T_2 domains. Here, the symbols with subscript S indicate the variables or samples are for the source domain, and those with T are for the target domain. For simplicity, we also denote the mixed target domain as T.

To extract modality-invariant features, we first employ VAE to transform both source domain and the mixed target domain into a common feature space, whose distributions are close to standard normal one $\mathcal{N}(0, I)$ for each domain. As demonstrated in [13], the VAE loss for each domain can be formulated as follows,

$$\mathcal{L}_{VAE}^S = E_{p_S(x)}[KL(q_S(z|x||\mathcal{N}(0,I))] - E_{q_S(z|x)}[p_S(x|z)], \qquad (1)$$

and

$$\mathcal{L}_{VAE}^T = E_{p_T(x)}[KL(q_T(z|x||\mathcal{N}(0,I))] - E_{q_T(z|x)}[p_T(x|z)], \qquad (2)$$

where $KL(\cdot||\cdot)$ denotes the Kullback-Leibler (KL) divergence between two distributions, z is the latent feature variable, and p, q are the domain distributions and their approximation, respectively. In this work, we assume that $q_{S/T}(z|x) = \mathcal{N}(u_{S/T}, \Sigma_{S/T})$, where $u_{S/T} = (u_{S/T}^1, \cdots, u_{S/T}^n) \in \mathbb{R}^n$, $\Sigma_{S/T} = \text{diag}(\lambda_{S/T}^1, \cdots, \lambda_{S/T}^n) \in \mathbb{R}^{n \times n}$.

To reduce the distribution discrepancy between z_S and z_T, we estimate their distribution distance with L_2 norm as follows,

$$\mathcal{L}_{DD} = \int [q_S(z) - q_T(z)]^2 dz \approx \frac{1}{M^2} \sum_{i=1}^M \sum_{j=1}^M \left[k(x_S^i, x_S^j) + k(x_T^i, x_T^j) - 2k(x_S^i, x_T^j) \right], \quad (3)$$

where $k(x_S^i, x_T^j) = \dfrac{e^{-\frac{1}{2}\sum_{l=1}^n \frac{(u_{S_l}^i - u_{T_l}^j)^2}{\lambda_{S_l}^i + \lambda_{T_l}^j}}}{(2\pi)^{\frac{n}{2}} \cdot (\prod_{l=1}^n (\lambda_{S_l}^i + \lambda_{T_l}^j))^{\frac{1}{2}}}$, and M is the sampling number in the stochastic gradient descent optimization process. Note that Eq. (3) is very similar with the form of Maximum Mean Discrepancy (MMD) [14].

Finally, we employ multi-class cross entropy loss (CELoss) for training of the *segmentor* in Fig. 1. We denote the loss as $\mathcal{L}_{seg}(\hat{y}_S, y_S)$, where $\hat{y}_S$ is the prediction.

Overall, the total loss for domain adaptation can be formulated as follows,

$$\mathcal{L} = \alpha_1 \cdot \mathcal{L}_{VAE}^S + \alpha_2 \cdot \mathcal{L}_{VAE}^T + \alpha_3 \cdot \mathcal{L}_{DD} + \alpha_4 \cdot \mathcal{L}_{seg}, \qquad (4)$$

where $\alpha_1, \alpha_2, \alpha_3$ and α_4 are the trade-off parameters.

It should be noted that mixing domains $T1$ and $T2$ do not always work. The relation of the label distribution, i.e., $p_S(y)$, $p_{T_1}(y)$ and $p_{T_2}(y)$, could affect this mixing strategy greatly. More specifically, when both $p_{T_1}(y)$ and $p_{T_2}(y)$ are similar to $p_S(y)$, mixing two domains with $T = (T_1, T_2)$ may make the input target images more complex, which challenges the ability of extracting modality-invariant features for the network of encoder, and leads to worse segmentation accuracy. However, mixing strategy can be useful for situation where $p_{T_1}(y)$ and $p_{T_2}(y)$ are complementary, which means that there exists large gap with $p_S(y)$ for each one, while the mixing distribution $p_T(y)$ is similar to $p_S(y)$, i.e., if the mixing operator is uniformly sampling from the two target domains, we have $p_S(y) \approx \frac{1}{2}(p_{T_1}(y) + p_{T_2}(y))$, and $KL(p_{T_1}(y)||p_{T_2}(y)) > \delta$, δ is large enough.

3   Experiments and Results

We validated the mixing strategy with the proposed framework in two situations: (1) The target label distributions were complementary, i.e., each target label distribution was very different from the source one, while their mixing label distribution was similar to it. We verified that the mixing strategy could benefit the segmentation of UDA. We refer this VAE-based framework with mixing strategy in this situation as mxVAEDA$^{\oplus}$; (2) Each target label distribution was similar with that of the source domain. We verified that the mixing operation might make the input data more complex, and thus degrade the segmentation of UDA. We refer the method in this situation as mxVAEDA$^{\ominus}$.

We validated the mixing strategy on MSCMR image (i.e., BSSFP CMR, LGE CMR and T2 CMR) with BSSFP CMR as the source domain and other two modality images as target domains.

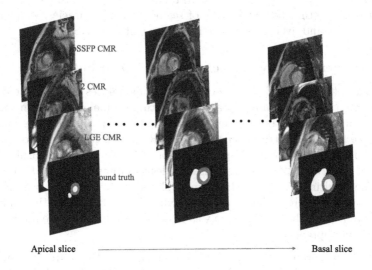

Apical slice ——————————————————————→ Basal slice

Fig. 2. Examples of *unpaired* CMR images from apical to basal slices: bSSFP CMR, T2 CMR, LGE CMR, and ground truth segmentation of the source image.

Dataset: The MICCIA 2019 Multi-Sequence Cardiac MR Segmentation (MS-CMRSeg) Challenge [15,16] provided 45 multi-sequence CMR images from patients who underwent cardiomyopathy. Each patient had been scanned using the three CMR sequences, i.e. the LGE, T2 and bSSFP. The task of this challenge was to segment the right ventricular cavity (RV), left ventricular cavity (LV) and left ventricular myocardium (MYO) from LGE CMR. To validate the method in the two aforementioned situations, we removed slices in each sample to make the label distributions of the three modalities to be similar. For evaluation, we randomly selected 10 LGE CMR cases for testing, 5 cases for validation. The training set included (1) all 45 bSSFP CMR with labels as the source data, (2) the remaining 30 LGE CMR as the T_1 unlabeled target data and (3) randomly selected T2 CMR as the T_2 unlabeled target data. We finally obtained 94 2D training slices for each domain. For experiments, all images were intensity normalized, resized to the same resolution of 1.6×1.6 mm and cropped with an ROI of 112×112 pixels. Figure 2 illustrates some samples of the three modalities.

Model Configuration: As shown in Fig. 1, we used U-Net as the encoder, which has been demonstrated to be effective and efficient in medical image segmentation tasks [17]. We also used the bottom features of the U-Net for slice classification, predicting the rough position from the center of the heart. This classification task was used for constraining the network. We further employed multi-scale segmentation strategy. We outputted segmentations on three scale, and fused them via a convolution layer as the final results. For parameter selection, we set $\alpha_1 = \alpha_2 = 1.0$, $\alpha_3 = 10^3$, and ranged α_4 in $[10^{-2}, 10^{-3}, 10^{-4}]$ for the best parameter setup. For evaluation of segmentation accuracies, we employed the Dice metric and average symmetric surface distance (ASSD).

3.1   Situation One: Complementary Label Distribution

We validated the proposed method on the three modalities with complementary label distributions on training data, i.e., each target label distribution was very different from the source one, while their mixing label distribution was similar to it. To make the label distribution of LGE CMR and T2 CMR to be complementary, we sampled slices around apical position and slices close to basal position from LGE CMR and T2 CMR training samples as T_1 and T_2 target data, respectively. The resulting 55 LGE CMR training slices and 39 T2 CMR training slices then had similar mixing label distribution with that of bSSFP CMR, while both were different from the latter, i.e., δ was large. For comparison, we implemented the following approaches:

- NoAdapt: We trained the encoder and segmentor on the source data without adaptation. The resulting model was applied directly to the LGE CMR testing data for segmentation.
- PnP-AdaNet [6]: This method employed a plug-and-play strategy, and used multi-layer feature maps to obtain the latent features. The domain adaptation was implemented on the feature level.
- VAEDA: The proposed method with bSSFP CMR as source data and LGE CMR as target data.
- mxVAEDA$^{\oplus}$: The proposed method with bSSFP CMR as source data, and mixed LGE CMR, T2 CMR as target data, where label distribution of the mixed LGE CMR and T2 CMR dataset was similar to that of bSSFP CMR, while different for each single modality. The method could benefit the segmentation of UDA in this situation.

As shown in Table 1, mixing complementary target data for domain adaptation training improved the accuracy, compared to the method without mixing, i.e., VAEDA. This was mainly due to the mixing label distribution close to that of the source domain. The improvement was particularly obvious on RV, as RV differed great from apical slice to basal slice, as illustrated in Fig. 2. Figure 3 visualized the segmentation results on the same images in Fig. 4 with different methods.

Table 1. Performance comparison of methods on LGE CMR images with complementary label distributions for the target domains, in which mxVAEDA$^\oplus$ benefited the segmentation of UDA. Note that both Dice and ASSD were computed based on the 3D images.

Methods	Dice (%)			ASSD (mm)		
	MYO	LV	RV	MYO	LV	RV
NoAdapt	60.9 ± 16.6	80.4 ± 13.6	67.7 ± 25.7	4.94 ± 3.96	6.76 ± 7.32	5.98 ± 6.85
PnP-AdaNet	52.4 ± 7.09	77.7 ± 8.69	57.7 ± 2.06	3.40 ± 1.89	6.61 ± 4.98	8.24 ± 9.30
VAEDA	63.8 ± 11.4	82.1 ± 9.63	68.8 ± 9.57	3.14 ± 1.54	3.98 ± 3.24	5.17 ± 1.97
mxVAEDA$^\oplus$	64.3 ± 11.1	83.2 ± 8.93	75.7 ± 9.13	3.81 ± 2.25	5.07 ± 2.74	3.85 ± 2.28

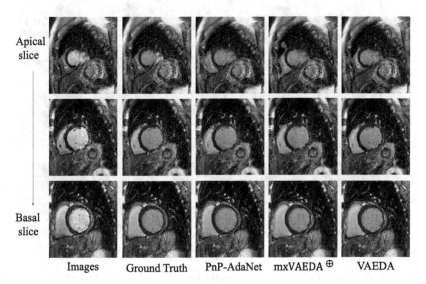

Apical slice

Basal slice

Images   Ground Truth   PnP-AdaNet   mxVAEDA $^\oplus$ VAEDA

Fig. 3. Visualization of segmentation results for different methods with complementary label distribution of target domains, in which mxVAEDA$^\oplus$ benefited the segmentation of UDA.

3.2   Situation Two: Similar Label Distribution

We validated the proposed method on the three modalities with similar label distributions for training data. For comparison, we also implemented methods including NoAdapt, PnP-AdaNet, VAEDA and mxVAEDA$^\ominus$.

As shown in Table 2, VAEDA performed well for segmentation task. By contrast, the mixing strategy made the domain adaptation harder, and degraded the performance greatly, especially on MYO, confirming that mixing strategy could degrade the segmentation of UDA in situation with similar label distributions. Although T2 CMR had similar label distribution with that of LGE CMR and bSSFP CMR, it made the mixed target domain more complex, and the network more difficult to extract modality-invariant features. Figure 4 illustrated some segmentation results of different methods.

Table 2. Performance comparison of methods on LGE CMR images with similar label distributions for the three domains, in which mxVAEDA$^\ominus$ degraded the segmentation of UDA. Note that both Dice and ASSD were computed based on the 3D images.

Methods	Dice (%)			ASSD (mm)		
	MYO	LV	RV	MYO	LV	RV
NoAdapt	60.9 ± 16.6	80.4 ± 13.6	67.7 ± 25.7	4.94 ± 3.96	6.76 ± 7.32	5.98 ± 6.85
PnP-AdaNet	68.8 ± 5.74	86.3 ± 6.05	75.6 ± 18.5	1.87 ± 0.672	2.44 ± 1.77	4.31 ± 4.57
VAEDA	70.3 ± 7.84	86.2 ± 8.34	76.9 ± 9.43	2.20 ± 1.19	2.52 ± 2.44	3.02 ± 1.78
mxVAEDA$^\ominus$	58.9 ± 12.5	78.6 ± 11.8	75.1 ± 8.61	3.73 ± 1.22	3.88 ± 2.37	3.18 ± 1.28

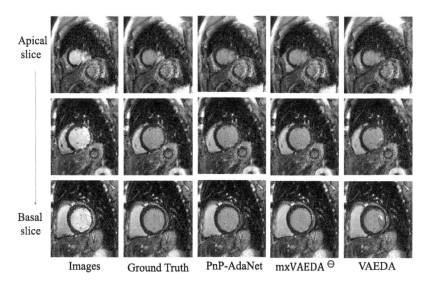

Apical slice

Basal slice

Images   Ground Truth   PnP-AdaNet   mxVAEDA$^\ominus$ VAEDA

Fig. 4. Visualization of segmentation results for different methods with similar label distribution of all domains, in which mxVAEDA$^\ominus$ degraded the segmentation of UDA.

4   Conclusion

Domain adaptation for multi-target domains could be complex. How to utilize the data from different domains to improve adaptation result is of great interest. In this work, we investigated the two situations of multiple domains for UDA. One is that the target domains have complementary label distributions. The other is that both target domains are similar to the source domain in terms of label distribution. We concluded that the mixing strategy could benefit the segmentation of UDA for the former situation, while degrade the performance in the latter situation. While we have validated these arguments on multi-sequence MR cardiac images (i.e., BSSFP CMR, LGE CMR and T2 CMR), more dataset should be validated and more complex situation for the discrepancy of the target domains (i.e., δ) should be studied in the future.

References

1. Karamitsos, T.D., Francis, J.M., Myerson, S., Selvanayagam, J.B., Neubauer, S.: The role of cardiovascular magnetic resonance imaging in heart failure. J. Am. Coll. Cardiol. **54**(15), 1407–1424 (2009)
2. Petitjean, C., Dacher, J.-N.: A review of segmentation methods in short axis cardiac MR images. Med. Image Anal. **15**(2), 169–184 (2011)
3. Raina, R., Battle, A., Lee, H., Packer, B., Ng, A.Y.: Self-taught learning: transfer learning from unlabeled data. In: Proceedings of the 24th International Conference on Machine Learning, pp. 759–766. ACM (2007)
4. Shimodaira, H.: Improving predictive inference under covariate shift by weighting the log-likelihood function. J. Stat. Plann. Inference **90**(2), 227–244 (2000)
5. Csurka, G.: A comprehensive survey on domain adaptation for visual applications. In: Csurka, G. (ed.) Domain Adaptation in Computer Vision Applications. ACVPR, pp. 1–35. Springer, Cham (2017). https://doi.org/10.1007/978-3-319-58347-1_1
6. Dou, Q., et al.: PnP-AdaNet: plug-and-play adversarial domain adaptation network at unpaired cross-modality cardiac segmentation. IEEE Access **7**, 99065–99076 (2019)
7. Zhang, Z., Yang, L., Zheng, Y.: Translating and segmenting multimodal medical volumes with cycle- and shape-consistency generative adversarial network. In: Proceedings of the IEEE Conference on Computer Vision and Pattern Recognition, pp. 9242–9251 (2018)
8. Goodfellow, I., et al.: Generative adversarial nets. In: Advances in Neural Information Processing Systems, pp. 2672–2680 (2014)
9. Arjovsky, M., Chintala, S., Bottou, L.: Wasserstein generative adversarial networks. In: Proceedings of the 34th International Conference on Machine Learning - Volume 70, ser. ICML'17, pp. 214–223. JMLR.org (2017)
10. Kamnitsas, K., et al.: Unsupervised domain adaptation in brain lesion segmentation with adversarial networks. In: Niethammer, M., et al. (eds.) IPMI 2017. LNCS, vol. 10265, pp. 597–609. Springer, Cham (2017). https://doi.org/10.1007/978-3-319-59050-9_47
11. Chen, C., Dou, Q., Chen, H., Qin, J., Heng, P.: Synergistic image and feature adaptation: towards cross-modality domain adaptation for medical image segmentation. In: The Thirty-Third AAAI Conference on Artificial Intelligence, pp. 865–872 (2019)
12. Ouyang, C., Kamnitsas, K., Biffi, C., Duan, J., Rueckert, D.: Data efficient unsupervised domain adaptation for cross-modality image segmentation. In: Shen, D., et al. (eds.) MICCAI 2019. LNCS, vol. 11765, pp. 669–677. Springer, Cham (2019). https://doi.org/10.1007/978-3-030-32245-8_74
13. Kingma, D.P., Welling, M.: Auto-encoding variational Bayes. CoRR arXiv:1312.6114 (2013)
14. Tzeng, E., Hoffman, J., Zhang, N., Saenko, K., Darrell, T.: Deep domain confusion: maximizing for domain invariance. CoRR arXiv:1412.3474 (2014)
15. Zhuang, X.: Multivariate mixture model for myocardial segmentation combining multi-source images. IEEE Trans. Pattern Anal. Mach. Intell. **41**(12), 2933–2946 (2019)

16. Zhuang, X.: Multivariate mixture model for cardiac segmentation from multi-sequence MRI. In: Ourselin, S., Joskowicz, L., Sabuncu, M.R., Unal, G., Wells, W. (eds.) MICCAI 2016. LNCS, vol. 9901, pp. 581–588. Springer, Cham (2016). https://doi.org/10.1007/978-3-319-46723-8_67
17. Ronneberger, O., Fischer, P., Brox, T.: U-Net: convolutional networks for biomedical image segmentation. In: Navab, N., Hornegger, J., Wells, W.M., Frangi, A.F. (eds.) MICCAI 2015. LNCS, vol. 9351, pp. 234–241. Springer, Cham (2015). https://doi.org/10.1007/978-3-319-24574-4_28

Uncertainty-Aware Training for Cardiac Resynchronisation Therapy Response Prediction

Tareen Dawood[1(✉)], Chen Chen[3], Robin Andlauer[1], Baldeep S. Sidhu[1,2],
Bram Ruijsink[1,2], Justin Gould[1,2], Bradley Porter[1,2], Mark Elliott[1,2],
Vishal Mehta[1,2], C. Aldo Rinaldi[1,2], Esther Puyol-Antón[1], Reza Razavi[1,2],
and Andrew P. King[1]

[1] School of Biomedical Engineering and Imaging Sciences, King's College London,
London, UK
tareen.dawood@kcl.ac.uk
[2] Guys and St Thomas' Hospital, London, UK
[3] BioMedIA Group, Department of Computing, Imperial College London,
London, UK

Abstract. Evaluation of predictive deep learning (DL) models beyond conventional performance metrics has become increasingly important for applications in sensitive environments like healthcare. Such models might have the capability to encode and analyse large sets of data but they often lack comprehensive interpretability methods, preventing clinical trust in predictive outcomes. Quantifying uncertainty of a prediction is one way to provide such interpretability and promote trust. However, relatively little attention has been paid to how to include such requirements into the training of the model. In this paper we: (i) quantify the data (aleatoric) and model (epistemic) uncertainty of a DL model for Cardiac Resynchronisation Therapy response prediction from cardiac magnetic resonance images, and (ii) propose and perform a preliminary investigation of an uncertainty-aware loss function that can be used to retrain an existing DL image-based classification model to encourage confidence in correct predictions and reduce confidence in incorrect predictions. Our initial results are promising, showing a significant increase in the (epistemic) confidence of true positive predictions, with some evidence of a reduction in false negative confidence.

Keywords: Uncertainty · Cardiac resynchronisation therapy · Reliability · Trust · Awareness

1 Introduction

Cardiac Resynchronisation Therapy (CRT) is an established treatment for advanced heart failure (HF) patients [1]. Current clinical guidelines for CRT patient selection use a limited set of characteristics: New York Heart Association classification, left ventricular ejection fraction, QRS duration, the type of

© Springer Nature Switzerland AG 2022
E. Puyol Antón et al. (Eds.): STACOM 2021, LNCS 13131, pp. 189–198, 2022.
https://doi.org/10.1007/978-3-030-93722-5_21

bundle branch block, etiology of cardiomyopathy and atrial rhythm [2]. However, despite this a significant number of patients (approximately 1 in 3) gain little or no symptomatic benefit from CRT [1]. The variability in CRT response has gathered clinical interest, with recent research identifying several factors that are associated with CRT response. For example, clinical studies have shown that apical rocking, septal flash and myocardial scarring are all associated with CRT response [3].

In parallel, other researchers have sought to leverage the power of machine learning (ML) techniques in predicting response to CRT. Peressutti *et al.* [4] used supervised multiple kernel learning (MKL) to develop a predictive CRT response model incorporating both motion (obtained from cardiac magnetic resonance (CMR) images) and non-motion data. Cikes *et al.* [5] showed that unsupervised MKL and multimodal data (clinical parameters and image data) can be used to phenotype HF patients and identify which groups benefit the most from CRT. Puyol-Antón *et al.* [6] developed a deep learning (DL)-based model based on CMR imaging that predicted CRT response and also provided explanatory concepts to aid interpretability.

In recent years, DL models have dominated medical research, but they are often developed without consideration of how the models will be used in clinical practice. In most cases where a DL model could potentially be used for complex predictive problems (such as CRT patient selection), it is unlikely that the model will be used as a stand-alone "black box" tool to replace clinicians. Rather, it will act as decision-support tool to aid clinical decision making. This consideration raises the important issue of clinical trust in the model [7].

One way to develop clinical trust in predictive outcomes is to quantify the confidence of a model in its automated decision. Two identified sources of uncertainty are aleatoric uncertainty, which is caused by noisy data inputs and epistemic uncertainty, which refers to uncertainties inherent in the model itself [8]. Abdar *et al.* [9] performed a comprehensive review of the techniques, applications and challenges of uncertainty quantification in DL. Interestingly, the literature indicates that uncertainty quantification has been predominantly applied to segmentation applications and less so for disease diagnosis or treatment response predictions. We believe that uncertainty information is an important measure to include when providing cardiologists with an automated decision-support tool, in order for them to start developing trust and effectively utilising the tool to aid their daily clinical work.

To date, whereas a number of works have quantified uncertainty in DL models, relatively little attention has been paid to including knowledge of uncertainty in the training of such models. For example, in CRT response prediction, confident incorrect predictions should be avoided, whereas less confident incorrect predictions are less of a problem. Likewise, for correct predictions it is preferred that they should be confident rather than less confident. It would be desirable for a predictive model to take these requirements into account, rather than purely focusing on prediction accuracy.

Related work to this idea includes Geifman and El-Yaniv [10] who proposed a method for 'selective' image classification, in which only confident cases are attempted based on a user-defined risk level. Recent work by Ding *et al.* proposed uncertainty-aware training for segmentation tasks by incorporating an additional loss term that aimed to maximise performance on confident outputs [11]. In this paper we draw inspiration from Ding *et al.* and propose a novel uncertainty-aware training loss for a classification problem, and investigate its effect on the aleatoric and epistemic uncertainties of a DL-based CRT response prediction model.

Our key contributions in this paper are: (1) for the first time, we quantify the epistemic and aleatoric uncertainties of a state-of-the-art DL-based CRT response prediction model, and (2) to the best of our knowledge this is the first investigation of an uncertainty-aware training method for a classification problem in medical imaging.

2   Materials

We used two databases to train and evaluate our uncertainty-aware CRT prediction model. First, the CMR short-axis (SA) stacks of 10,000 subjects (a mix of healthy and cardiovascular patients) from the UK Biobank dataset [12] were used for pre-training of the segmentation and variational autoencoder (VAE) models described in Sect. 3. A second database from the clinical imaging system of Guy's and St Thomas' NHS Foundation Trust (GSTFT) Hospital was also used, consisting of 20 heart failure (HF) patients and 73 CRT patients. All 73 CRT patients met the conventional criteria for CRT patient selection. The HF patients were used to fine tune the VAE. The 73 CRT patients were used to train and evaluate the baseline and uncertainty-aware CRT prediction models (i.e. VAE and classifier, see Sect. 3.)

For the CRT patients, CMR imaging was performed prior to CRT and the CMR multi-slice SA stack was used in this study. The Siemens Aera 1.5T, Siemens Biograph mMR 3T, Philips 1.5T Ingenia and Philips 1.5T and 3T Achieva scanners were used to perform CMR imaging. The typical slice thickness was 8–10 mm, in-plane resolution was between $0.94 \times 0.94 \, mm^2$ and $1.5 \times 1.5 \, mm^2$ and the temporal resolution was 13–31 ms/frame. Using post-CRT echocardiography images (at 6 month follow up), a positive response was defined as a 15% reduction in left ventricular (LV) end-systolic volume.

Institutional ethics approval was obtained for use of the clinical data and all patients consented to the research and for the use of their data. All relevant protocols were adhered to in order to retrieve and store the patient data and related images. For all CMR datasets, the top three slices of the SA stack were employed as the input to the model described in Sect. 3. All slices were spatially re-sampled to 80×80 pixels and temporally re-sampled to $T = 25$ time samples before being used for training/evaluation of the models.

3   Methods

As a baseline CRT response prediction model we used the state-of-the-art DL-based approach proposed by Puyol-Antón *et al.* [6]. This consists of a segmentation model to automatically segment the LV blood pool, LV myocardium and right ventricle (RV) blood pool from the 3 CMR SA slices at all time frames over the cardiac cycle. These segmentations are then used as input to a VAE with a classification network to predict CRT response from a concatenation of the latent representations for each time frame. Details of training the segmentation and VAE/classifier models are provided in Sect. 3.3. Note that we do not include the explanatory concept classifier proposed in [6] in our baseline as we wish to concentrate our analysis on CRT response prediction confidence in this paper (i.e. we use the baseline model only from [6]). The architecture of this model is illustrated in Fig. 1.

3.1   Uncertainty-Aware Loss Function

The baseline CRT prediction model utilised a loss function that combined the standard reconstruction and Kullback-Leibler terms for the VAE and a term for the CRT response classifier. In this work we propose an additional term for uncertainty-aware training, inspired by [11]. The final loss function used was:

$$\mathcal{L}_{\text{total}} = \mathcal{L}_{\text{re}} + \beta\mathcal{L}_{\text{KL}} + \gamma\mathcal{L}_{\text{C}} + \alpha\mathcal{L}_{\text{U}} \tag{1}$$

where $\mathcal{L}_{\text{re}}$ is the cross-entropy between the input segmentations to the VAE and the output reconstructions, $\mathcal{L}_{\text{KL}}$ is the Kullback-Leibler divergence between the latent variables and a unit Gaussian, $\mathcal{L}_{\text{C}}$ is the binary cross entropy loss for the CRT response classification task and $\mathcal{L}_{\text{U}}$ is the uncertainty-aware loss function. β, γ and α are used to weight the level of influence each term has to the total loss.

The uncertainty-aware loss term is defined as:

$$\mathcal{L}_{\text{U}} = \frac{1}{N_{FP}} \sum_{i\in\text{FP}, j\in\text{TP}} \max(\mathcal{P}_i^{+ve} - \mathcal{P}_j^{+ve} + m, 0)$$
$$+ \frac{1}{N_{FN}} \sum_{i\in\text{FN}, j\in\text{TN}} \max(\mathcal{P}_i^{-ve} - \mathcal{P}_j^{-ve} + m, 0) \tag{2}$$

Here, $\mathcal{FP}$, $\mathcal{TP}$, $\mathcal{FN}$ and $\mathcal{TN}$ represent sets of samples from a training batch which are classified as false positives, true positives, false negatives and true negatives respectively, and N_{FP} and N_{FN} are counts of the number of false positives and false negatives in the batch. $\mathcal{P}^{+ve}$ and $\mathcal{P}^{-ve}$ represent the class probabilities of the classifier (after the Softmax layer) for positive and negative response, i.e. $\mathcal{P}^{+ve} = 1 - \mathcal{P}^{-ve}$. The subscripts for these terms represent the sample used as input. Intuitively, Eq. (2) evaluates all pairs of correct/incorrect positive/negative predictions in a training batch, and the terms will be positive when the incorrect prediction (i) has a higher confidence than the correct one (j). If the correct one has a higher confidence than the incorrect one by a margin of the hyperparameter m or more it will be zero.

3.2   Uncertainty Quantification

To evaluate the uncertainty characteristics of the baseline CRT prediction model and assess the impact of our uncertainty-aware loss function we estimate the aleatoric and epistemic uncertainties. The specific points at which uncertainty was estimated are illustrated in Fig. 1. To estimate the aleatoric uncertainty of the CRT response prediction model we generated multiple plausible segmentation inputs to the VAE using inference-time drop out in the segmentation model with probability $= 0.5$ in the decoder layers. Aleatoric uncertainty was then estimated using the CRT response prediction of the original data's segmentations and those from 19 additional segmentation sets generated in this way, i.e. the original and 19 additional segmentations were propagated through the VAE and CRT classifier.

The epistemic uncertainty of the CRT response prediction model was estimated using random sampling in the latent space of the VAE. Again, the original embedding together with 19 additional random samples were used for estimating epistemic uncertainty. Increasing the number of samples from the latent space did not have a statistically significant difference on the estimate but did adversely affect simulation time, therefore just 20 samples were used for epistemic uncertainty estimation. For both types of uncertainty, the CRT response predictions were made for the 20 samples and used to compute a response prediction confidence as the percentage of positive responses out of the 20.

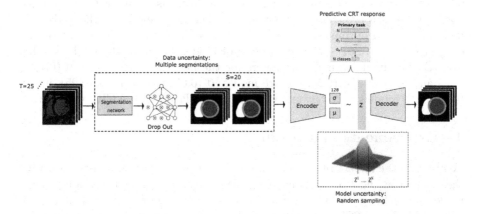

Fig. 1. Diagram showing the architecture of the baseline VAE/classification model developed by Puyol-Antón *et al.* [6] and the stages at which uncertainty estimates were made. The CMR SA images are segmented using a DL-based model, and these segmentations act as the inputs to the VAE. The classification is performed in the latent space of the VAE and points at which aleatoric uncertainty and epistemic uncertainty were estimated are shown in the dotted blocks.

3.3 Tools, Implementation and Evaluation

We implemented all methods using the Python programming language, the PyCharm IDE and the open source PyTorch tensor library. All models were trained on a NVIDIA A6000 48 GB using an Adam optimiser with a faster learning rate for the VAE and a slower rate for the CRT classifier (10^{-2}) to (10^{-8}), with a batch size of 8.

The segmentation model was pre-trained using the 10,000 UK Biobank subjects [12]. We then fine-tuned the model using 300 slices (multiple slices/time points from 20 CMR scans) from the HF patients. The VAE was pre-trained using the UK Biobank data and fined tuned with the HF data and then both the VAE and CRT classifier were trained with the CRT patient data for 300 epochs. For training the uncertainty-aware model, the fine-tuned VAE model was used, the uncertainty-aware loss function introduced and then both the VAE and CRT classifier trained for 300 epochs using the 73 CRT patients.

Both the baseline model and the uncertainty-aware model were trained and evaluated using a 5-fold nested cross validation. For each of the 5 outer folds, an inner 2-fold cross validation was performed to optimise hyperparameters using the training data. In these inner folds, the set of hyperparameters yielding the highest accuracy was selected. The optimal hyperparameters were used to train a model (using all training data for that fold) which was then applied to the held-out (outer) fold. This process was repeated for all outer folds. In this way, hyperparameter optimisation was performed using training data and the model was always applied to completely unseen data. Note also that the CRT data had not been used in pre-training either the segmentation model or the VAE. The hyperparameters optimised in this way were (see Eq. (1)): β ($0.001-2$), γ ($0-2$), α ($0.01-2$), m ($0.01-1$) and the size of hidden layers in the CRT classifier ($32, 64$). The final optimal parameters for the baseline model were, β (0.1), γ (0.6) and for the uncertainty-aware model, β (0.001), γ (0.5), α (0.05) and m (0.6). Both models were optimal with 32 hidden layers in the CRT classifier.

4 Experiments and Results

To evaluate the impact of our uncertainty-aware loss function, we first measured the overall classification accuracy of the baseline and then the uncertainty-aware model and achieved a balanced accuracy (i.e. average of sensitivity and specificity) of 70% and 68% respectively. We note that this was lower than the value reported in [6], as a result of implementing nested cross validation, which prevents leakage of data between the training and testing sets.

We further summarise the epistemic and aleatoric uncertainties of the outputs (estimated as described in Sect. 3.2). The results are shown in Tables 1 and 2 for epistemic and aleatoric uncertainty respectively. These tables both show counts of the numbers of true positives (TP), false negatives (FN), false positives (FP) and true negatives (TN) in 3 confidence bands. Note that these bands represent confidence in the predicted response, i.e. for the subjects predicted as responders it is the percentage confidence in positive response, and for subjects predicted

Table 1. Summary of epistemic (model) uncertainty for the baseline and uncertainty-aware models. (GT = Ground Truth, TP = True positive, FN = False Negative, FP = False positive, TN = True negative)

Baseline model				
Confidence	GT +ve		GT −ve	
	Pred. +ve (TP)	Pred. −ve (FN)	Pred. +ve (FP)	Pred. −ve (TN)
0–30	1	3	10	6
31–70	11	5	2	7
71–100	23	2	0	3
Uncertainty-aware model				
0–30	3	5	14	4
31–70	6	1	0	7
71–100	27	3	0	3

Table 2. Summary of aleatoric (data) uncertainty for the baseline and uncertainty-aware models. (GT = Ground Truth, TP = True positive, FN = False Negative, FP = False positive, TN = True negative)

Baseline model				
Confidence	GT +ve		GT −ve	
	Pred. +ve (TP)	Pred. −ve (FN)	Pred. +ve (FP)	Pred. −ve (TN)
0–30	3	4	11	6
31–70	6	0	1	4
71–100	26	6	0	6
Uncertainty-awareness trained model				
0–30	3	5	13	7
31–70	3	0	0	4
71–100	30	4	1	3

as non-responders it is the percentage confidence in negative response (which is 100 minus the positive response confidence).

The results show that, for epistemic and aleatoric uncertainty, the uncertainty-aware model encouraged confidence in correct predictions (i.e. see the true positives column). There is also some evidence of reduced confidence in incorrect predictions for the non-responders for epistemic uncertainty. However, overall significant changes were not as evident for false negatives for both aleatoric and epistemic uncertainty. Further analysis is required to develop a more robust method to improve the outcomes, however, as a preliminary investigation the results are encouraging and suggest the need to include uncertainty awareness training when building prediction models.

5   Discussion and Conclusion

In this paper we have proposed a novel uncertainty-aware loss term for classification problems, in which we envisage the model being used as a decision-support tool for clinicians. In such situations, to promote clinical trust and to maximise the utility of the tool, it is imperative that confident incorrect predictions are minimised. To illustrate the effects of including the developed uncertainty-aware loss function, aleatoric and epistemic uncertainty were estimated for a baseline CRT response prediction model and the model that utilised the new loss term.

The results indicate an increase in confident true positive predictions for both epistemic and aleatoric uncertainty. There is also some evidence of a reduction in confidence of incorrect predictions, particularly in epistemic uncertainty. However, despite these promising signs the method requires further investigation and refinement. One weakness in the proposed framework is that our uncertainty-aware loss term is based on the class probabilities of the classifier network which are estimates of epistemic uncertainty and it is known that these probabilities can overestimate confidence in predictions [13]. It is not possible to directly include uncertainties estimated using sampling based approaches such as inference-time drop out due to them not being differentiable. In future work we will address this issue by seeking to incorporate more reliable and differentiable estimates of uncertainty, such as the use of soft labels, to reduce overestimation of confidence. Further investigation into uncertainty estimation approaches during training of DL models should be explored to develop a more robust and accurate method for uncertainty-aware training.

Future work will also investigate methods to improve the aleatoric uncertainty outcomes, such as the implementation of a probabilistic U-net segmentation model [14] instead of using the inference-time drop out technique, which may produce more realistic segmentations and improved estimates of aleatoric uncertainty. Additionally, by incorporating electrical and mechanical parameters with the existing model similar to [15] the CRT prediction accuracy could be improved. In their paper, Albatat *et al.* [15] used mesh generation to develop an electromechanical model that could be used for patient specific CRT prediction. This approach alongside investigating alternate classification model architectures, and potentially expanding the size of the data set, may improve the performance achieved with the developed uncertainty-aware loss function, to improve the accuracy and reliability of CRT predictions.

We believe that the preliminary investigation and framework we have proposed represents an important step on the road towards clinical translation of DL models for complex predictive tasks, such as CRT patient selection. Although we demonstrated the framework in this paper for CRT response prediction we believe it will be applicable to a range of other complex diagnostic/prognostic tasks for which DL models will likely be used for decision-support. In future work, we aim to evaluate the impact of our confidence estimates on clinical decision making with more robust methodologies.

Acknowledgements. This work was supported by the NIHR Guys and St Thomas Biomedical Research Centre and the Kings DRIVE Health CDT for Data-Driven Health. This research has been conducted using the UK Biobank Resource under Application Number 17806. The work was also supported by the EPSRC through the SmartHeart Programme Grant (EP/P001009/1).

References

1. Katbeh, A., Van Camp, G., Barbato, E., et al.: Cardiac resynchronization therapy optimization: a comprehensive approach. Cardiology **142**, 116–128 (2019)
2. Authors/Task Force Members, Brignole, M., Auricchio, A., Baron-Esquivias, G., et al.: 2013 ESC guidelines on cardiac pacing and cardiac resynchronization therapy: the task force on cardiac pacing and resynchronization therapy of the European Society of Cardiology (ESC). Developed in collaboration with the European Heart Rhythm Association (EHRA). Eur. Heart J. **34**(29), 2281–2329 (2013)
3. Stătescu, C., Ureche, C., Enachi, S., Radu, R., Sascău, R.A.: Cardiac resynchronization therapy in non-ischemic cardiomyopathy: role of multimodality imaging. Diagnostics **11**(4), 625 (2021)
4. Peressutti, D., Sinclair, M., Bai, W., et al.: A framework for combining a motion atlas with non-motion information to learn clinically useful biomarkers: application to cardiac resynchronisation therapy response prediction. Med. Image Anal. **35**, 669–684 (2017)
5. Cikes, M., Sanchez-Martinez, S., Claggett, B., et al.: Machine learning-based phenogrouping in heart failure to identify responders to cardiac resynchronization therapy. Eur. J. Heart Fail. **21**(1), 74–85 (2019)
6. Puyol-Antón, E., et al.: Interpretable deep models for cardiac resynchronisation therapy response prediction. In: Martel, A.L., et al. (eds.) MICCAI 2020. LNCS, vol. 12261, pp. 284–293. Springer, Cham (2020). https://doi.org/10.1007/978-3-030-59710-8_28
7. Elton, D.C.: Self-explaining AI as an alternative to interpretable AI. In: Goertzel, B., Panov, A.I., Potapov, A., Yampolskiy, R. (eds.) AGI 2020. LNCS (LNAI), vol. 12177, pp. 95–106. Springer, Cham (2020). https://doi.org/10.1007/978-3-030-52152-3_10
8. Hüllermeier, E., Waegeman, W.: Aleatoric and epistemic uncertainty in machine learning: an introduction to concepts and methods. Mach. Learn. **110**(3), 457–506 (2021)
9. Abdar, M., Pourpanah, F., Hussain, S., et al.: A review of uncertainty quantification in deep learning: techniques, applications and challenges. Inf. Fusion **76**, 243–297 (2021)
10. Geifman, Y., El-Yaniv, R.: Selective classification for deep neural networks. In: Advances in Neural Information Processing Systems, pp. 4878–4887 (2017)
11. Ding, Y., Liu, J., Xu, X., et al.: Uncertainty-aware training of neural networks for selective medical image segmentation. In: Medical Imaging with Deep Learning, pp. 156–173. PMLR (2020)
12. Petersen, S.E., Matthews, P.M., Francis, J.M., et al.: UK Biobank's cardiovascular magnetic resonance protocol. J. Cardiovasc. Magn. Reson. **18**(1), 8 (2015)
13. Hendrycks, D., Gimpel, K.: A baseline for detecting misclassified and out-of-distribution examples in neural networks. arXiv preprint arXiv:1610.02136 (2016)

14. Kohl, S.A.A., et al.: A probabilistic U-Net for segmentation of ambiguous images. arXiv preprint arXiv:1806.05034 (2018)
15. Albatat, M., et al.: Electromechanical model to predict cardiac resynchronization therapy. In: 2018 40th Annual International Conference of the IEEE Engineering in Medicine and Biology Society (EMBC), pp. 5446–5459. IEEE (2018)

Cross-domain Artefact Correction
of Cardiac MRI

Caner Özer[1(✉)] and İlkay Öksüz[1,2]

[1] Department of Computer Engineering, Istanbul Technical University,
İstanbul, Turkey
{ozerc,oksuzilkay}@itu.edu.tr
[2] School of Biomedical Engineering and Imaging Sciences, King's College London,
London, UK

Abstract. Artefacts constitute a paramount issue in medical imaging where the prevalence of artefacts may severely impact the clinical diagnosis accuracy. Specifically, the mistriggering and motion family of artefacts during the cardiac MR image acquisition would eventually damage the visibility of certain tissues, such as the left ventricular, right ventricular, and myocardium. This would cause the ejection fraction to be incorrectly estimated and the patient's heart condition to be incorrectly evaluated. Despite much research on medical image reconstruction, relatively little work has been done for cardiac MRI artefact correction. In this work, inspired by the image reconstruction literature, we propose to use an auto-encoder-guided mistriggering artefact correction method, which not only corrects the artefacts in the image domain but also in the k-space domain with the introduction of an enhanced structure. We conduct a variety of experiments on photos and medical images to compare the performances of different network architectures under mistriggering artefacts and gaussian noise. We demonstrate the superiority of the cross-domain network in the case of k-space-related artefacts.

Keywords: Cardiac MRI · Image denoising · Image artefacts · Auto-encoders

1 Introduction

Magnetic Resonance Imaging (MRI) is a widely used non-invasive imaging technique for constructing a detailed picture of different tissue types in a variety of anatomical regions [13]. This brings a significant advantage over different modalities including computed tomography (CT) in terms of fine-grained resolution with nonexistent exposure to ionizing radiation. However, MRI can be hindered with artefacts due to motion in the case of dynamic organs such as the heart [3].

Although there are a couple of artefact correction methodologies available in the literature, works reducing the artefacts on cardiac MRI are mostly focused on brain MRI volumes. Kidoh et al. converts the brain MRI volumes to discrete cosine transformation (DCT) coefficients prior to the use of convolutional

© Springer Nature Switzerland AG 2022
E. Puyol Antón et al. (Eds.): STACOM 2021, LNCS 13131, pp. 199–207, 2022.
https://doi.org/10.1007/978-3-030-93722-5_22

blocks [4]. Liu et al. uses multiple dense U-Net-like blocks in order to obtain artefact-free brain MRI volumes [6]. For reducing artefacts on dynamic-contrast enhanced MRI of the liver, a patch-wise training scheme is proposed [10]. Finally, 3D Convolutional Neural Networks are used to undermine the effect of motion artefacts on the mid-ventricular slices of cardiac MRI volumes [8]. However, none of these related works use a cross-domain learning strategy that combines a joint image and frequency-space learning.

Depending on the nurture of the artefact or noise, it is critical to use cross-domain architectures in order to handle the problem more efficiently. For instance, a two-stream cross-domain architecture is used for image demoireing, which is the process of removing Moire patterns on an image, where a discrete cosine transformation is used within the frequency network alongside an image network [11]. Also, this kind of architecture design can be seen in KIKI-net [2] where the goal is to reconstruct the MR image given the undersampled k-space acquisitions. In this regard, we adopt the idea of cross-domain-based training of KIKI-net for artefact correction. We hereby refer to the image and frequency domains with our cross-domain definition.

In this paper, we propose to use cross-domain auto-encoders for the mistriggering artefact correction and image denoising tasks. We have two contributions to the current literature. First, we demonstrate that the correction in the k-space domain is essential due to the nature of the image artefact. We show that for the mistriggering artefact correction task, a more stable training can be performed, and a better-corrected image can be obtained on the CMR short-axis view, where we verify this claim through visual and numeric analysis. Second, we show that the nature of the artefact is a key component in the success of different correction architectures. We validate our findings both on computer vision and cardiac MRI data.

2   Method

We propose to use two different network architectures for correcting artefacts, which resemble U-Net [9] and DnCNN [14] architectures as shown in Fig. 1a and 1b, respectively. The first model involves feature aggregations from previous levels as of the same spatial resolution, whereas the second model is simply composed of stacked layers. The first model also uses max-pooling and upscaling layers in order to change the spatial resolution of intermediate-level feature represensations. In this work, we investigate the use of different block structures which we describe in the following paragraphs.

For both of these network architectures, the main purpose is to estimate the noise pattern within an image, which can be due to the presence of Gaussian noise or mistriggering artefacts. The estimated noise pattern will be subtracted from the image, which lets to obtain a denoised or corrected image. Although the noise pattern is estimated by transforming the input into feature maps by passing through the convolutional blocks, the convolutional block choice is critical depending on the source of the noise. Also, these network architectures will work on independent 2D image slices.

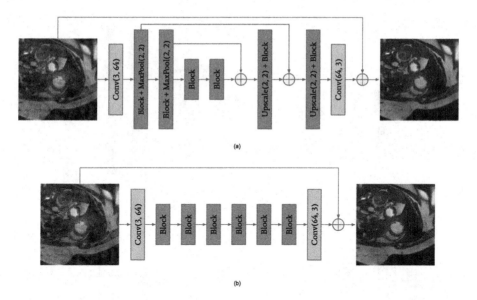

Fig. 1. Network architectures. (a) U-Net-like architecture. (b) DnCNN architecture.

We utilize four different block configuration settings which are used in the network architectures. First of all, as shown in Fig. 2a, we use convolutional, batch normalization, and activation layers in a sequential order which we define as Baseline Convolutional Block working on Image domain (BCB_I). We also construct its frequency domain counterpart as in Fig. 2b, which calculates the Fast Fourier Transform (FFT) of the latent representation, processes the real and imaginary parts of the representation through 2 branches of convolutional, batch normalization and activation layers, and converts the latent representation back to spatial domain through Inverse Fast Fourier Transform (IFFT). We name this block Baseline Convolutional Block working on Frequency domain (BCB_F). Then, we obtain Enhanced Convolutional Block (ECB) by expanding BCB_I with a two-branch BCB_I, which is inspired by [12]. However, different than the aforementioned work, we also average the output of these two branches at the end as in Fig. 2c. Finally, since our motivation is also to improve the quality of the images having mistriggering artefacts that occur on k-space, we propose our cross-domain approach where we adopt BCB_I and BCB_F that the latter has Fourier and inverse Fourier transformation layers right at the start and end of the branches. We aim to perform learning on both image and frequency domains sequentially in this architecture inspired from the image reconstruction literature [2]. After passing the latent representation through BCB_I, the complex-valued numbers (generated by FFT) layer will be handled by a two-branch mechanism, where the upper branch would handle real values while the bottom would handle imaginary values. In this respect, we use the output values of these two branches in an Inverse Fast Fourier Transformation (IFFT) layer to make the switch from

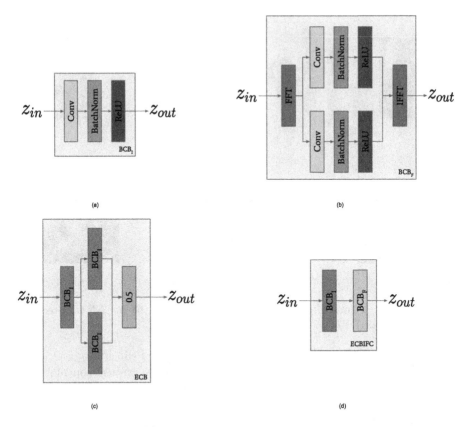

Fig. 2. Building blocks. (a) Baseline Convolutional Block working on Image domain. (b) Baseline Convolutional Block working on Frequency domain. (c) Enhanced Convolutional Block. (d) Enhanced Convolutional Block with Image and Frequency Components

frequency to image space as in Fig. 2d, and name this as Enhanced Convolutional Block with Image and Frequency Components (ECBIFC).

Within each of these blocks, we define the filter and kernel sizes of the convolutional layers as 64 and 3 × 3. We use these blocks within the network architectures and as a result, there exists 0.226 million parameters in the network architectures that use the BCB_I whereas the number of parameters becomes 0.448 million for the models that use BCB_F and 0.671 million for ECB or ECB-IFC, which both have the same number of parameters. However, the number of floating point operations is reduced to its one-third as a result of introducing residual connections and max-pool/upscaling layers in the U-Net model.

3   Experimental Results

We adopted an image denoising repository[1] in order to build our methodology. We conduct 2 experiments on 2 datasets, namely, BSDS300 [7], a color image dataset, and ACDC challenge dataset [1], a 3D+time cardiac MRI dataset. For the BSDS300 dataset, we use 200 images for training and 100 for testing, whereas for the ACDC dataset, we use 100 patient volumes for training and 50 for testing. We train all of these networks for 200 epochs with a mini-batch size of 64 where the parameters of the networks are optimized through ADAM [5] by using the mean squared error loss function. The code and experiments will be made available on this repository[2].

We evaluate the network architectures on both datasets for the image denoising objective, whereas we only use the ACDC dataset for the mistriggering artefact correction objective. First, we evaluate all of the possible scenarios on an image denoising task, where we inject additive white Gaussian noise with a standard deviation of 30 for both of these datasets. Then, we generate synthetic mistriggering artefacts on the ACDC dataset with a corruption ratio of $C = 0.15$. We test all of the possible network architecture and building block combinations to see the effect of improvement on artefact correction. Finally, we evaluate these two scenarios quantitatively in terms of Peak Signal-to-Noise (PSNR) metric.

3.1   Mistriggering Artefact Generation

We generate synthetic mistriggering artefacts [8] by using information from high-quality images in the cardiac cycle. We alter the k-space representation of a cine sequence by swapping the lines inside the k-space representation of adjacent video frames. The percentage of the lines that have been changed is controlled by a hyperparameter, C, which is called the corruption ratio. Increasing the factor C makes the mistriggering artefacts more apparent, and we take $C = 0.15$ in order to create more realistic image artefacts.

Suppose that the number of lines in an image slice and the number of total adjacent slices is denoted by N and T, respectively. Given a 2D+time patient image, I, the pseudocode for k-space-based mistriggering artefact generation is described in Algorithm 1.

In this algorithm, we first define the total number of lines that will be changed, namely, n'. After that, we draw n' lines from a Uniform distribution that is defined between 0 to $N - 1$. Then, we form another discrete standard normal distribution whose mean value is aligned to the current time-slice t. We set the probability mass function of t to 0 in this distribution. In this way, it is guaranteed that values different than the current time-slice, which we call the source time-frames, will be drawn from the distribution. Finally, we apply these changes by transferring the k-space lines from source to target, so we have the images with mistriggering artefacts.

---

[1] https://github.com/lychengr3x/Image-Denoising-with-Deep-CNNs.
[2] https://github.com/canerozer/xdomain-ac.

Algorithm 1: k-space based randomized mistriggering artefact generation procedure

---

Result: I', Images with mistriggering artefact
$I, N, T, P, t, t', n', I'$
while $t \leq T$-1 **do**
  $n' := NP$
  select *random* n' *lines from a Uniform distribution*
  form *a discrete standard Normal distribution*
  align *x-axis of the distribution according to t*
  select *random* t' *lines from that distribution*
  apply *changes on* I' *using I, n' and t'*
  $t := t + 1$
end

3.2 Preprocessing

For ACDC Dataset, we independently process 2D slices of each systole and diastole phase of the CMR volumes. Then, we resize the images to 256×256 and train the network architectures by randomly cropping with a size of 192×192 regardless of the dataset. For test time evaluation, we take the center crop of the image which has the same input size as the training. After rescaling the pixel intensity values between $[-1, 1]$, we will be injecting Gaussian noise if the objective is image denoising.

3.3 Quantitative Analysis

In order to measure the denoising and mistriggering artefact correction performances of the network architectures, we use Peak Signal-to-Noise Ratio (PSNR) metric, which divides the square of the dynamic range of the image to the Mean Square Error (MSE) of corrected output, $\tilde{x}$, and clean image, x and takes its logarithm as shown below:

$$PSNR = 10log_{10}\left(\frac{4}{MSE(\tilde{x}, x)}\right). \tag{1}$$

In Table 1, we illustrate PSNR values for all of the datasets and objectives. We also provide the PSNR values for the scenario that when no correction is applied to the images. The DnCNN model (Model 1) is generally better than the one that resembles the U-Net architecture (Model 2) in terms of PSNR. We may relate this to the use of max-pool layers, which are present for reducing the computational complexity of the U-Net-like model. In addition, we see a mostly positive impact of using the ECB except for the denoising task for the BSDS dataset. However, when we involve processing on the k-space domain, the performance slightly deteriorates for the denoising on both BSDS and ACDC datasets. We relate this degradation in performance to the fact that Gaussian noise is present on both frequency and image domains as Gaussian distribution.

Nevertheless, we see the strength of involving learning on the frequency space on the artefact reduction task. Even though models with ECB and ECBIFC blocks have the same number of parameters, since we introduce the synthetic parameters on the k-space, we observe a positive impact on the PSNR metric on both models due to the underlying source of the artefacts (e.g., k-space). We see that using the cross-domain-based training scheme has improved the mistriggering artefact correction performance by 1.42 and 0.39 dB for Model 1 and 2, respectively. Interestingly, we see a fall in the PSNR metric from 25.64 to 23.45 when we use Model 1 with BCB_I blocks for the artefact reduction task on the ACDC dataset. In addition, the improvement caused by using BCB_F is somehow limited. We relate these performance limitations with inadequate model capacity and lacking residual connections within the architecture and even improve by stacking image and frequency baseline blocks.

We also provide qualitative results to underscore the network architectures' image denoising and artefact correction capabilities as in Fig. 3. We use Model 1 with BCB_I for the first row (BSDS Denoising), Model 1 with EBC for the second row (ACDC Denoising), and Model 1 with ECBIFC for the third row (ACDC Artefact Reduction). In all three cases, it is clear that the networks are able to suppress the noise and artefact factors, despite the smoothing. Nevertheless, we can see the details of the images more clearly (i.e., the marching band in the first row or cardiac anatomical structures for the last two rows), which also support our claim for image quality improvement as a result of using block structures, depending on the problem setting.

Table 1. PSNR results of 2 network architectures when different block structures are used. The experiments are performed on the BSDS and ACDC datasets for denoising (D) and mistriggering artefact reduction (AR) tasks.

		DnCNN				U-Net			
Dataset	No Correction	BCB_I	BCB_I	ECB	ECBIFC	BCB_I	BCB_F	ECB	ECBIFC
BSDS (D)	18.58	**29.13**	26.46	28.75	25.11	28.40	26.07	28.60	27.76
ACDC (D)	18.58	33.93	33.55	**34.06**	33.67	32.49	31.08	32.99	32.28
ACDC (AR)	25.64	23.45	26.06	27.01	**28.43**	27.26	25.60	27.79	28.18

4 Discussion

In this paper, we propose a method for correcting cardiac MR image artefacts using a cross-domain network. The inclusion of cross-domain knowledge improves the image quality in case of mistriggering artefacts by +1.42 and +0.39 dB of increment after comparing ECB with ECBIFC. When white Gaussian noise is applied to degrade image quality, it is observed that cross-domain network performs slightly worse than single domain counterpart. This study underlines the significance of utilizing cross-domain information in the case the source of the

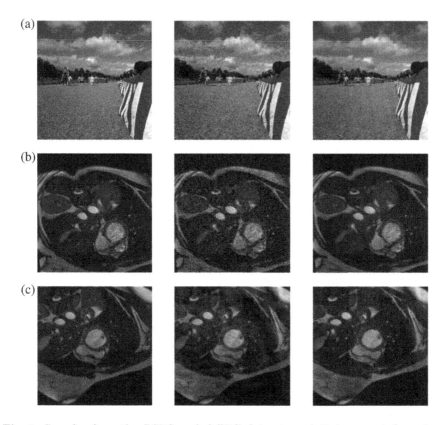

Fig. 3. Samples from the BSDS and ACDC datasets and their corrected versions. Columns from left to right: (i) Clean image, (ii) Noisy image, (iii) Corrected/Denoised Image. Rows from up to bottom: (a) Sample from the noisy BSDS dataset, using Model 1 with BCB_I (b) Sample from the noisy ACDC dataset, using Model 1 with ECB (c) Sample from the ACDC dataset with mistriggering artefacts, using Model 1 with ECBIFC.

artefact is in k-space. Future work will include applying the idea of utilizing cross-domain knowledge on state-of-the-art image denoising networks with multiple sources of image quality deteriorations. A global image correction algorithm design will be essential to enable the clinical translation of these tools.

Acknowledgements. This paper has been produced benefiting from the 2232 International Fellowship for Outstanding Researchers Program of TUBITAK (Project No: 118C353). However, the entire responsibility of the publication/paper belongs to the owner of the paper. The financial support received from TUBITAK does not mean that the content of the publication is approved in a scientific sense by TUBITAK.

References

1. Bernard, O., et al.: Deep learning techniques for automatic MRI cardiac multi-structures segmentation and diagnosis: Is the problem solved? IEEE TMI **37**(11), 2514–2525 (2018)
2. Eo, T., Jun, Y., Kim, T., Jang, J., Lee, H.J., Hwang, D.: KIKI-net: cross-domain convolutional neural networks for reconstructing undersampled magnetic resonance images. Magn. Reson. Med. **80**(5), 2188–2201 (2018)
3. Ferreira, P.F., Gatehouse, P.D., Mohiaddin, R.H., Firmin, D.N.: Cardiovascular magnetic resonance artefacts. JCMR **15**(1), 41 (2013)
4. Kidoh, M., et al.: Deep learning based noise reduction for brain MR imaging: tests on phantoms and healthy volunteers. Magn. Reson. Med. Sci. **19**(3), 195–206 (2020)
5. Kingma, D.P., Ba, J.: Adam: a method for stochastic optimization. In: Bengio, Y., LeCun, Y. (eds.) 3rd International Conference on Learning Representations, ICLR 2015, San Diego, CA, USA, 7–9 May 2015, Conference Track Proceedings (2015). http://arxiv.org/abs/1412.6980
6. Liu, J., Kocak, M., Supanich, M., Deng, J.: Motion artifacts reduction in brain MRI by means of a deep residual network with densely connected multi-resolution blocks (DRN-DCMB). Magn. Reson. Imaging **71**, 69–79 (2020)
7. Martin, D., Fowlkes, C., Tal, D., Malik, J.: A database of human segmented natural images and its application to evaluating segmentation algorithms and measuring ecological statistics. In: ICCV, vol. 2, pp. 416–423, July 2001
8. Oksuz, I., et al.: Deep learning using K-space based data augmentation for automated cardiac MR motion artefact detection. In: Frangi, A.F., Schnabel, J.A., Davatzikos, C., Alberola-López, C., Fichtinger, G. (eds.) MICCAI 2018. LNCS, vol. 11070, pp. 250–258. Springer, Cham (2018). https://doi.org/10.1007/978-3-030-00928-1_29
9. Ronneberger, O., Fischer, P., Brox, T.: U-net: convolutional networks for biomedical image segmentation. In: Navab, N., Hornegger, J., Wells, W.M., Frangi, A.F. (eds.) MICCAI 2015. LNCS, vol. 9351, pp. 234–241. Springer, Cham (2015). https://doi.org/10.1007/978-3-319-24574-4_28
10. Tamada, D., Kromrey, M.L., Ichikawa, S., Onishi, H., Motosugi, U.: Motion artifact reduction using a convolutional neural network for dynamic contrast enhanced MR imaging of the liver. Magn. Reson. Med. Sci. **19**(1), 64–76 (2020)
11. Vien, A.G., Park, H., Lee, C.: Dual-domain deep convolutional neural networks for image demoireing. In: CVPRW, June 2020
12. Xie, S., Girshick, R.B., Dollár, P., Tu, Z., He, K.: Aggregated residual transformations for deep neural networks. In: CVPR, pp. 5987–5995 (2017)
13. Zbontar, J., et al.: fastMRI: an open dataset and benchmarks for accelerated MRI (2018)
14. Zhang, K., Zuo, W., Chen, Y., Meng, D., Zhang, L.: Beyond a Gaussian denoiser: residual learning of deep CNN for image denoising. IEEE TIP **26**(7), 3142–3155 (2017)

Detection and Classification of Coronary Artery Plaques in Coronary Computed Tomography Angiography Using 3D CNN

Jun-Ting Chen[1], Yu-Cheng Huang[2], Holger Roth[3], Dong Yang[3], Chih-Kuo Lee[4], Wen-Jeng Lee[2], Tzung-Dau Wang[5], Cheng-Ying Chou[6], and Weichung Wang[7(✉)]

[1] Graduate Program of Data Science, National Taiwan University and Academia Sinica, Taipei, Taiwan
r08946015@ntu.edu.tw

[2] Department of Medical Imaging, National Taiwan University Hospital, Taipei 100, Taiwan

[3] NVIDIA, Bethesda, Maryland, USA
{hroth,dongy}@nvidia.com

[4] Department of Internal Medicine, National Taiwan University Hospital Hsin-Chu Branch, Hsinchu, Taiwan

[5] Cardiovascular Center and Divisions of Cardiology and Hospital Medicine, Department of Internal Medicine, National Taiwan University Hospital, Taipei, Taiwan
tdwang@ntu.edu.tw

[6] Department of Biomechatronics Engineering, National Taiwan University, Taipei, Taiwan
chengying@ntu.edu.tw

[7] Institute of Applied Mathematical Sciences, National Taiwan University, Taipei, Taiwan
wwang@ntu.edu.tw

Abstract. Measuring the existence of coronary artery plaques and stenoses is a standard way of evaluating the risk of cardiovascular diseases. Coronary Computed Tomography Angiography (CCTA) is one of the most common assessments of cardiovascular disease. Manually detecting stenoses over the complete coronary artery tree of each patient takes an expert about 10 min. Previous works about automated detection of plaques required carefully controlling parameters of region-based segmentation algorithms or hand-craft features extraction for machine learning algorithms. Current works presented the feasibility of deep learning algorithms for plaque and stenosis detection. This study aims to develop an efficient workflow for detecting plaques and classification using a 3D convolutional neural network (CNN). Based on a fully convolutional structure, our model has better stability than recurrent layers at the training stage. Furthermore, our method is efficient since the model takes only cross-sectional planes (CSP) images as the input, rather than including other views containing replicate information. For each patient, it takes less

E. Puyol Antón et al. (Eds.): STACOM 2021, LNCS 13131, pp. 208–218, 2022.
https://doi.org/10.1007/978-3-030-93722-5_23

than 25 s to perform the whole workflow. The first step of our workflow is to extract CSP images on each control point of a given set of coronary artery centerlines. We then stack several contiguous CSP images together to form a 3D volume, which is the input of the CNN model. The model will predict the type of plaque (soft, mixed, or calcified) at each control point. With this information, we can not only detect the precise location of plaques but also summarize the existence of plaque in each artery segment. In addition, we consider the lumen radius estimation as the next step to approach stenosis detection.

Keywords: Coronary plaque · Coronary Computed Tomography Angiography (CCTA) · Convolutional neural network (CNN)

1 Introduction

1.1 Clinical Significance

Cardiovascular diseases (CVDs) can cause damage to the heart and blood vessels and are the world's leading cause of mortality. In 2019, it claimed about 18.6 million lives [10]. Out of all CVDs, coronary artery diseases account for most deaths. Severe stenosis in the coronary artery is related to the occurrence of subsequent acute myocardial infarction(AMI), which is also called a heart attack. AMI may result in a high mortality rate and significant morbidity in those who survived. Early detection and risk factor modification were shown to reduce mortality and morbidity rate in patients at risk of AMI. Coronary artery catheter angiography is used as a reference standard in detecting coronary artery stenosis and plaque. However, it is an invasive process that might result in severe complications in rare occurrences. State-of-the-art coronary CT angiography (CCTA) is shown to have high sensitivity and fair specificity in detecting coronary artery stenosis, while maintaining a low radiation dose and minimal invasiveness. Aside from stenosis detection and grading, CCTA offers extra information regarding the characteristics and composition of the atherosclerotic plaque. Certain plaque types and characteristics are shown to be independent predicting factors of death and morbidity. Thus CCTA is included as a part in the management guideline of cardiac atherosclerotic disease by the European Society of Cardiology and American Heart Association. However, reading of CCTA is technically demanding. It has to be read by sub-specialty trained experts on dedicated coronary CTA workstations with lengthy manual post-processing and reading time [9]. Therefore, we aim to create an efficient workflow for detecting plaque and filtering out health regions.

1.2   Related Work

Previously, intensity-based thresholding methods were used for automatic detection of calcified plaque [11] but they would cause many false positives (FP) and mostly were insufficient for detecting soft plaques. With automatic vessel segmentation, work on soft plaque detection was proposed [6], however, it requires carefully controlling parameters as well as previous works do. Several works [14, 16] on both detection and characterization were proposed based on SVM, and requiring hand-craft features as inputs. More recent studies [2, 15] invest in both plaque characterization and stenosis detection using neural networks. Reik et al. [15] design a 3D CNN combined with recurrent layers to extract features from segments cubes in multi-planar reformation (MPR) images, wherein the training phase takes manually annotated negative (health) segments. Another study [2] also processes CCTA volumes into MPR images and designs a deeper 3D CNN evaluated on a large private dataset of 493 patients in total. It is not clear if such deep networks can perform effectively on smaller datasets which are more widely available. Other work [8] on severe stenosis detection, which is highly related to plaque detection, is also based on CNN but requires both CSP and MPR as input and is therefore more inefficient.

1.3   Contribution

In this paper, we propose an efficient workflow for detecting plaque and classifying its type along the full length of the coronary arteries. The overview of the workflow is presented in Fig. 1. We first extract a cross-sectional plane (CSP) at each point of a set of centerlines representing the coronary arteries. Since the plaque is a consequent structure in the coronary artery, the plaque type information is related to the characterization at neighboring points. We then formulate this task as a time series problem, with each control point as a time stamp. The extracted contiguous CSP will be stacked together to enter the 3D CNN for predicting. (1) Our method is efficient since we only take CSP views of coronary arteries, without requiring other views, such as curved planar reformatted views. For each patient, parsing the image volume and meta information from Dicom files and generating cross-sectional planes cost about 20 s. It then takes less than 3 s to perform prediction with extracted CSPs. Therefore it takes less than 25 s to run the workflow for each patient. (2) Also, instead of manually splitting the segments from the whole vessel, we apply the automatic sliding process to traverse the whole arteries. To perform point-wise classification, every point will be assigned with the same annotation as the plaque it belongs to. Therefore, our method and training process can easily transfer to every private dataset with annotation containing type, start, and endpoint information of each plaque. The result shows that over 90% of the plaques are detected.

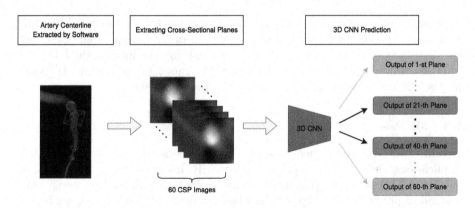

Fig. 1. Overview of our method. First, we use the vessel centerline extracted by software to generate cross-sectional planes. We then collect a sequence of 60 contiguous CSP images for the inputs of the 3D CNN and got the output of central 20 CSP images. The outputs of 20 CSP images at both end sides will be abandoned, and only preserved at the training stage for computing loss values.

2 Methodology

2.1 Cross-Sectional Plane Extraction

The vessel centerlines were first automatically generated by Philips IntelliSpace Portal (Philips, Amsterdam, Netherlands) software, and manually modified and corrected. Associated centerline information, including control point coordinates, tangent vectors, and normal vectors, is then generated automatically according to the final annotated centerline. We then utilize this information to generate cross-sectional views of coronary arteries. The cross-sectional planes are extracted from every control point of the whole coronary arteries. The average spacing between each two nearby control points is about 0.38 mm, which is also the spacing between each two nearby cross-sectional planes. Therefore the extracted cross-sectional planes would not miss any important information contained in the curved-planar reconstruction view. The pixel space of the CSP is set to 0.1 × 0.1 in millimeters to get smooth enough images. To cover the lumen and vessel voxels within each plane, each image has a size of 79 × 79 pixels. Therefore each extracted plane will cover a 7.9 × 7.9 area in millimeters, surpassing most areas of the artery vessel.

2.2 Piecewise Linear Image Normalization

Unlike usual RGB images with pixel values ranging from 0 to 255, the Hounsfield Units (HU) value of CT volumes used in this study range widely from −1024 to 3071. According to Fig. 2, we plot the histogram of all collected CSP images and found that most HU values are in the range $[-200, 600]$, and a small portion of values are in the range $[600, 1200]$. Adapting from min-max normalization and applying value clipping, we create a piece-wise linear transformation $T(X)$ that compresses the HU values to the range $[0, 1]$. In order to approximately match the distribution of pixel intensities when normalizing the images, the HU value in $[-200, 600]$ will be linearly transformed to the range $[0, 0.9]$, and the HU value in $[600, 1200]$ will be linearly transformed to the range $[0.9, 1]$.

2.3   3D Resnet

Recently, 3D CNNs are generally applied on time series tasks and medical image applications, such as organ segmentation on MRI images. Our model is based on the **R(2+1)D** structure [13], a 3D Resnet variant for video action recognition. Instead of predicting the type of the whole sequence of CSP images, we let the model predict the types at each CSP image. As we mentioned previously, the plaque type of each CSP image is related to neighboring images. We formulate the task as a time-series problem, with each CSP image in a sequence as a video frame. Thus we replace the spatio-temporal pooling with a 1×1 convolution layer and an output of size $L \times 4 \times 1 \times 1$. Here, 4 is the number of plaque types and L is the length of each CSP image sequence. The network often has poor prediction near the two end sides of each sequence, hence for the sequence of $L = 60$ CSP images, we only take the prediction output corresponding to the central 20 CSP images, with 20 CSP images prediction at both end sides abandoned. To traverse all the points of each patient's artery tree, we use the sliding window movement for selecting the CSP sequences. Starting from the proximal end of each vessel, we take 60 contiguous CSP images as the input sequence for the model and output the prediction of the central 20 images, and then move forward the distal end with 20 points, repeating the same approach and so on until approaching the distal end.

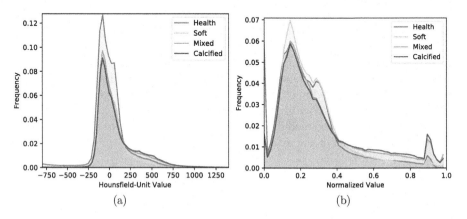

Fig. 2. Histogram of (a) HU value (b) normalized values distribution, of all collected cross-sectional images, according to each plaque types.

2.4   Training Strategy

We found that the target data type of CCTA is unbalanced: Plaque only occupies a small portion of the whole arterial tree, and the soft type only accounts for a small part among all three types of plaque. Plain training will cause significant overfitting and will worsen the performance on the soft type. Therefore, we apply the following technique to avoid the above issues.

Weighted Loss. Class specified loss weight has been a potential solution for class-imbalanced classification tasks. By calculating the point number of each type, reported in Table 1, in all $173,781$ extracted planes, over 90% of planes belong to the health type. We then set the rounded values of the "Sqrt. Ratio" to be the class weight $c = \langle 1, 11, 5, 7 \rangle$ for cross-entropy loss. In addition to the class weights, we also introduce a sequence weight to the loss. Despite we throw the outputs regarding several planes at both end sides in the prediction stage, we include these outputs into loss calculation for backward propagation in the training stage. The loss values of the planes near both end sides play the role of auxiliary loss, hence required down-weighting. As Fig. 3 shows, the planes toward both end sides have the weights exponentially decreasing.

Epoch Data Sampling. Data resampling has been considered an efficient solution to solve data imbalanced classification problems [1]. Recent work [4] showed this strategy also works for deep learning algorithms on highly imbalanced datasets. In this work, we simply apply a down-sampling strategy to the

Table 1. Collection information of each type. The first column indicates the number of planes of each type. The second column indicates the ratio, taking the number of all planes 173,781 divided by that of each type. The third column presents the squared root value of the ratio shown in the second row.

Plaque type	Number of control points	Ratio	Sqrt of ratio
Health (Non plaque)	161,032	1.079	1.039
Soft	1,350	128.727	11.346
Mixed	7,339	23.679	4.866
Calcified	4,060	42.803	6.542

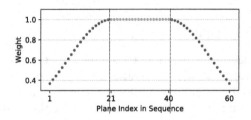

Fig. 3. Sequential weight for a sequence with length $L = 60$.

sequence data before each epoch starts. Given a keep ratio $k = \langle k_0, k_1, k_2, k_3 \rangle$, for the i-th type with N_i sequences, a number $\text{floor}(N_i \cdot k_i)$ of sequences will be randomly sampled from the N_i sequences. Then each epoch would contain $\sum_{i=0}^{3} \text{floor}(N_i \cdot k_i)$ sequences. Also adapted from the "Sqrt. Ratio" reported in 1, starting with the keep ratio $k = \langle 0.1, 1, 0.5, 0.7 \rangle$, we performed a grid search to find a proper selection for the keep ratio k. Finally, we chose $k = \langle 0.1, 1, 0.7, 0.9 \rangle$.

Optimizer and Learning Rate Decay Schedule. Recently, the AdamW algorithm [7] has become a popular choice of optimization for backward propagation due to the regularization effect. Since the scale of 3D Resnet is not that deep or wide, we simply pick a larger regularization factor $\lambda = 10^{-3}$. Besides, the initial learning rate and decay schedule are critical for convergence. We chose cyclic learning rates [12] for the learning rate decay scheduler. In addition to the learning rate warmup scheme [3] to prevent instability, the cyclic schedule also performs the restart scheme to avoid fall into some improper local minimum. Moreover, to make the maximum bound (max_lr) deceasing, we chose the exp_range policy, with the exponential factor $gamma = 0.999$.

3   Evaluation

3.1   Dataset

In this study, a private dataset of 127 CCTA scans from a single tertiary referral center (National Taiwan University Hospital (NTUH)) is used to evaluate our proposed workflow. The dataset is composed of patients with intermediate cardiovascular risks who received out-patient CCTA to decide further deposition. CCTA scans are performed and reconstructed on state-of-the-art CT scanners with dedicated CCTA protocols. The cardiac phases with least motion artifact are chosen and initial post-processing and subsequent annotation are done on Philips IntelliSpace Portal (Philips, Amsterdam, Netherlands). The following are the common properties of the 127 CCTA volumes, showing in the form of the mean value $\pm$ standard deviation: 293.7 ± 82 slices, with the slice thickness 0.507 ± 0.098 mm, and each slice of shape 512×512 pixels with the pixel space $0.390 \times 0.390 \pm 0.018 \times 0.018$ mm. The 127 scans contain 97 soft, 241 mixed, 289 calcified plaques in total. The 127 cases are divided into 3 splits, 88 cases for training, 22 cases for validation, and 17 cases for testing.

3.2   Experimental Results

In the training stage, we utilize the **F1** scores of 4 types of plaque to monitor the learning status. In our observation, the historical curve of **F1** scores showed frequent oscillations. We consider the weighted average of **F1** scores on the validation set, with the weight $w = \langle 1, 9, 5, 7 \rangle$, as the monitoring metric for selecting model weights (checkpoints). The selected model has point-based accuracy of 0.8758. For further analysis, we define that a plaque region is **detected** if over 70% of its points are classified to one of 3 positive plaque types, and a plaque is correctly **characterized** if over 70% of its points are correctly classified. Table 2 represents the above information of the prediction. One of the test case predictions is visualized in Fig. 4, which presents that all the plaques are predicted, but with some region misclassified.

Table 2. Summary of prediction on 17 test cases showing the number of correctly detected and correctly characterized plaque types.

Plaque type	Soft	Mix	Calcified	Overall
Number of plaques	18	46	49	113
Detected	10 (55.6%)	46 (100%)	49 (100%)	105 (92.9%)
Correct characterized	7 (38.9%)	13 (28.2%)	29 (59.2%)	49 (43.4%)

Rotterdam Coronary Evaluation Dataset. Furthermore, to evaluate the generalization of our method, we include the prediction result on the Rotterdam Coronary Evaluation Dataset [5]. To extract cross-sectional planes, we first applied finite differential to estimate the tangent and normal vectors of the given centerline coordinations. Then the CSP images were extracted with the same setting mentioned in Sect. 2.1. Without fine-tuning on the Rotterdam dataset, we directly performed inference using the model only trained on NTUH dataset. One of the test case predictions is visualized in Fig. 5, where most of the plaques are predicted, but with some region misclassified. Table 3 shows the summary of directly prediction on Rotterdam dataset. The result is worse than the prediction of NTUH dataset, which is possibly caused by different image quality and reference standard between Rotterdam and NTUH dataset.

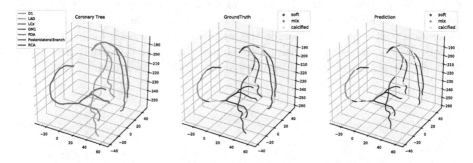

Fig. 4. Visualization of Prediction. Each curve is a centerline of vessels, and the regions in blue indicate health (non-plaque). (Color figure online)

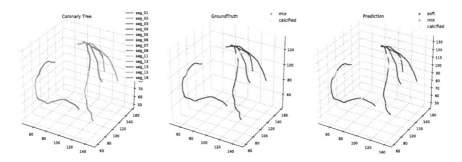

Fig. 5. Visualization of Prediction on the Rotterdam dataset. Each curve is a centerline of vessels, and the regions in blue indicate health (non-plaque). (Color figure online)

Table 3. Summary of prediction on 18 cases of Rotterdam dataset, showing the number of correctly detected and correctly characterized plaque types.

Plaque type	Soft	Mix	Calcified	Overall
Number of Plaques	33	31	39	103
Detected	12 (36.3%)	18 (58.0%)	22 (56.4%)	52 (50.4%)
Correct characterized	7 (21.2%)	3 (9.6%)	12 (30.7%)	22 (21.3%)

4   Discussion and Conclusions

In this paper, we proposed an efficient workflow for detecting plaque in coronary arteries using 3D Resnet. We apply several strategies for reducing data imbalance and overfitting issues. Our results show that most of the plaques can be automatically detected. But there are still some misclassified regions, especially for the soft type plaque. Furthermore, false-positive predictions tend to exist in bifurcation and the distal end of vessels. The current result might be improved with adjustment on the length of the input sequences. Moreover, the evaluation part requires to include the significance of stenosis for each plaque. In other words, the plaques with severe stenosis are more important than those with slight stenosis, hence require more accurate characterization.

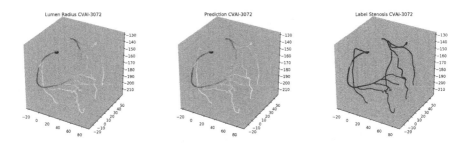

Fig. 6. A visualization of lumen radius estimation.

On the other hand, our future work is applying the proposed workflow on stenosis detection. Lumen radius estimation is a prerequisite approach for severe stenosis detection. Our preliminary work on lumen radius estimation is visualized in Fig. 6. The preliminary result shows the possibility of stenosis detection using our workflow.

Acknowledgments. This study was supported by Ministry of Science and Technology (MOST; MOST108-3011-F002-004). The funding source had no role in study design, data collection, analysis or interpretation, report writing or the decision to submit this paper for publication.

References

1. Branco, P., Torgo, L., Ribeiro, R.P.: A survey of predictive modeling on imbalanced domains. ACM Comput. Surv. (CSUR) **49**(2), 1–50 (2016)
2. Candemir, S., White, R.D., Demirer, M., Gupta, V., Bigelow, M.T., Prevedello, L.M., et al.: Automated coronary artery atherosclerosis detection and weakly supervised localization on coronary CT angiography with a deep 3-dimensional convolutional neural network. Comput. Med. Imaging Graph. **83**, 101721 (2020)
3. Goyal, P., et al.: Accurate, large minibatch SGD: training ImageNet in 1 hour (2018)
4. Johnson, J.M., Khoshgoftaar, T.M.: The effects of data sampling with deep learning and highly imbalanced big data. Inf. Syst. Front. **22**(5), 1113–1131 (2020)
5. Kirişli, H., et al.: Standardized evaluation framework for evaluating coronary artery stenosis detection, stenosis quantification and lumen segmentation algorithms in computed tomography angiography. Med. Image Anal. **17**(8), 859–876 (2013)
6. Lankton, S., Stillman, A., Raggi, P., Tannenbaum, A.R.: Soft plaque detection and automatic vessel segmentation. Georgia Institute of Technology (2009)
7. Loshchilov, I., Hutter, F.: Decoupled weight decay regularization. In: International Conference on Learning Representations (2018)
8. Tejero-de-Pablos, A., et al.: Texture-based classification of significant stenosis in CCTA multi-view images of coronary arteries. In: Shen, D., et al. (eds.) MICCAI 2019. LNCS, vol. 11765, pp. 732–740. Springer, Cham (2019). https://doi.org/10.1007/978-3-030-32245-8_81
9. Pugliese, F., et al.: Learning curve for coronary CT angiography: what constitutes sufficient training? Radiology **251**(2), 359–368 (2009). https://doi.org/10.1148/radiol.251208038410.1148/radiol.2512080384. PubMed https://www.ncbi.nlm.nih.gov/pubmed/1940157019401570
10. Roth, G.A., et al.: Global burden of cardiovascular diseases and risk factors, 1990–2019: update from the GBD 2019 study. J. Am. Coll. Cardiol. **76**(25), 2982–3021 (2020). PubMed Central https://www.ncbi.nlm.nih.gov/pmc/articles/PMC7755038PMC7755038. https://doi.org/10.1016/j.jacc.2020.11.01010.1016/j.jacc.2020.11.010. PubMed https://www.ncbi.nlm.nih.gov/pubmed/3306932633069326
11. Saur, S.C., Alkadhi, H., Desbiolles, L., Székely, G., Cattin, P.C.: Automatic detection of calcified coronary plaques in computed tomography data sets. In: Metaxas, D., Axel, L., Fichtinger, G., Székely, G. (eds.) MICCAI 2008. LNCS, vol. 5241, pp. 170–177. Springer, Heidelberg (2008). https://doi.org/10.1007/978-3-540-85988-8_21

12. Smith, L.N.: Cyclical learning rates for training neural networks. In: 2017 IEEE Winter Conference on Applications of Computer Vision (WACV), pp. 464–472. IEEE (2017)
13. Tran, D., Wang, H., Torresani, L., Ray, J., LeCun, Y., Paluri, M.: A closer look at spatiotemporal convolutions for action recognition. In: 2018 IEEE/CVF Conference on Computer Vision and Pattern Recognition, pp. 6450–6459 (2018). https://doi.org/10.1109/CVPR.2018.00675
14. Zhao, F., Wu, B., Chen, F., Cao, X., Yi, H., Hou, Y., et al.: An automatic multi-class coronary atherosclerosis plaque detection and classification framework. Med. Biol. Eng. Comput. **57**(1), 245–257 (2019)
15. Zreik, M., Van Hamersvelt, R.W., Wolterink, J.M., Leiner, T., Viergever, M.A., Išgum, I.: A recurrent CNN for automatic detection and classification of coronary artery plaque and stenosis in coronary CT angiography. IEEE Trans. Med. Imaging **38**(7), 1588–1598 (2018)
16. Zuluaga, M.A., Hush, D., Delgado Leyton, E.J.F., Hoyos, M.H., Orkisz, M.: Learning from only positive and unlabeled data to detect lesions in vascular CT images. In: Fichtinger, G., Martel, A., Peters, T. (eds.) MICCAI 2011. LNCS, vol. 6893, pp. 9–16. Springer, Heidelberg (2011). https://doi.org/10.1007/978-3-642-23626-6_2

Predicting 3D Cardiac Deformations with Point Cloud Autoencoders

Marcel Beetz[1]($\boxtimes$), Julius Ossenberg-Engels[1], Abhirup Banerjee[1,2],
and Vicente Grau[1]

[1] Institute of Biomedical Engineering, Department of Engineering Science,
University of Oxford, Oxford OX3 7DQ, UK
marcel.beetz@eng.ox.ac.uk
[2] Division of Cardiovascular Medicine, Radcliffe Department of Medicine,
University of Oxford, Oxford OX3 9DU, UK
abhirup.banerjee@cardiov.ox.ac.uk

Abstract. Mechanical contraction and relaxation of the heart play an important role in evaluating healthy and diseased cardiac function. Mechanical patterns consist of complex non-linear 3D deformations that vary considerably between subjects and are difficult to observe on 2D images, which impacts the prediction accuracy of cardiac outcomes. In this work, we aim to capture 3D biventricular deformations at the end-diastolic (ED) and end-systolic (ES) phases of the cardiac cycle with a novel geometric deep learning approach. Our network consists of an encoder-decoder structure that works directly with light-weight point cloud data. We initially train our network on pairs of ED and ES point clouds stemming from a mixed population of subjects with the aim of accurately predicting ED outputs from ES inputs as well as ES outputs from ED inputs. We validate our network's performance using the Chamfer distance (CD) and find that ED and ES predictions can be achieved with an average CD of 1.66 ± 0.62 mm on a dataset derived from the UK Biobank cohort with an underlying voxel size of $1.8 \times 1.8 \times 8.0$ mm [8]. We derive structural and functional clinical metrics such as myocardial mass, ventricular volume, ejection fraction, and stroke volume from the predictions and find an average mean deviation from their respective gold standards of 1.6% and comparable standard deviations. Finally, we show our method's ability to capture deformation differences between specific subpopulations in the dataset.

Keywords: Cardiac deformation prediction · Point cloud autoencoders · Cardiac anatomy reconstruction · Cardiac contraction · Geometric deep learning · Cardiac MRI

1 Introduction

Cardiovascular diseases are the most common cause of death in the world, accounting for 32% of all annual fatalities in 2019 [10]. A major driver behind this is the often insufficient understanding of cardiac pathologies, particularly

© Springer Nature Switzerland AG 2022
E. Puyol Antón et al. (Eds.): STACOM 2021, LNCS 13131, pp. 219–228, 2022.
https://doi.org/10.1007/978-3-030-93722-5_24

the relationship between structural changes and cardiac function. Furthermore, the differentiation between subject subpopulations in terms of cardiac function is often not taken into account. All of these affect predictions of clinical outcomes and consequently complicate the application of personalised treatment plans. In an attempt to address this issue, we aim to gain a more comprehensive understanding of the structure-function interactions in the heart. To this end, we propose a novel point cloud autoencoder network (PCN) to capture 3D cardiac deformation and model subpopulation-specific cardiac mechanical function. While deep learning on point cloud data has recently been applied to multiple problems in cardiac image analysis, such as segmentation [12], disease classification [4], and surface reconstruction [1], this work proposes, to the best of our knowledge, the first point cloud-based deep learning approach for cardiac deformation modelling. Grid-based deep learning techniques have previously been used for survival prediction [2], image registration [6], and motion-modelling for image sequences [5], but have not investigated subpopulation differences. We improve upon the closest prior work in this area [7] in several ways. Firstly, we utilize 3D instead of 2D data which enables us to capture real cardiac deformation and allows us to calculate clinical metrics based on volumetric changes in the cardiac cycle, such as ejection fraction (EF) and stroke volume (SV). Secondly, we work directly with point cloud data, rather than pixel information, which is more memory-efficient and has a straighforward expansion to 3D. Thirdly, the PCN is more efficient at working with surface models than grid-based networks, allowing us to scale to considerably higher surface densities. Finally, we employ the PCN to predict both phase directions, i.e. end-systolic (ES) cardiac shape conditioned on structural end-diastolic (ED) inputs as well as ED outputs from ES inputs, allowing us to model both cardiac contraction and relaxation.

2   Methods

We first provide a brief overview of our dataset before explaining the network architecture and training procedure of our proposed method.

2.1   Dataset

Our dataset consists of ∼500 female and ∼500 male subjects that were randomly chosen from the UK Biobank study [8]. We select cine MRI acquisitions at both ED and ES phases of the cardiac cycle for each subject and use the multi-step pipeline described in [1] to obtain corresponding 3D point cloud representations of the biventricular anatomy.

2.2   Network Architecture

Our method takes as input a multi-class point cloud that represents the biventricular anatomy at either end of the cardiac cycle (ED or ES) and is tasked to predict the corresponding deformed anatomical surface at the other extreme phase of the cardiac cycle. The network architecture consists of an encoder-decoder

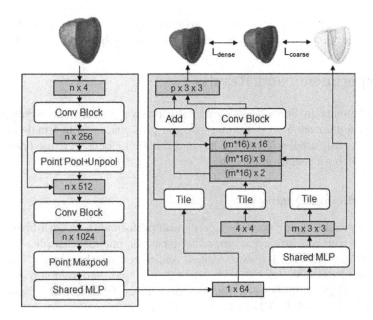

Fig. 1. Architecture of the proposed point cloud autoencoder network. The input point cloud consists of n points, each of which is stored as a 4-dimensional vector (x, y, z point coordinates and a class label). The network outputs both a coarse low-density point cloud and a dense high-resolution point cloud with separate 3D coordinates to represent class information.

structure inspired by the Point Completion Network [13] and its application to cardiac image analysis [1] (Fig. 1).

We use an extended version of the PointNet++ architecture [9] as our encoder by adding a conditional input to each point to indicate its cardiac substructure and multiple fully connected layers to facilitate the processing of the complex multi-class biventricular anatomy point clouds. The decoder first uses a shared multi-layer perceptron (MLP) to predict a low-resolution multi-class point cloud (coarse output) with the aim of capturing the global anatomical structure. This is followed by a FoldingNet [11] step, which outputs a high resolution multi-class point cloud (dense output) with accurate shapes both on a local and global level.

2.3 Training

The loss function of our network consists of two terms, one acting on the coarse prediction (L_{coarse}) and the other on the dense output (L_{dense}). In both cases, the predictions are compared to the gold standard point cloud using the Chamfer distance. A weighting parameter α is multiplied with the dense loss in the combined loss term to put more emphasis on generating a globally realistic shape at the beginning of training, while encouraging both high local and global accuracy in later training stages. Accordingly, we start training with a low α of 0.01 and then gradually increase it during training until it reaches a value of 5.0.

3   Experiments and Results

For experiments in Sects. 3.1–3.3, all networks were trained on ~800 pairs of ED and ES point clouds, stemming in equal proportions from female and male subjects. At test time, ~75 pairs of ED and ES point clouds were used as separate test sets for both female and male subjects.

3.1   Prediction Quality

In order to evaluate our network's prediction performance of both biventricular contraction and relaxation, two separate networks, one for each phase direction (ED to ES and ES to ED), were trained and tested jointly on both female and male subject data. We visualize the input, predicted, and gold standard point clouds of multiple sample cases from the unseen test dataset of the ES prediction task in Fig. 2 and of the ED prediction task in Fig. 3.

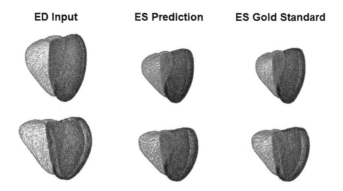

Fig. 2. ES prediction results of our method on two sample cases of the unseen test dataset.

Next, we quantify the performance of both the ED prediction and the ES prediction networks by calculating the Chamfer distances between the network predictions and the corresponding gold standard point clouds in the test set. We report the results separately for each sex and for the three ventricular sub-structures (left ventricular (LV) endocardium, LV epicardium, and right ventricular (RV) endocardium) to evaluate both overall and subpopulation-specific prediction quality, as well as enable a more localized assessment of deformation accuracy (Table 1).

Our method achieves Chamfer distances similar to the voxel resolution of the original MRI acquisition across all classes, sexes, and phases, while LV and ES predictions show slightly better results than RV and ED predictions.

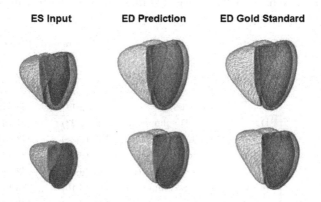

Fig. 3. ED prediction results of our method on two sample cases of the unseen test dataset.

Table 1. Prediction results of the proposed method on the test dataset.

Sex	Input phase	Predicted phase	Class	Chamfer distance (mm)
Female	ED	ES	LV endocardium	1.35 (±0.65)
			LV epicardium	1.39 (±0.57)
			RV endocardium	1.69 (±0.77)
	ES	ED	LV endocardium	1.64 (±0.73)
			LV epicardium	1.61 (±0.81)
			RV endocardium	2.04 (±0.86)
Male	ED	ES	LV endocardium	1.37 (±0.47)
			LV epicardium	1.38 (±0.35)
			RV endocardium	1.75 (±0.50)
	ES	ED	LV endocardium	1.89 (±0.70)
			LV epicardium	1.68 (±0.43)
			RV endocardium	2.16 (±0.62)

Values represent mean (± standard deviation (SD)).

3.2   Cardiac Anatomy Prediction

In order to assess the predictive ability of our method from a clinical perspective, we first convert both the point clouds predicted by our method in Sect. 3.1 and the gold standard point clouds of the test dataset to multi-class surface meshes using the Ball Pivoting Algorithm [3]. We then calculate the LV volume, LV mass, and RV volume for the predicted and gold standard meshes for both phase directions and report the results split by sex in Table 2.

Table 2. Cardiac volume metrics calculated from meshed point clouds predicted by our method.

Sex	Input phase	Predicted phase	Clinical metric	Gold standard	Prediction	% Diff
Female	ED	ES	LV ES volume (ml)	49 (±11)	49 (±9)	0.0
			RV ES volume (ml)	62 (±15)	62 (±12)	0.0
			LV mass (g)	86 (±14)	88 (±14)	3.4
	ES	ED	LV ED volume (ml)	123 (±23)	121 (±22)	1.6
			RV ED volume (ml)	146 (±25)	144 (±23)	1.4
			LV mass (g)	86 (±14)	85 (±14)	1.2
Male	ED	ES	LV ES volume (ml)	64 (±14)	63 (±11)	1.6
			RV ES volume (ml)	89 (±16)	86 (±14)	3.4
			LV mass (g)	121 (±24)	120 (±20)	0.8
	ES	ED	LV ED volume (ml)	153 (±29)	150 (±24)	2.0
			RV ED volume (ml)	187 (±28)	183 (±25)	2.1
			LV mass (g)	121 (±24)	121 (±22)	0.0

Values represent mean (± SD) in all cases.

Overall, we find good alignment between the predicted and gold standard scores in both means and standard deviations across all metrics, prediction tasks, and sexes. Differences tend to be greater for male hearts and ED predictions.

3.3 Cardiac Function Analysis

Since cardiac deformation is assessed using function-specific metrics in clinical practice, we also calculate ejection fraction (EF) and stroke volume (SV) of all test predictions and gold standard data for both chambers and phase directions, separating male and female subjects, to provide further validation of our method's performance in modelling biventricular deformations (Table 3).

We observe an average difference between the gold standard and predicted values of 1.72% indicating good overall prediction performance. Similar to the clinical anatomy metrics reported in Table 2, female hearts and ES predictions show slightly better scores.

Table 3. Cardiac function metrics calculated from meshed point clouds predicted by our method.

Sex	Clinical metric	Gold standard	ES prediction		ED prediction	
			Ours	% Diff	Ours	% Diff
Female	LV EF (%)	60 (±7)	60 (±6)	0.0	60 (±5)	0.0
	LV SV (ml)	75 (±17)	75 (±18)	0.0	73 (±14)	2.7
	RV EF (%)	65 (±23)	65 (±22)	0.0	63 (±19)	3.1
	RV SV (ml)	85 (±19)	84 (±19)	1.2	85 (±23)	0.0
Male	LV EF (%)	58 (±6)	59 (±6)	1.7	57 (±6)	1.7
	LV SV (ml)	90 (±21)	90 (±21)	0.0	86 (±17)	4.4
	RV EF (%)	59 (±18)	58 (±16)	1.7	56 (±19)	5.1
	RV SV (ml)	102 (±31)	104 (±18)	2.0	98 (±30)	3.9

Values represent mean (± SD) in all cases.

3.4   Subpopulation-Specific Cardiac Deformations

After evaluating our network's prediction performance on the entire test set, we also aim to analyze whether the PCN is able to extract features specific to certain subgroups in the data, enabling subpopulation-specific deformation modelling. To test this hypothesis, two separate networks were first trained on point cloud data pairs from exclusively female subjects, one in the ED to ES and the other in the ES to ED direction. They were then tested on data from both female and male subjects to investigate potential sex-specific differences in cardiac deformation prediction. Results are presented in Fig. 4 displaying the respective histograms of the CDs for each sex (different colors) and for both directions of deformation prediction (different subfigures). We find statistically significant differences in the prediction performance between male and female hearts for both prediction tasks (Kolmogorov-Smirnov (KS) test: p-value < 0.001).

4   Discussion

The results in Table 1 demonstrate that our method achieves high prediction accuracy with average CDs comparable to the voxel size of the underlying images (1.8 × 1.8 × 8.0 mm) [8]. Low standard deviations of approximately 0.5 mm indicate a robust prediction performance and show that our method is capable of accurately capturing the highly variable and complex non-linear 3D biventricular contraction and relaxation patterns. This is further corroborated by the qualitative prediction results in Fig. 2 and 3, which show good alignment between prediction and gold standard on both a global and local scale for both phase directions. Since we work with lightweight point cloud data instead of the grid-based data structures used in previous approaches, we are able to achieve these results on surface representations with higher resolution and hence greater anatomical accuracy. This is only made possible by the PCN architecture which

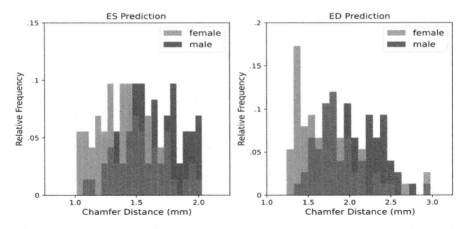

Fig. 4. Comparison of prediction performance on both female and male subjects using networks trained on only female cases. The Chamfer distances between prediction and gold standard point clouds were calculated for all female (green color) and male (blue color) cases in the test dataset. The resulting distributions of the Chamfer distances for each sex are presented as two histograms. The relative frequency (y-axis) indicates how often CD values fall within each of the pre-defined distance ranges (x-axis). Results are depicted for both ES prediction (left) and ED prediction (right). (Color figure online)

allows for fast and memory-efficient point cloud processing compared to previous deep learning approaches requiring regular grid structures. We observe that results are slightly better for predictions of ES from ED than vice versa and for female test subjects compared to males. We suspect that these lower overall CDs are likely caused by the fact that female hearts and hearts at ES are of smaller average size than male hearts and hearts at ED, while still retaining the same point cloud resolution. The difficulty in precisely locating the ES frame in an image sequence could also affect these results. When considering both volumetric cardiac anatomy and cardiac function metrics, our method's predictions result in very similar values to the reference values. This provides further evidence of the accuracy of the predicted shapes, both on their own and in relation to their respective input point clouds, and is crucial for the acceptance of more complex data-driven methods in clinical practice, as the results are in line with established clinical image-based biomarkers. From the histograms in Fig. 4, it can be seen that when training PCNs on just female point cloud data pairs and then testing on a mixed-sex population, the CDs for female subjects are significantly lower than those for male subjects. To better compare the distributions of CDs between test groups, the KS test was applied to assess histogram similarity. We find p-values below 0.001 in both phase directions, which indicate statistically significant differences between the distributions. This supports the hypothesis that cardiac shape deformation is subpopulation-specific. When given one subpopulation to train on, the network extracts deformation patterns that are unique to that subpopulation in the data and cannot be found in the other, thereby performing better during test time.

5   Conclusion

We have shown in this work that point cloud autoencoder networks are highly capable of modelling global and local non-linear cardiac deformations and predicting relevant clinical metrics with a high level of accuracy. Their memory-sparse set-up makes them more efficient than grid-based methods, easily scalable to higher resolutions, and suitable for use in larger network architectures. Furthermore, we have observed that PCNs can be used to extract subpopulation-specific cardiac deformations, which opens up a range of future research avenues.

Acknowledgments. This research has been conducted using the UK Biobank Resource under Application Number '40161'. The authors express no conflict of interest. The work of M. Beetz was supported by the Stiftung der Deutschen Wirtschaft (Foundation of German Business). The work of J. Ossenberg-Engels was supported by the Engineering and Physical Sciences Research Council (EPSRC) and Medical Research Council (MRC) [grant number EP/L016052/1]. The work of A. Banerjee was supported by the British Heart Foundation (BHF) Project under Grant HSR01230. The work of V. Grau was supported by the CompBioMed 2 Centre of Excellence in Computational Biomedicine (European Commission Horizon 2020 research and innovation programme, grant agreement No. 823712).

References

1. Beetz, M., Banerjee, A., Grau, V.: Biventricular surface reconstruction from cine MRI contours using point completion networks. In: 2021 IEEE 18th International Symposium on Biomedical Imaging (ISBI), pp. 105–109 (2021)
2. Bello, G.A., et al.: Deep-learning cardiac motion analysis for human survival prediction. Nat. Mach. Intell. **1**(2), 95–104 (2019)
3. Bernardini, F., Mittleman, J., Rushmeier, H., Silva, C., Taubin, G.: The ball-pivoting algorithm for surface reconstruction. IEEE Trans. Visual Comput. Graphics **5**(4), 349–359 (1999)
4. Chang, Y., Jung, C.: Automatic cardiac MRI segmentation and permutation-invariant pathology classification using deep neural networks and point clouds. Neurocomputing **418**, 270–279 (2020)
5. Krebs, J., Mansi, T., Ayache, N., Delingette, H.: Probabilistic motion modeling from medical image sequences: application to cardiac cine-MRI. In: Pop, M., et al. (eds.) STACOM 2019. LNCS, vol. 12009, pp. 176–185. Springer, Cham (2020). https://doi.org/10.1007/978-3-030-39074-7_19
6. Krebs, J., Mansi, T., Mailhé, B., Ayache, N., Delingette, H.: Unsupervised probabilistic deformation modeling for robust diffeomorphic registration. In: Stoyanov, D., et al. (eds.) DLMIA/ML-CDS -2018. LNCS, vol. 11045, pp. 101–109. Springer, Cham (2018). https://doi.org/10.1007/978-3-030-00889-5_12
7. Ossenberg-Engels, J., Grau, V.: Conditional generative adversarial networks for the prediction of cardiac contraction from individual frames. In: Pop, M., et al. (eds.) STACOM 2019. LNCS, vol. 12009, pp. 109–118. Springer, Cham (2020). https://doi.org/10.1007/978-3-030-39074-7_12
8. Petersen, S.E., et al.: UK Biobank's cardiovascular magnetic resonance protocol. J. Cardiovasc. Magn. Reson. **18**(1), 1–7 (2015)

9. Qi, C.R., Yi, L., Su, H., Guibas, L.J.: PointNet++: deep hierarchical feature learning on point sets in a metric space. In: Advances in Neural Information Processing Systems, pp. 5099–5108 (2017)

10. WHO: Cardiovascular disease death rate (2019). https://www.who.int/en/newsroom/fact-sheets/detail/cardiovascular-diseases-(cvds)

11. Yang, Y., Feng, C., Shen, Y., Tian, D.: FoldingNet: interpretable unsupervised learning on 3D point clouds. arXiv preprint arXiv:1712.07262 (2017)

12. Ye, M., et al.: PC-U net: learning to jointly reconstruct and segment the cardiac walls in 3D from CT data. In: Puyol Anton, E., et al. (eds.) STACOM 2020. LNCS, vol. 12592, pp. 117–126. Springer, Cham (2021). https://doi.org/10.1007/978-3-030-68107-4_12

13. Yuan, W., Khot, T., Held, D., Mertz, C., Hebert, M.: PCN: point completion network. In: 2018 International Conference on 3D Vision (3DV), pp. 728–737 (2018)

Influence of Morphometric and Mechanical Factors in Thoracic Aorta Finite Element Modeling

Ruifen Zhang[(✉)], Monica Sigovan, and Patrick Clarysse

Univ Lyon, CNRS UMR 5220, Inserm U1294, INSA Lyon,
Université Claude Bernard Lyon 1, Creatis, Villeurbanne, France
Ruifen.zhang@creatis.insa-lyon.fr

Abstract. Evaluation of mechanical properties from thoracic aorta finite element (FE) modeling can help to better stratify patients in need of intervention. This paper assesses the influence of various factors in the configuration of a FE model focused on the ascending aorta. Essential factors including morphological descriptors, material elasticity, boundary conditions (BCs), and load are investigated on a synthetic aorta model. Based on representative values retrieved from literature, the impact of these factors is studied using statistical hypothesis testing (one-way ANOVA).

Our results show that wall thickness, blood pressure, ascending aorta diameter, and material elasticity have the most dominant impacts onto the aortic stress distributions. Such an information is useful to properly configure patient-specific thoracic aorta FE modeling.

Keywords: Finite element modeling · Ascending thoracic aorta · Sensitivity analysis

1 Introduction

In the cardiovascular disease landscape, the ascending thoracic aorta aneurysm (ATAA) is one of the major pathologies of the aorta. ATTA is associated with a high morbidity and high mortality rate in case of rupture [1]. The vascular diameter, measured radio-graphically, remains the conventional criterion for surgical repair and primary determinant for risk stratification [2]. However, prior studies have shown that a significant number of patients developed adverse consequences below the standard diameter criteria of 5.0 cm, whereas some patients presented mean diameters > 5.5 cm without dissection or rupture. Clearly, additional markers or measurements are needed for the diagnosis and monitoring of ATAA in view of preventing complications [3].

Promoted from advances in medical imaging, computational patient-specific Finite Element (FE) modeling of thoracic aorta mechanics can provide unique insights into vessel integrity in both healthy and diseased states. It is believed that aortic dissection or rupture may occur when the stress in the aorta wall is

E. Puyol Antón et al. (Eds.): STACOM 2021, LNCS 13131, pp. 229–238, 2022.
https://doi.org/10.1007/978-3-030-93722-5_25

high enough to damage the tissue itself [4]. Consequently, wall stress distribution from a FE analysis may present a useful information for risk assessment. However, to our knowledge, there is no comprehensive survey on the factors influencing aorta FE modeling and the resulting wall stress distributions.

In this paper, the influence of several geometrical and mechanical factors on the distribution of wall stress is investigated in an idealized but realistic FE aorta model.

2   Methodology

The study of the impact of factors is based on the analysis of the response of synthetic aorta models to various morphometric and mechanical factors. Statistical analysis of the response considers mean, mean minus or plus standard deviation values of the factors.

2.1   Constructions of Synthetic Aorta Geometric Models

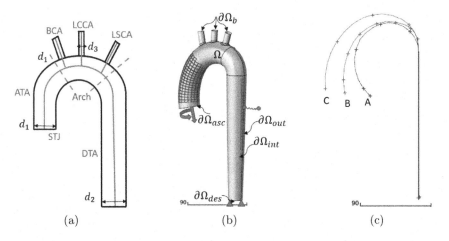

(a)   (b)   (c)

Fig. 1. Configuration of the synthetic thoracic aorta FE modeling. (a) Schematic view of the synthetic aorta geometry with lumen centerline in blue. (b) Synthetic aorta geometry constructed in FreeCAD. The configuration of BCs in aorta FE modeling comprises: stretch in the longitudinal direction and twist at the ascending aortic root section $\partial\Omega_{asc}$; fixed displacement restriction on the descending section $\partial\Omega_{des}$ and branch sections $\partial\Omega_b$ in all degrees of freedom; support by spine modeled with a spring on the spine contact region (green surfacic elements). The systolic blood pressure is applied to the internal lumen surface $\partial\Omega_{int}$. (c) Centerlines A, B, and C sketched in the 2D-plane. (BCA: branchiocephalic; LCCA: left common carotid; LSCA: left subclavian artery.) (Color figure online)

Based on the general schematic thoracic aortic geometry in Fig. 1(a), the synthetic aorta models were constructed in FreeCAD software by sweeping the sino-tubular junction (STJ) cross-section along a curved centerline down to the diaphragm level (Fig. 1(b)). To integrate shape variations, three centerlines were sketched in a 2D plane that correspond to typical ascending aorta shapes from normal to less or more curved arch (Fig. 1(c)). Line A presents highest curvature on the ascending part, line B has moderate curvature (healthy case), and line C is the most opened aorta with smaller curvature. The lengths for ascending, aortic arch, and descending parts are 89, 64 and 202 mm, respectively (obtained from one patient in our dataset). Mean and standard deviation (SD) of the ascending thoracic aorta diameter (ATD), descending thoracic aorta diameter (DTD) and aortic wall thickness (T) were set from population data available in the scientific literature (Table 2) according to

$$\overline{X} = \frac{\sum N_i \overline{X}_i}{\sum N_i}, \quad s(X) = \sqrt{\frac{\sum N_i((\overline{X}_i - \overline{X})^2 + s(X_i)^2)}{\sum N_i}} \tag{1}$$

where $\overline{X}_i$, $s_i(X)$, N_i are the mean, standard deviation and number of subjects in each data set indexed with i. Due to the limitations in spatial resolution hampering estimation of wall thickness in medical imaging, the measurements of diameters and wall thickness from *in-vitro* samples were included. The statistics of the parameters are: ATD $= 32.8 \pm 4.4$ mm, DTD $= 22.3 \pm 2.6$ mm, and $T = 2.2 \pm 0.5$ mm. Supra-aortic branches of branchiocephalic (BCA), left common carotid (LCCA), and left subclavian artery (LSCA) originating from the top of the aortic arch were attributed the same diameter of 10 mm. The generated geometries correspond to a geometry at rest, often named zero-pressure geometry.

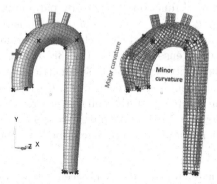

Fig. 2. The non-dilated control idealized mesh in yellow and the dilated pathological idealized mesh in grey after RBF morphing. One control point (red) in the middle of the ascending major curvature part controls the morphing of the aneurysm. Other control points in blue impose restriction onto the mesh. (Color figure online)

In order to introduce pathological aorta shapes into the study, virtual ascending aortic aneurysms were created by a localized dilation of the healthy FE mesh using a radial basis functions (RBF) based algorithm with a Gaussian kernel [5,13]. The major curvature part smoothly protrudes outward by 20% and 40%

of its normal diameter representing moderate to more severe aneurysms with sectional radius of 40 and 46 mm, respectively ((Fig. 2, right)). One control point in the middle of the ascending major curvature domain was selected to pilot the aneurysm deformation in the outward direction, as well as several control points around the arch and the descending parts to constrain the mesh. The weights and the polynomial coefficients for RBF morphing function were calculated from coordinates of the control points in the reference and deformed frames.

2.2   Material Law for Aorta Wall

The constitutive law (i.e. strain energy function SEF Ψ) proposed by Gasser et al. [14] is adopted for aorta FE modeling; it is based on the histological information in aortic wall, referred as GOH law (Eq. (2)) in the sequel. The SEF is formulated as follows

$$
\begin{aligned}
\Psi &= \Psi_{\mathrm{vol}}(J) + \overline{\Psi}_{\mathrm{g}} + \sum_{i=1,2} \overline{\Psi}_{\mathrm{fi}} \\
&= \Psi_{\mathrm{vol}}(J) + \frac{1}{2}c_{10}(\overline{I}_1 - 3) + \sum_{i=1,2} \frac{k_1}{2k_2}[exp(k_2\overline{E}_i^2) - 1],
\end{aligned}
\tag{2}
$$

where $\Psi_{\mathrm{vol}}(J)$ denotes a Lagrange contribution, which enforces the associated volume incompressibility kinematical constraint; $\overline{\Psi}_g$ a ground-matrix term and $\overline{\Psi}_{\mathrm{fi}}$ a collagen fiber reinforcement term for the purely isochoric contribution; $\overline{\Psi}_g$ is modeled by an isotropic Neo-Hookean law with coefficient c_{10} (Pa); $\overline{\Psi}_{\mathrm{fi}}$ is modeled by an exponential function weighted by parameter k_1 (Pa) and dimensionless parameter k_2; $\overline{E}_i = \kappa\overline{I}_1 + (1 - 3\kappa)\overline{I}_{4i} - 1$, where $\overline{I}_1$, $\overline{I}_{4i}$ are the first and the fourth invariants of the right Cauchy-Green strain tensor; the angle γ in a generalized structure tensor $\mathbf{H} = \kappa\mathbf{I} + (1 - 3\kappa)\mathbf{a}_0 \otimes \mathbf{a}_0$ represents the mean orientations of two family fibers, which are symmetric about the circumferential direction and $\mathbf{a}_0$ is the orientation direction; $0 < \kappa < 1/3$ accounts for the fiber orientation dispersion. With $\kappa = 0$, the collagen fibers are aligned to the mean direction γ (no dispersion). On the contrary, $\kappa = 1/3$ characterizes an isotropic distribution of the orientations of the collagen fibers around γ.

Table 1. GOH material parameters for three levels of material elasticity.

Elasticity	c_{10} [MPa]	k_1 [MPa]	k_2 [-]	κ	$\gamma°$	Refs
Stiff	0.06	1	5	0	35	[18]
Mid	0.0095	5.15	8.64	0.24	38.8	[19]
Compliant	0.014	0.81	12.42	0.18	39.1	[19]

Three levels of material elasticity were identified from literature based on GOH biaxial stretch-stress experiments (Table 1).

2.3 Physiological Conditions for the Synthetic Aorta Wall

Applied loads and BCs arise primarily from three sources: the hemodynamic load that acts on the luminal surface, the spine constraint on the aorta, and an inherent pre-stretch of the aorta tissue. It is generally observed that the ascending aorta moves up-and-down in the longitudinal direction following the systolic-diastolic motion of the heart. Aortic root stretch of 11.6 ± 2.9 mm [15] and twists of $0°$, $6°$, $12°$ were applied to mimic this behavior at the aortic root section $\partial\Omega_{asc}$ (Fig. 1b). A spring with stiffness $k = 10^8$ N/mm, a dashpot of damping coefficient $c = 10^5$ N/(mm/s) [16] were imposed to the spine contact region. The aortic descending $\partial\Omega_{des}$ and branch $\partial\Omega_b$ sections were kept fixed (Fig. 1(b)). Aortic lumen $\partial\Omega_{int}$ was submitted to blood pressure to deform the aorta from the zero-pressure geometry to the end-systolic (ES) state.

2.4 Statistical Analysis of FE Modeling Factors

Table 2. Three levels of each biomechanical factor in synthetic aortic FE modeling.

Factors	Metrics	−SD	Mean	+SD	#Subjects	References
Aortic geometry	ATD [mm]	28.55	32.83	37.11	1520	[6–8,12]
	DTD [mm]	19.7	22.3	24.6	38	[7,9]
	Thickness T [mm]	1.7	2.2	2.7	427	[9–12]
	Curvature [-]	A	B	C	–	
	Aneurysm [%]	0	20	40	–	
Load	Systolic BP [mmHg]	100	120	140	–	
BCs	Stretch [mm]	8.7	11.6	14.5	73	[15]
	Twist [degrees]	0	6	12	–	
Elasticity	GOH law	stiff	mid	compliant	–	[18,19]

A statistical hypothesis testing was designed to answer two questions: (i) Do different values of factors have significant impact onto the stress distribution? (ii) Which factors have the most impact onto FE modeling? In this test, the independent variables are the 9 factors in Table 2 where we consider three value levels: mean, mean minus and plus SD. The dependent variable is the von Mises stress value distribution on the ascending part of the synthetic aorta model. One-way analysis of variance (one-way ANOVA) was selected for this multi-group testing, as the independent variables are categorical and the dependent variable is numerical. The null hypothesis H_o of this one-way ANOVA is that the mean values of von Mises stress for each level are same $x_1 = x_2 = x_3$. If the F-statistic p-value is lower than critical value $\alpha = 0.05$, then at least one of the levels is statistically significantly different from the others, and the null hypothesis is rejected. On the contrary, subgroup means are fairly close to the overall mean and/or distributions overlap and are hard to distinguish. Thus the first question was addressed through the p values. The importance of the 9 factors was assessed based on the ranking of the effect size η^2. η^2 is the percentage of variance in the dependent variable accounting for independent variables. The higher the values, the more influential the factors.

3   Results

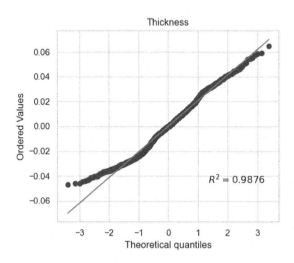

Fig. 3. QQ-plot for the thickness parameter to validate the normal distribution assumption for one-way ANOVA testing.

Value of von Mises stress at the hexahedra element first Gaussian integration points distributed over the ascending aorta were collected. With a meshing seed size of 5 mm, there were around 630 elements and therefore about 630 × 3 (levels) data per case. As a precondition to one-way ANOVA, the normality assumption of sample distributions was validated by the correlation value R^2 of the ANOVA model's residual probability with Gaussian distribution (e.g.quantile-quantile, QQ-plot). For example, normal assumption for the wall thickness was confirmed with residuals closely distributed along the 45°C line with $R^2 = 0.98$, as shown in Fig. 3. In addition, the homogeneity assumption of variances was checked with Bartlett's test.

The results of the ANOVA in Table 3 shows that except for twist and DTD, all the other investigated factors show significant differences between their three levels with p-values <0.05. In addition to this p-value which tells if at least one level is significantly different from the others, 'post-hoc' Tukey's honestly significantly difference (HSD) tests were conducted to identify the level that is significantly different from the others. For example in Fig. 4, for twist, large p-values indicate no difference between the three level pairs. In contrast, significant differences exists between all the three pairs for thickness (p-values<0.01).

Table 3. Summary table of one-way ANOVA tests, including residuals (Res.), sum of square (SS), degrees of freedom (df), mean square (MS), F-statistic, p value and effect size η^2.

Factors	SS	df	MS	F-statistic	p-value	η^2
ATD	0.428	2.0	0.214	489.48	<0.001	0.335
Res.	0.847	1939	4.37e$-$4			
Pressure	1.053	2.0	0.526	1310.48	0.0	0.576
Res.	0.775	1929	4.02e$-$4			
Material	0.348	2.0	0.174	6732.14	<0.001	0.294
Res.	0.839	1929	4.35e$-$4			
Stretch	0.004	2.0	0.00208	5.19	0.0056	0.0054
Res.	0.773	1929	4.01e$-$4			
Twist	1.13e$-$29	2.0	5.65e$-$30	1.44e$-$26	1.0	1.496e$-$29
Res.	7.54e$-$01	1929	3.91e$-$4			
Thickness	2.242	2.0	1.12	2586	0.0	0.723
Res.	0.858	1979	4.34e$-$4			
DTD	4.19e$-$4	2.0	2.09e$-$4	0.53	0.589	5.88e$-$4
Res.	0.711	1796	3.96e$-$4			
Curvature	0.0187	2.0	9.37e$-$3	11.69	<0.001	0.0107
Res.	1.735	2164	8.02e$-$4			
Aneurysm	0.038	2.0	0.0192	29.9	<0.001	0.0301
Res.	1.241	1929	6.43e$-$4			

These two illustrated cases are representative observations for non-prominent and prominent factors.

The effect size η^2 ranking shows that wall thickness, blood pressure, ascending diameter (ATD), and material elasticity have dominant impacts onto the aortic stress distributions, with respectively $\eta^2 = 0.723, 0.576, 0.335, 0.294$.

4   Discussion

This synthetic study gives attention points for the proper configuration of patient specific FE based aorta modeling. From the conclusions of the above analysis, FE modeling was performed on one patient (43 years old, male) with a bicuspid aortic valve (BAV) and congenital disease represented by a narrowing of the aorta (i.e. Co-arctation). Approval was obtained from the local ethics committee (ID-RCB: 2018-A01742-53). Contrast-enhanced CT imaging with 0.58 * 0.58 * 0.45 mm resolution was performed on a dual-energy CT scanner (IQon, Philips, Best, The Netherlands). Aorta lumen was segmented from CT images at end-diastole (ED) with in-house software CreaTools, and meshed with a uniform wall

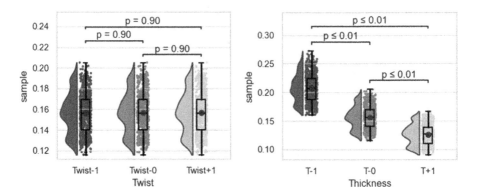

Fig. 4. Raincloud plots of sample's histograms together with boxplot. P-values are from Tukey's pair comparison for one-way ANOVA 'post-hoc' testing.

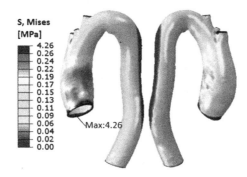

Fig. 5. Spatial distribution of von Mises stress in MPa from a patient-specific aorta FE modeling.

thickness of 1.5 mm. FE simulation was performed within Abaqus/Standard, fixing the descending and branches sections and imposing ED to ES displacements at the aortic root section (estimated from 3D image registration), and ES blood pressure of 120 mmHg. In this simulation, mechanical GOH parameters were estimated from an inverse modeling simulation ($c_{10} = 0.445$ MPa, $k_1 = 0.206$ MPa, $k_2 = 5$, $\kappa = 0.091$, $\gamma = 25.1°$). Several parameters therefore remain not subject based. The obtained spatial distribution for von Mises stress is shown in Fig. 5. Apart unrealistic stress values at the aortic root section, and focusing on the ascending part, high stress values located on the minor curvature region and range of stress values were comparable to other works [20,21].

The importance of wall thickness in aorta FE modeling has been recognized, notably in studies on abdominal aorta [22,23]. Usually, and as in this work, a uniform thickness is assumed, which can be unrealistic, in particular in presence of a pathology. As demonstrated recently by our group [17], the black blood dual CT technique produces new contrast allowing to identify the vascular wall, therefore allowing the integration of patient-specific wall thickness. This is one aspect we would like to investigate. Blood pressure appears also as an essential factor. Relative aortic blood pressure may be assessed from 4D flow Magnetic Resonance Imaging [24], another imaging techniques that can also be used for patient specific FE aorta modeling. The elasticity of the aortic wall material that we studied above is to treat the five GOH parameters as a single set, the global sensitivity study for the influence of each individual parameter on the aortic FE model will be published later in our inverse model. Material parameters reveal also as influential but remain difficult to assess for individuals. This is the object of inverse modeling approaches that aim at estimating patient specific biomechanical wall properties. Several works have shown the challenge to uniquely determine material law parameters from available data. This study can help to set up such modeling and even extended to analyze the sensitivity to all material law parameters.

Acknowledgements. This study was conducted within the framework of the LABEX PRIMES (ANR-11-LABX-0063) project of the University of Lyon, within the "Investissements d'Avenir" (ANR-11-IDEX-0 0 07) program operated by the French National Research Agency (ANR) and the International Research Project METISLAB from the CNRS. It is supported by China Scholarship Council (CSC), and the QUANTAAS ANR (ANR-18-CE19-0025-01) project.

References

1. Mokashi, S.A., Svensson, L.G.: Guidelines for the management of thoracic aortic disease in 2017. Gen. Thorac. Cardiovasc. Surg. **67**(1), 59–65 (2019)
2. Hiratzka, L.F., et al.: American College of Cardiology Foundation, American Heart Association Task Force on Practice Guidelines, American Association for Thoracic Surgery, American College of Radiology, American Stroke Association, Society of Cardiovascular Anesthesiologists, Society for Cardiovascular Angiography and Interventions, Society of Interventional Radiology, Society of Thoracic Surgeons, Society for Vascular Medicine and North American Society for Cardiovascular Imaging: 2010 ACCF/AHA/AATS/ACR/ASA/SCA/SCAI/SIR/STS/SVM guidelines for the diagnosis and management of patients with thoracic aortic disease. J. Am. Coll. Cardiol. **55**(14), e27–e129 (2010)
3. Evangelista, A., et al.: Insights from the international registry of acute aortic dissection: a 20-year experience of collaborative clinical research. Circulation **137**(17), 1846–1860 (2018)
4. Wang, D.H., Makaroun, M.S., Webster, M.W., Vorp, D.A.: Effect of intraluminal thrombus on wall stress in patient-specific models of abdominal aortic aneurysm. J. Vasc. Surg. **36**(3), 598–604 (2002)
5. Buhmann, M.D.: Radial Basis Functions: Theory and Implementations, vol. 12. Cambridge University Press (2003)
6. Mao, S.S., et al.: Normal thoracic aorta diameter on cardiac computed tomography in healthy asymptomatic adults: impact of age and gender. Acad. Radiol. **15**(7), 827–834 (2008)
7. Morrison, T.M., Choi, G., Zarins, C.K., Taylor, C.A.: Circumferential and longitudinal cyclic strain of the human thoracic aorta: age-related changes. J. Vasc. Surg. **49**(4), 1029–1036 (2009)

8. Doyle, B.J., Norman, P.E., Hoskins, P.R., Newby, D.E., Dweck, M.R.: Wall stress and geometry of the thoracic aorta in patients with aortic valve disease. Ann. Thorac. Surg. **105**(4), 1077–1085 (2018)

9. García-Herrera, C.M., et al.: Mechanical characterisation of the human thoracic descending aorta: experiments and modelling. Comput. Methods Biomech. Biomed. Eng. **15**(2), 185–193 (2012)

10. Mousavi, S.J., Avril, S.: Patient-specific stress analyses in the ascending thoracic aorta using a finite-element implementation of the constrained mixture theory. Biomech. Model. Mechanobiol. **16**(5), 1765–1777 (2017)

11. Di Giuseppe, M., et al.: Identification of circumferential regional heterogeneity of ascending thoracic aneurysmal aorta by biaxial mechanical testing. J. Mol. Cell. Cardiol. **130**, 205–215 (2019)

12. Cosentino, F., et al.: On the role of material properties in ascending thoracic aortic aneurysms. Comput. Biol. Med. **109**, 70–78 (2019)

13. Capellini, K., et al.: Computational fluid dynamic study for aTAA hemodynamics: an integrated image-based and radial basis functions mesh morphing approach. J. Biomech. Eng. **140**(11), 111007 (2018)

14. Gasser, T.C., Ogden, R.W., Holzapfel, G.A.: Hyperelastic modelling of arterial layers with distributed collagen fibre orientations. J. R. Soc. Interface **3**(6), 15–35 (2006)

15. Plonek, T., et al.: The evaluation of the aortic annulus displacement during cardiac cycle using magnetic resonance imaging. BMC Cardiovasc. Disord. **18**(1), 1–6 (2018)

16. Pagoulatou, S.Z., et al.: The effect of the elongation of the proximal aorta on the estimation of the aortic wall distensibility. Biomech. Model. Mechanobiol. **20**(1), 107–119 (2021)

17. Rotzinger, D.C., Si-Mohamed, S.A., Shapira, N., Douek, P.C., Meuli, R.A., Boussel, L.: "Dark-blood" dual-energy computed tomography angiography for thoracic aortic wall imaging. Eur. Radiol. **30**(1), 425–431 (2020)

18. Farzaneh, S., Trabelsi, O., Avril, S.: Inverse identification of local stiffness across ascending thoracic aortic aneurysms. Biomech. Model. Mechanobiol. **18**(1), 137–153 (2019)

19. Weisbecker, H., Pierce, D.M., Regitnig, P., Holzapfel, G.A.: Layer-specific damage experiments and modeling of human thoracic and abdominal aortas with non-atherosclerotic intimal thickening. J. Mech. Behav. Biomed. Mater. **12**, 93–106 (2012)

20. Wisneski, A.D., et al.: Patient-specific finite element analysis of ascending thoracic aortic aneurysm. J. Heart Valve Dis. **23**(6), 765 (2014)

21. Nathan, D.P., et al.: Increased ascending aortic wall stress in patients with bicuspid aortic valves. Ann. Thorac. Surg. **92**(4), 1384–1389 (2011)

22. Shang, E.K., et al.: Local wall thickness in finite element models improves prediction of abdominal aortic aneurysm growth. J. Vasc. Surg. **61**(1), 217–223 (2015)

23. Joldes, G.R., Miller, K., Wittek, A., Doyle, B.: A simple, effective and clinically applicable method to compute abdominal aortic aneurysm wall stress. J. Mech. Behav. Biomed. Mater. **58**, 139–148 (2016)

24. Bouaou, K., et al.: Analysis of aortic pressure fields from 4D flow MRI in healthy volunteers: associations with age and left ventricular remodeling. J. Magn. Reson. Imaging **50**(3), 982–993 (2019)

Multi-disease, Multi-view and Multi-center Right Ventricular Segmentation in Cardiac MRI Challenge (M&Ms-2)

Right Ventricle Segmentation via Registration and Multi-input Modalities in Cardiac Magnetic Resonance Imaging from Multi-disease, Multi-view and Multi-center

Xiaowu Sun[⊠], Li-Hsin Cheng, and Rob J. van der Geest

Division of Image Processing, Department of Radiology, Leiden University Medical Center, Leiden, The Netherlands
x.sun@lumc.nl

Abstract. Quantitative assessment of cardiac function requires accurate segmentation of cardiac structures. Convolutional Neural Networks (CNNs) have achieved immense success in automatic segmentation in cardiac magnetic resonance imaging (cMRI) given sufficient training data. However, the performance of CNN models greatly degrade when the testing data is from different vendors or different centers. In this paper, we introduce the use of image registration to propagate annotation masks from labeled images to unlabeled images as to enlarge the training dataset. Furthermore, we investigated various input modalities including 3D volume, single-channel 2D image, multi-channel 2D image constructed from spatial and temporal stack to extract more features to improve domain generalization in cMRI segmentation. We evaluated our method in M&Ms-2 challenge testing data (https://www.ub.edu/mnms-2/), achieving averaged Dice scores of 0.925, 0.919 and Hausdorff Distance of 10.587 mm, 6.045 mm in right ventricular segmentation in short-axis view and long-axis view respectively.

Keywords: Cardiac MRI · Deep learning · Label propagation · Input modality · Generalization

1 Introduction

In clinical routine, cardiac magnetic resonance imaging (cMRI) is considered a standard reference for the diagnosis of cardiac disease. Accurate segmentation of cardiac structures such as left ventricle (LV), myocardium and right ventricle (RV) is essential to quantitatively assess the cardiac function. Traditional manual segmentation method not only is time-consuming but also prone to inter-rater experience.

In recent years, deep learning based automatic segmentation approaches have been achieved immense success in cardiac segmentation. Tran et al. was the first to employ the fully convolutional neural (FCN) network for LV and RV segmentation in short-axis MRI [1]. Poudel proposed a recurrent FCN network ensembling the spatial information for LV segmentation [2]. However, the performance of most of those deep learning

© Springer Nature Switzerland AG 2022
E. Puyol Antón et al. (Eds.): STACOM 2021, LNCS 13131, pp. 241–249, 2022.
https://doi.org/10.1007/978-3-030-93722-5_26

based models degrade dramatically when the trained model is applied directly on other unseen datasets from different centers or vendors. Differences in image protocols, disease characteristics, scanner-specific bias and the other factors remain even after careful pre-processing [3]. In addition, the RV has a more complex shape and border characteristics compared to the LV. Hence, the M&Ms-2 challenge is motivated to build a method to segment the RV using multi-center, multi-disease and multi-view cMRI data.

The most straight forward approach to tackle this problem is to collect and annotate data from multiple centers, vendors and patient patologies. Tao used a large heterogeneous data with 41,593 images from different centers and different vendors to train a CNN model and achieved a good generalization [4]. Chen demonstrated that applying data augmentation strategies on a single-site single-scanner dataset could improve the performance on an unseen dataset across different sites or scanners [5]. Based on those studies, we hypothesize that a large-scaled pooling data from different domains could improve a model's performance on an unseen dataset. Additionally, in the conventional CNN models, the information derived from the neighboring images is usually ignored. Hence, we introduced two stack model to extract the spatiotemporal features to improve the performance.

In this paper, given limited data, we investigated several methods to generate more training data and extract more features including 1): The use of image registration to propagate annotation masks to unlabeled phases 2): Introducing the spatial and temporal neighboring images to construct a multi-channel 2D image to integrate more spatiotemporal information for the RV segmentation task.

2 Materials and Methods

2.1 Data

Table 1. Description of training, validation and testing dataset

Pathology	Number of training	Number of validation	Number of testing
Normal subjects	40	5	30
Dilated left ventricle	30	5	25
Hypertrophic cardiomyopathy	30	5	25
Congenital arrhythmogenesis	20	5	10
Tetralogy of fallot	20	5	10
Interatrial communication	20	5	10
Dilated right ventricle	0	5	25
Tricuspid valve regurgitation	0	5	25
Total	160	40	160

The M&Ms-2 challenge provides 360 cases (160 for training, 40 for validation and 160 for testing) in both short-axis (SA) and long-axis (LA) views from four different centers, acquired with three vendors (General Electric, Philips and Siemens). As shown in Table 1, except the normal subjects, there are five pathologies in the training dataset, two pathologies are not present in the training dataset but only in the validation and testing dataset. In addition, only end-diastolic (ED) and end-systolic (ES) phases in the training data are annotated by experienced experts, including LV, RV and left ventricular myocardium (MYO). Although this challenge focused on the RV segmentation, in our experiments, LV and MYO annotations were also used to constrain the RV segmentation.

2.2 Registration

In the available dataset, only the ED and ES phases are labeled, while the other phases are continuous in time consistent with the ED and ES phases. All the phases from the same case have an almost identical intensity distribution, which will alleviate the errors caused by inter-subject variability [6]. Hence, we used intensity-based registration method to propagate the labels, regarding the ED and ES as the template. The progress is described in Fig. 1. Given three phases (ED, ES and unlabeled), the ED and ES are firstly registered to the unlabeled phase, generating two geometric transformation matrixes named ES-Tf and ED-Tf, then the transformation matrix with smaller norm was used to propagate the mask. Matlab inbuilt functions *imregtform* and *imwarp* were used to implement the registration [11]. Mean square error (MSE) and Affine were set as the similarity metric and transformation type.

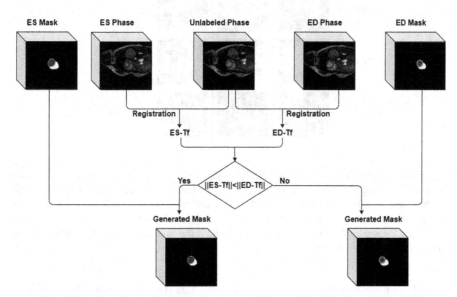

Fig. 1. Registration method to generate a mask for a unlabeled phase in SA view.

2.3 Input Modality of Network

As illustrated in Fig. 2 a short-axis cine MRI scan contains multiple slices and multiple phases. Images from the same slice level describe a cardiac cycle, while images from the same phase describe the complete heart structure. In conventional methods [4, 5], each single-channel 2D image or a 3D volume with the whole images from the same phase is usually considered as the input of a network. Although using a single-channel 2D image as the input could enlarge the training dataset, a 3D volume can provide more spatial information for the segmentation than using a 2D image. As a compromise, a spatial stack or temporal stack model, as proposed in our previous work [7], can be used to build a multi-channel 2D image which can provide accompanying spatial or temporal information respectively.

Spatial Multi-Channel 2D Image (SMI). The slices from the same phase are used to construct a SMI. As illustrated in Fig. 2, three 2D images from Phase 9 Slice 5, 6, 7 are used to build a 3-channels SMI for the image of Slice 6 Phase 9.

Temporal Multi-Channel 2D Image (TMI). In a similar way, a TMI consists of several neighboring phases of a particular slice. As shown in Fig. 2, images from Slice 6 Phase 8, 9, 10 are used to construct a 3-channels TMI for the image of Slice 6 Phase 9.

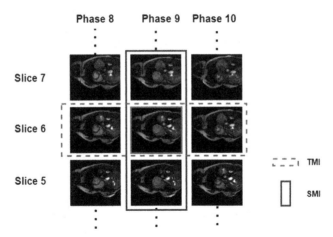

Fig. 2. An example of constructing a spatial multi-channel 2D image and temporal multi-channel 2D image. The image in the red box is the target image which will be segmented. The three spatial neighboring images in the blue box is called an SMI with three channels, where the top one is the first channel, the middle one is the second channel and so on. The TMI consists of three temporal neighboring images in the dash-line box, the left one is the first channel and the right is the last channel. (Color figure online)

Table 2 shows a brief summary of the training data size in SA view after combining the registration, SCI (single-channel 2D image), SMI and TMI. The original MnMS-2 dataset contains 320 3D volumes and 2,704 2D images with labeled annotation for training, when applying the registration approach to propagate the annotation masks, the data size of 3D volume and single-channel 2D image increased to 4,152 and 32,330 respectively. The LA-view images were acquired as single slice, resulting in the LA images being multi-phase single-slice. The 3D volumes and SMI can not be constructed in LA view. Hence, the model achieving the best performance in SA view was used as the pre-trained model for the LA view instead of using different input modalities.

Table 2. Training dataset description in SA MRI. **SCI**: single-channel 2D image. The number of channel in SMI and TMI is set to 5.

Data modality	Used registration	Training data size
SCI	No	2,704
SCI	Yes	32,330
3D volume	No	320
3D volume	Yes	4,152
SMI	No	2,704
SMI	Yes	32,330
TMI	No	2,704
TMI	Yes	32,330

2.4 Network Architecture

nnUNet [8] based on the U-Net architecture is a fully automatic and out-of-the-box medical image segmentation framework. To improve the robustness of domain shift in cardiac MRI, nnUNet_MMS [9] was specially designed by investigating various data augmentation techniques and ranked first at the first edition of M&Ms [10]. Hence, we introduced nnUNet as the baseline, and built our method upon nnUNet_MMS. All the models in this study are based on a 2D network. The data augmentation methods are the same in nnUNet_MMS model.

Since the propagated masks are not as accurate as the manual masks, those pseudo data was used to pre-train the model and the manually labeled data was applied to fine-tune the pre-trained network. The results are reported using Dice and Hausdorff Distance (HD). All experiments were executed on an NVIDIA Quadro RTX 6000 GPU with 24 GB internal memory.

3 Experiments and Results

3.1 Validation Set Results

We first evaluated the performance of different networks with different input modalities in the SA view from the validation dataset. Then we compared the results in the LA view with or without pre-training from SA view.

The Table 3 shows that using 3D volume without registration processing as the input, nnUNet_MMS achieved a slightly better dice than nnUNet, but yielded worse HD. However, when the registration method is applied to generate more 3D volume data to pre-train nnUNet_MMS, it achieved the best performance with a Dice of 0.922 and HD of 9.472 mm. It also can be observed that the segmentation results derived from the two stack models (SMI and TMI) are better than that from SCI, which confirmed that SMI and TMI could provide more spatial and temporal information for the segmentation task.

The results in LA views presented in Table 4 illustrates that the performance increased by 0.01 and 0.7 in terms of Dice and HD as a result of transferring the pre-trained model from SA view to LA view. Figures 3 and 4 show some segmentation examples derived from the best model.

Table 3. Segmentation results generated from different networks with different input modalities in the validation dataset in SA view.

Network	Input	Used registration data to pre-train network	Dice	HD (mm)
nnUNet(Baseline)	3D volume	No	0.912	10.318
nnUNet_MMS	3D volume	No	0.915	10.475
	3D volume	Yes	**0.922**	**9.472**
	SMI	No	0.919	9.577
	SMI	Yes	0.920	9.539
	TMI	No	0.917	10.343
	TMI	Yes	0.914	12.221
	SCI	No	0.915	11.354
	SCI	Yes	0.914	10.515

Table 4. Segmentation results in the validation dataset in LA view.

Network	Transfer from SA	Dice	HD
nnUnet(Baseline)	No	0.910	6.004
nnUNet_MMS	No	0.908	6.081
	Yes	**0.920**	**5.343**

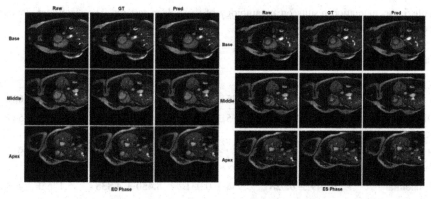

Fig. 3. A visual example from the apex, middle and base levels at ED (left) and ES (right) phases in SA view.

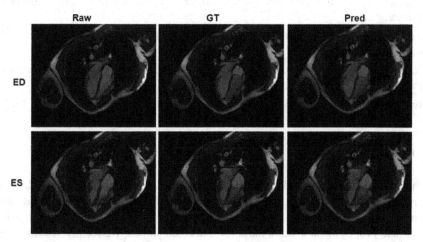

Fig. 4. A visual example at ED and ES phase in LA view.

3.2 Testing Set Results

We chose the model which performs best in the validation data as the final model. As the testing dataset is hidden by the organizer, we submitted our final model to the organizer and evaluated the performance online. Table 5 shows the details of our method on the hidden test data. In the SA view our method performed best in congenital arrhythmogenesis yielding 0.949, 8.45 mm for Dice and HD. The best results in LA view are generated from the normal subjects with 0.935 and 5.006 mm for Dice and HD. In addition, two pathologies (dilated right ventricle and tricuspid valve regurgitation) are not present in the training data but only in the testing data. The results on those two pathologies reveal that our approach obtains promising performance on an unseen pathology.

Table 5. Segmentation results on 8 pathology of the hidden test set. The mean and standard deviation are reported.

Pathology	Dice		HD (mm)	
	SA	LA	SA	LA
Normal subjects	0.922 ± 0.050	**0.935 ± 0.035**	8.999 ± 4.540	**5.006 ± 2.657**
Dilated left ventricle	0.922 ± 0.084	0.915 ± 0.052	13.257 ± 13.134	5.944 ± 3.547
Hypertrophic cardiomyopathy	0.934 ± 0.057	0.932 ± 0.033	10.214 ± 5.842	5.343 ± 2.916
Congenital arrhythmogenesis	**0.949 ± 0.028**	0.934 ± 0.031	**8.450 ± 4.838**	5.125 ± 1.738
Tetralogy of fallot	0.920 ± 0.034	0.914 ± 0.037	14.157 ± 8.232	7.404 ± 3.673
Interatrial communication	0.910 ± 0.048	0.906 ± 0.066	12.045 ± 4.189	8.021 ± 6.089
Dilated right ventricle	0.924 ± 0.045	0.897 ± 0.121	10.397 ± 5.223	7.064 ± 5.091
Tricuspidal regurgitation	0.923 ± 0.040	0.914 ± 0.039	9.236 ± 3.675	6.112 ± 3.349
Overall	0.925 ± 0.055	0.919 ± 0.063	10.587 ± 7.241	6.045 ± 3.824

4 Conclusion

In this paper, we investigated label propagation and multiple input modalities to increase the robustness in right ventricle segmentation from multi-disease, multi-view and multi-center cMRI data. To enlarge the training dataset, we exploited the use of image registration to propagate annotation masks to unlabeled phases. We further systematically investigated the effect of using different input modalities including 3D volumes, single-channel 2D image, spatial stack and temporal stack. The results illustrate that spatial stack and temporal stack provide more information for the segmentation task, and using 3D volume with label propagation could further improve the generalization ability in a unseen dataset.

Declaration. The authors of this paper declare that the segmentation methods implemented in this challenge has not used any pre-trained models nor additional MRI datasets other than those provided by the organizers.

References

1. Tran, P.V.: fully convolutional neural network for cardiac segmentation in short-axis MRI. arXiv preprint arXiv:1604.00494 (2016)

2. Poudel, R.P.K., Lamata, P., Montana, G.: Recurrent fully convolutional neural networks for multi-slice MRI cardiac segmentation. In: Zuluaga, M.A., Bhatia, K., Kainz, B., Moghari, M.H., Pace, D.F. (eds.) Reconstruction, segmentation, and analysis of medical images, pp. 83–94. Springer International Publishing, Cham (2017). https://doi.org/10.1007/978-3-319-522 80-7_8

3. Glocker, B., et al.: Machine learning with multi-site imaging data: an empirical study on the impact of scanner effects. arXiv preprint arXiv:1910.04597 (2019)

4. Tao, Q., et al.: Deep learning-based method for fully automatic quantification of left ventricle function from cine MR images: a multivendor, multicenter study. Radiology **290**(1), 81–88 (2019)

5. Chen, C., et al.: Improving the generalizability of convolutional neural network-based segmentation on CMR images. Front. Cardiovas. Med. **7**, 105 (2020)

6. Zhang, Y., et al.: Semi-supervised cardiac image segmentation via label propagation and style transfer. In: Puyol Anton, E., et al. (eds.) STACOM 2020. LNCS, vol. 12592, pp. 219–227. Springer, Cham (2021). https://doi.org/10.1007/978-3-030-68107-4_22

7. Sun, X., et al.: SAUN: Stack attention U-Net for left ventricle segmentation from cardiac cine magnetic resonance imaging. Med. Phys. **48**(4), 1750–1763 (2021)

8. Isensee, F., et al.: nnU-Net: a self-configuring method for deep learning-based biomedical image segmentation. Nat. Methods **18**(2), 203–211 (2021)

9. Full, P.M., Isensee, F., Jäger, P.F., Maier-Hein, K.: Studying robustness of semantic segmentation under domain shift in cardiac MRI. In: Puyol Anton, E., et al. (eds.) STACOM 2020. LNCS, vol. 12592, pp. 238–249. Springer, Cham (2021). https://doi.org/10.1007/978-3-030-68107-4_24

10. Campello, V.M., et al.: Multi-centre, multi-vendor and multi-disease cardiac segmentation: The M&Ms challenge. In: IEEE Transactions on Medical Imaging (2021)

11. Intensity-based automatic image registration - MATLAB& Simulink. https://www.mathwo rks.com/help/images/intensity-based-automatic-image-registration.html

Using MRI-specific Data Augmentation to Enhance the Segmentation of Right Ventricle in Multi-disease, Multi-center and Multi-view Cardiac MRI

Tewodros Weldebirhan Arega$^{(\boxtimes)}$, François Legrand, Stéphanie Bricq, and Fabrice Meriaudeau

ImViA Laboratory, Université Bourgogne Franche-Comté, Dijon, France

Abstract. Accurate segmentation of right ventricle (RV) from cardiac MRI is essential to evaluate the structure and function of the RV and to further study cardiac disorders. However, it is a difficult task due to its complex crescent shape and the presence of wall irregularities in its cavity. As part of the multi-disease, multi-center, and multi-view RV segmentation in cardiac MRI challenge (M&Ms-2), we propose to solve the problem using a fully automatic deep learning method that employs different data augmentation techniques. More specifically, we applied MRI-specific based, intensity and spatial data augmentation techniques to reduce the variation among the multi-center images with various cardiac pathologies. MRI-specific data augmentation are transformations that simulate image artifacts specific to MRI such as random bias field, random ghosting and random motion artifacts. We evaluate the proposed method in the validation set of the challenge. Among the data augmentation techniques applied, the MRI-specific based data augmentation enhanced the segmentation results of both long-axis and short-axis images in terms of Dice coefficient and Hausdorff Distance (HD). From the experiments, it shows us that the usage of MRI-specific transformations alongside intensity and spatial transformations in cardiac MRI can increase the variety of the training dataset and further help to improve the generalization capabilities of the models in multi-center, multi-disease cardiac MRI images. The proposed method ranked second at the M&Ms-2 challenge.

Keywords: Cardiac MRI · Multi-view · Right ventricle · Segmentation · Deep learning · Data augmentation

1 Introduction

Cardiac magnetic resonance (CMR) is a set of magnetic resonance imaging (MRI) utilized for the diagnosis and prognosis of cardiac diseases. It provides anatomical and functional information of the heart particularly of the left ventricle (LV) and right ventricle (RV).

E. Puyol Antón et al. (Eds.): STACOM 2021, LNCS 13131, pp. 250–258, 2022.
https://doi.org/10.1007/978-3-030-93722-5_27

Compared to LV, RV segmentation is a less commonly investigated field of research, mostly due to the difficulty of the task. RV has a complex crescent shape which varies with the imaging slice level. It also has wall irregularities in the cavity and its structure is more prone to strong remodeling in presence of cardiac diseases or lesions [4,15].

Before the introduction of deep learning to the medical image analysis domain, few works were focusing on the RV segmentation task. A single publicly available dataset existed at the time for the task, known as the RVSC (Right Ventricle Segmentation Challenge) contest [15]. The solutions of the challenge mainly used atlas-based and prior based methods as well as image-driven approaches that use the cardiac motion [15].

Recently, deep learning based solutions have been proposed to segment RV from CMR. Luo *et al.* proposed two-stage cascaded deep learning method that first localizes the bi-ventricular as region of interest (ROI) and then segments the RV myocardium from the localized region of the CMR [11]. In addition, several works [5,6,9,12,13,17] were proposed to segment the RV alongside LV and myocardium from the publicly available CMR images provided by the ACDC 2017 challenge [2] and the first M&Ms challenge (2020) [3]. However, these previous challenges and methods focused only on short-axis (SA) segmentation. RV segmentation from long-axis (LA) segmentation is relatively less studied area of research. Few existing works were performed in the context of multi-view CMR segmentation [1,16].

In this paper, we proposed a fully automatic deep learning method that employs various data augmentation techniques to segment RV from multi-disease, multi-center and multi-view CMR. MRI-specific based, intensity and spatial data augmentations were applied to increase the training dataset variety and improve the model generalization capability. MRI-specific data augmentations are transformations related to the k-space that simulate image artifacts specific to MRI. In this paper, we used random k-space motion artifact, random k-space ghosting artifact and random bias field artifact which are more related to cardiac MR. We evaluated our method on the validation dataset of the challenge. Most of the applied data augmentation techniques improved the segmentation result. Particularly, the MRI-specific data augmentation enhanced the result of both LA and SA images.

2 Dataset

The Multi-Disease, Multi-View and Multi-Center Right Ventricular Segmentation in Cardiac MRI (M&Ms-2)[1] challenge dataset consists of 360 healthy and pathological subjects which have 7 different RV and LV pathologies. All subjects were scanned in three clinical centres from Spain using three different magnetic resonance scanner vendors (Siemens, General Electric and Philips). The training data contains 160 annotated images. Another 40 cases, which contain 5 cases for each pathology, were used as a validation set to create a public leaderboard. The

---

[1] https://www.ub.edu/mnms-2/.

test set consists of 160 healthy and pathological subjects. For each subject, short-axis and long-axis view CMR acquired at End-Systolic (ES) and End-Diastolic (ED) time frames are provided. In addition to the contours of RV, the LV and myocardium annotations are also given. Note that two pathologies (Tricuspidal Regurgitation and Congenital Arrhythmogenesis) are only present in the validation and testing sets but not in the training set to evaluate generalization to unseen pathologies [3]. We declare that the segmentation method implemented for participation in the M&Ms 2 challenge has not used any pre-trained models nor additional MRI datasets other than those provided by the organizers.

3   Methods

3.1   Network Architecture

To segment the RV from SA and LA CMR images, we employed separate models for each of them. Since the LA images are 2 dimensional, we used a 2D segmentation model to segment the RV from the LA CMR. The network architecture is based on 2D nnU-Net [8]. We modified the default nnU-Net network architecture. Dropout is incorporated at the middle layers of the network to reduce overfitting and enhance the generalization as can be seen from Fig. 1. Inspired by [5], we replaced the instance normalization by batch normalization (BN) to improve the network's generalizability.

The U-Net's encoder and decoder consists of 12 convolutional layers where each convolution is followed by batch normalization and Leaky ReLU (negative slope of 0.01) activation function. The kernel size of the convolution is 3×3. During pre-processing, all images were resampled to have a pixel resolution of $1.25\,\mathrm{mm} \times 1.25\,\mathrm{mm}$ and the intensity of every image was normalized to have zero-mean and unit-variance.

To segment RV from the SA CMR images, a 3D nnU-Net [8] network was employed. Similar to the modification applied to the 2D nnU-Net, we used batch normalization instead of instance normalization and added dropout to the network architecture. For the pre-processing, we resampled all the volumes to $1.2\,\mathrm{mm} \times 1.2\,\mathrm{mm} \times 9.6\,\mathrm{mm}$ and applied Z-score to normalize each volume. A connected component analysis was used as a post-processing step to remove small false positive pixels for both SA and LA images.

3.2   Data Augmentation

To improve the generalization and robustness of the models in the multi-center and multi-disease dataset, we used various data augmentations, including intensity based data augmentation, spatial data augmentation and MRI-specific data augmentation. From the intensity based augmentation, we applied brightness, contrast, Gaussian noise, Gaussian blur and gamma. Spatial augmentations such as random rotation, random scaling, random flipping, elastic deformation and low resolution simulation are used. MRI specific data augmentations [14] such as

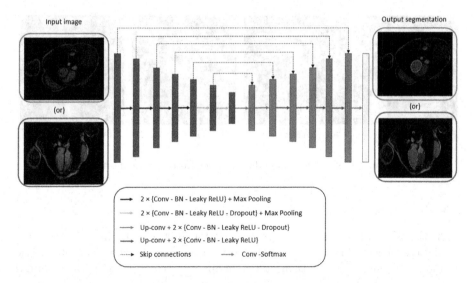

Fig. 1. Overview of the network architecture.

random bias field, random k-space motion artifact and random k-space ghosting were applied to simulate MRI-specific image artifacts.

Bias field is a low frequency and smooth signal that corrupts MRI images due to the magnetic field inhomogeneities of the MRI scanner. It changes the intensity values of the pixels so that the same tissue has different intensity distributions [10]. Random ghosting and random motion are artifacts that are created due to patient's motion during MRI acquisition, and they affect the k-space. The k-space is a temporary image space in which signals from the MRI scanner are stored during data acquisition. Then the MR image is generated by calculating the inverse Fourier transform of the k-space. During the cardiac cycle, the heart moves generating a cardiac pulsation. This motion can generate motion and ghosting artifacts along the phase-encoding direction. For random k-space ghosting artifact, it is generated by removing every nth plane of the k-space. For the random motion, it is simulated by filling the k-space with arbitrary rigidly-transformed versions of the images and then calculating the inverse transform of the compound k-space [14]. Some examples of the MRI-specific data augmentations are shown in Fig. 2.

The intensity and spatial augmentations were applied on the fly during training using batchgenerators library[2] while the MRI specific augmentations were generated using TorchIO library[3] before training on a randomly selected images and were added to the training dataset.

---

[2] https://github.com/MIC-DKFZ/batchgenerators.

[3] https://github.com/fepegar/torchio.

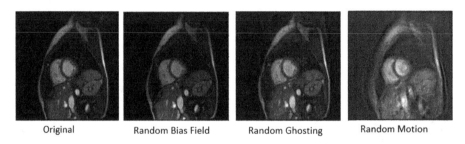

Original   Random Bias Field   Random Ghosting   Random Motion

Fig. 2. Examples of MRI-specific data augmentations on a short-axis CMR.

3.3 Training

Both the 2D and 3D models were trained for 1000 epochs in a 5 fold cross validation scheme. The weights of the network were optimized using Stochastic gradient descent (SGD) with nesterov momentum ($\mu = 0.99$) with an initial learning rate of 0.01. The learning rate was decayed using the "poly" learning rate policy [8]. The mini-batch size was 2 for the 3D model and 14 for the 2D model. The loss function is the sum of Dice loss and cross-entropy loss. The training was done on NVIDIA Tesla V100 GPUs using Pytorch deep learning framework based on nnU-Net implementation [8].

4 Results and Discussion

For the evaluation of segmentation results, Dice coefficient and Hausdorff distance (HD) metrics were mainly used. The baseline method is the default nnU-Net network. It uses light data augmentation that includes rotation, scaling, Gaussian blur and noise. The *Baseline + Dropout (BD)* modifies the baseline model by adding Dropout at the middle layers of the segmentation network as mentioned in Sect. 3.1. The *Baseline + Dropout + Batch Normalization (BDB)* uses batch normalization instead of the instance normalization compared to the *BD* network.

Dice coefficient measures the similarity of two images. It is calculated as the size of the overlap between segmented image and ground truth divided by the total size of the two images. Hausdorff distance is the greatest of all distances from a point in one set to the closest point in the other set [7]. This metrics focuses on outliers.

Regarding the data augmentation, we divided the experiments into three categories: heavy data augmentation, moderate data augmentation and MRI-specific augmentation. During the comparison of these three augmentations, we used the same network that is *Baseline + Dropout + Batch Normalization (BDB)*. The heavy data augmentation uses elastic deformation (α : $(0., 1300.), \sigma$: $(9., 15.)$, rotation $(-180, +180°)$, scaling $(0.65, 1.6)$, mirroring, additive brightness (μ : $0, \sigma$: 0.2), contrast, gamma $(0.6, 2.0)$, Gaussian noise and blurring. The moderate augmentation uses similar data augmentation as the heavy but the probabilities of applying the transformations are reduced. Besides, it does not use contrast. The MRI-specific augmentation employs random bias field, random ghosting and random motion artifacts with the default parameter values of the TorchIO functions.

Table 1 shows the comparison of the different network architectures and the various data augmentation types. This comparison was done on the validation set provided by the challenge organizers. It can be seen that the addition of dropout and batch normalization to the baseline network improved the Dice coefficient and HD of both SA and LA images.

Comparing the data augmentation experiments' performance in Table 1, it can be observed that the heavy data augmentation yielded the worst result in terms of the mean Dice and HD on both SA and LA images. This can be due to the fact that using heavy data augmentation that has many transformations with high probability generates unrealistic cardiac MR images that are more deviated from the original image. This may lead to poor segmentation performance as the network sees more unrealistically transformed images during training. The moderate data augmentation's performance for SA and LA was not similar. For LA images, it improved the segmentation result very well. Applying moderate data augmentation on LA increased the mean Dice by 0.32% and reduced the mean HD from 5.57 to 5.28 mm. However, for the SA, it yielded a worse result.

Table 1. Comparison of RV segmentation performance on various network architectures and data augmentation techniques on validation set of the challenge. The bold values are the best.

Method	Short-axis		Long-axis	
	DSC(%)	HD(mm)	DSC(%)	HD(mm)
Baseline	92.26	9.79	91.64	5.96
Baseline + Dropout (BD)	92.25	9.86	91.71	5.91
Baseline + Dropout + BN (BDB)	92.36	8.99	91.97	5.57
BDB + Heavy Augmentation	92.08	9.47	91.16	5.78
BDB + Moderate Augmentation	92.12	9.56	**92.29**	**5.28**
BDB + MRI-specific Augmentation	**92.53**	**8.98**	92.12	5.33

The MRI-specific data augmentation enhanced the segmentation result in both SA and LA images (Table 1). It outperformed the other data augmentation types for the SA by achieving the highest Dice coefficient (%) of 92.53 and the lowest HD of 8.98 *mm*. For the LA, it improved both the mean Dice and mean HD compared to the baseline method. However, its segmentation performance was slightly worse compared to the moderate data augmentation. This tells us that the MRI-specific augmentations applied are more relevant to the SA than to the LA.

Figure 3 shows the visual segmentation results of SA and LA images from the five-fold cross validation result. From the qualitative comparison, it can be observed that the MRI-specific data augmentation has slightly more robust results compared to the other methods in both LA and SA images.

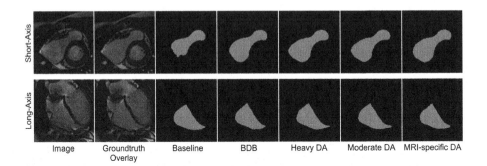

Fig. 3. Visual examples of the segmentation results. RV is marked in red. DA is data augmentation. (Color figure online)

For the final submission, we used an ensemble of two models for SA (BDB + MRI-specific augmentation model and BDB model) and a single model (BDB + Moderate augmentation model) for LA. The test result is summarized in Table 2. From the table, one can see that the proposed method is generalizable to unseen pathologies (Dilated RV and Tricuspidal Regurgitation) having Dice scores and HDs consistent with the rest of the dataset. Our method ranked in second place among 15 participating teams.

Table 2. Test set results. Interatrial C.: Interatrial Comunication, HCM: Hypertrophic Cardiomyopathy, Congenital H.: Congenital Arrhytmogenesis, Tricuspidal R.: Tricuspidal Regurgitation

Pathology	Short-axis		Long-axis		Cases
	DSC(%)	HD(mm)	DSC(%)	HD(mm)	
Normal	92.21 ± 5.28	8.67 ± 4.51	93.36 ± 3.49	5.24 ± 2.47	30
Tetralogy of Fallot	92.87 ± 3.07	10.66 ± 2.25	91.69 ± 3.91	7.42 ± 3.89	10
Congenital A.	93.88 ± 3.50	8.50 ± 3.96	93.06 ± 3.17	5.52 ± 2.05	10
Interatrial C.	84.89 ± 23.1	14.5 ± 11.0	91.31 ± 5.45	7.41 ± 4.96	10
Dilated LV	93.21 ± 4.95	10.4 ± 7.86	91.59 ± 4.94	6.30 ± 5.42	25
HCM	93.49 ± 5.98	8.81 ± 5.02	93.01 ± 3.45	5.28 ± 2.83	25
Dilated RV	*91.64 ± 4.65*	*10.67 ± 3.59*	*90.03 ± 9.27*	*7.04 ± 5.27*	*25*
Tricuspidal R.	*91.59 ± 4.94*	*9.86 ± 4.49*	*91.61 ± 3.97*	*6.01 ± 3.22*	*25*
Average	**92.07 ± 7.69**	**9.94 ± 5.81**	**91.98 ± 4.99**	**6.10 ± 4.03**	**160**

5 Conclusion

In this paper, we proposed a fully automatic deep learning method that leverages various data augmentation techniques to segment RV from multi-center and multi-view CMR images which have RV related pathologies. Intensity based, spatial and MRI-specific data augmentations were used to increase the size and variety of the training dataset. The MRI-specific data augmentations such as random motion, random ghosting and random bias field were applied to simulate cardiac MR artifacts. From the quantitative and qualitative results, we can say that using a moderate level spatial and intensity data augmentation improves segmentation result and applying MRI-specific data augmentations can further improve the results of long-axis and short-axis images and enhance the model's generalization capability.

Acknowledgements. This work was supported by the French National Research Agency (ANR), with reference ANR-19-CE45-0001-01-ACCECIT. Calculations were performed using HPC resources from DNUM CCUB (Centre de Calcul de l'Université de Bourgogne). We also thank the Mesocentre of Franche-Comté for the computing facilities.

References

1. Bai, W., et al.: Self-supervised learning for cardiac MR image segmentation by anatomical position prediction. In: Shen, D., et al. (eds.) MICCAI 2019. LNCS, vol. 11765, pp. 541–549. Springer, Cham (2019). https://doi.org/10.1007/978-3-030-32245-8_60

2. Bernard, O., et al.: Deep learning techniques for automatic MRI cardiac multi-structures segmentation and diagnosis: is the problem solved? IEEE Trans. Med. Imaging **37**, 2514–2525 (2018)

3. Campello, V.M., et. al: Multi-centre, multi-vendor and multi-disease cardiac segmentation: the M&Ms challenge. IEEE Trans. Med. Imaging **40**, 3543–3554 (2021)

4. Caudron, J., Fares, J., Vivier, P., Lefebvre, V., Petitjean, C., Dacher, J.: Diagnostic accuracy and variability of three semi-quantitative methods for assessing right ventricular systolic function from cardiac MRI in patients with acquired heart disease. Eur. Radiol. **21**, 2111–2120 (2011)

5. Full, P.M., Isensee, F., Jäger, P.F., Maier-Hein, K.: Studying robustness of semantic segmentation under domain shift in cardiac MRI. In: Puyol Anton, E., et al. (eds.) STACOM 2020. LNCS, vol. 12592, pp. 238–249. Springer, Cham (2021). https://doi.org/10.1007/978-3-030-68107-4_24

6. Girum, K.B., Créhange, G., Lalande, A.: Learning with context feedback loop for robust medical image segmentation. IEEE Trans. Med. Imaging **40**, 1542–1554 (2021)

7. Huttenlocher, D., Klanderman, G., Rucklidge, W.: Comparing images using the Hausdorff distance. IEEE Trans. Pattern Anal. Mach. Intell. **15**, 850–863 (1993)

8. Isensee, F., Jaeger, P., Kohl, S., Petersen, J., Maier-Hein, K.: NNU-Net: a self-configuring method for deep learning-based biomedical image segmentation. Nat. Methods **18**, 1–9 (2021). https://doi.org/10.1038/s41592-020-01008-z

9. Isensee, F., Jaeger, P.F., Full, P.M., Wolf, I., Engelhardt, S., Maier-Hein, K.H.: Automatic cardiac disease assessment on cine-MRI via time-series segmentation and domain specific features. In: Pop, M., et al. (eds.) STACOM 2017. LNCS, vol. 10663, pp. 120–129. Springer, Cham (2018). https://doi.org/10.1007/978-3-319-75541-0_13

10. Juntu, J., Sijbers, J., Dyck, D., Gielen, J.: Bias Field Correction for MRI Images. In: Kurzyski, M., Puchała, E., Woźniak, M., Żołnierek, A. (eds.) Computer Recognition Systems. Advances in Soft Computing, vol. 30. Springer, Heidelberg (2005). https://doi.org/10.1007/3-540-32390-2_64

11. Luo, G., An, R., Wang, K., Dong, S., Zhang, H.: A deep learning network for right ventricle segmentation in short-axis MRI. In: 2016 Computing in Cardiology Conference (CinC), pp. 485–488 (2016)

12. Ma, J.: Histogram matching augmentation for domain adaptation with application to multi-centre, multi-vendor and multi-disease cardiac image segmentation. In: Puyol Anton, E. (ed.) STACOM 2020. LNCS, vol. 12592, pp. 177–186. Springer, Cham (2021). https://doi.org/10.1007/978-3-030-68107-4_18

13. Painchaud, N., Skandarani, Y., Judge, T., Bernard, O., Lalande, A., Jodoin, P.-M.: Cardiac MRI segmentation with strong anatomical guarantees. In: Shen, D., et al. (eds.) MICCAI 2019. LNCS, vol. 11765, pp. 632–640. Springer, Cham (2019). https://doi.org/10.1007/978-3-030-32245-8_70

14. Pérez-García, F., Sparks, R., Ourselin, S.: TorchIO: a Python library for efficient loading, preprocessing, augmentation and patch-based sampling of medical images in deep learning. ArXiv abs/2003.04696 (2020)

15. Petitjean, C., et al.: Right ventricle segmentation from cardiac MRI: a collation study. Med. Image Anal. **19**(1), 187–202 (2015)

16. Vigneault, D., Xie, W., Ho, C., Bluemke, D., Noble, J.: Ω-Net (Omega-Net): fully automatic, multi-view cardiac MR detection, orientation, and segmentation with deep neural networks. Med. Image Anal. **48**, 95–106 (2018)

17. Zhang, Y., et al.: Semi-supervised cardiac image segmentation via label propagation and style transfer. In: Puyol Anton, E., et al. (eds.) STACOM 2020. LNCS, vol. 12592, pp. 219–227. Springer, Cham (2021). https://doi.org/10.1007/978-3-030-68107-4_22

Right Ventricular Segmentation from Short- and Long-Axis MRIs via Information Transition

Lei Li[1,2], Wangbin Ding[3], Liqin Huang[3], and Xiahai Zhuang[1(✉)]

[1] School of Data Science, Fudan University, Shanghai, China
zxh@fudan.edu.cn
[2] School of Biomedical Engineering, Shanghai Jiao Tong University, Shanghai, China
[3] College of Physics and Information Engineering, Fuzhou University, Fuzhou, China

Abstract. Right ventricular (RV) segmentation from magnetic resonance imaging (MRI) is a crucial step for cardiac morphology and function analysis. However, automatic RV segmentation from MRI is still challenging, mainly due to the heterogeneous intensity, the complex variable shapes, and the unclear RV boundary. Moreover, current methods for the RV segmentation tend to suffer from performance degradation at the basal and apical slices of MRI. In this work, we propose an automatic RV segmentation framework, where the information from long-axis (LA) views is utilized to assist the segmentation of short-axis (SA) views via information transition. Specifically, we employed the transformed segmentation from LA views as a prior information, to extract the ROI from SA views for better segmentation. The information transition aims to remove the surrounding ambiguous regions in the SA views. We tested our model on a public dataset with 360 multi-center, multi-vendor and multi-disease subjects that consist of both LA and SA MRIs. Our experimental results show that including LA views can be effective to improve the accuracy of the SA segmentation. Our model is publicly available at https://github.com/NanYoMy/MMs-2.

Keywords: RV segmentation · Short-axis and long-axis MRI · Information transition

1 Introduction

The segmentation of right ventricular (RV) is an essential preprocessing step for the cardiac functional assessment, such as the volume of ventricles, regional wall thickness, and ejection fraction. Manual delineations of the RV from short-axis (SA) and long-axis (LA) MRIs can be subjective and labor-intensive. However, automatic RV segmentation remains challenging, mainly due to the heterogeneous intensity, the complex variable shapes, and the unclear boundary of RV [4].

L. Li and W. Ding—The two authors have equal contributions to the paper.

© Springer Nature Switzerland AG 2022
E. Puyol Antón et al. (Eds.): STACOM 2021, LNCS 13131, pp. 259–267, 2022.
https://doi.org/10.1007/978-3-030-93722-5_28

In literature, most methods jointly segment both ventricles, and only a few methods focus exclusively on RV segmentation [2,15,16]. The joint segmentation of ventricles aims to employ the similar gray levels in their blood cavities and the relatively stable positions of two ventricles. Therefore, conventional atlas-based methods and model-based approaches combining with prior anatomical knowledge, are commonly used in these joint optimization methods [12]. Recently, with the development of deep learning (DL) in medical image computing, several DL-based algorithms have been proposed for automatic RV segmentation [8,9,13]. The superiority of employing both the SA and LA images instead of only the SA images has been demonstrated [7]. However, most current RV segmentation studies mainly focus on the algorithms solely using SA cardiac MRI [1]. Instead, the research on employing other views of MRIs to guide the segmentation of SA especially on the apical and basal slices, is rather rare. Moreover, due to the scarce of multi-center and multi-disease clinical dataset, the challenges of RV segmentation on the data from different centers and pathologies are rarely considered.

In this work, we propose a multi-view (LA and SA view) segmentation framework to delineate RV from multi-center and multi-disease MRIs. The framework is consists of a 2D and a 3D nnU-Net [5], which aim to segment RV from LA and SA views in successive. The nnU-Net has the advantage of self-automatic configuration, and therefore alleviates the burden of manual effort in the network configuration. Moreover, LA views can provide comprehensive information for the apical and basal slices of SA views, and also visualize atria clearly. We therefore employ an information transition scheme to assist the SA view segmentation via the corresponding LA view.

Related Literature. For the literature of the RV segmentation, one could refer to the review paper [1], where over forty research papers were evaluated. The review paper showed that current RV segmentation methods still can not properly solve all existing RV challenging issues. For the simultaneous SA and LA MRI segmentation, Koikkalainen et al. [7] employed both SA and LA MRI to segment ventricles and atria by transforming them into a same coordinate system. Vigneault et al. [13] proposed an Ω-net to segment ventricles and atria from MRIs with SA, four-chamber and two-chamber views. They simultaneously transformed all these views into a canonical orientation, and then performed the segmentation on the transformed images. Oghli et al. [11] assumed that RV cavity is continuous in the LA direction, and then transited the seed point of region growing method along the LA direction. Chen et al. [3] segmented left ventricular (LV) myocardium from SA views by combining the learned anatomical shape priors from various views. It is still an open question about how to effectively employ LA views for the RV segmentation of SA views.

2   Methodology

2.1   Segmentation Framework

Figure 1 presents the proposed segmentation framework, where SA and LA images are segmented via 3D and 2D nnU-Net [5], separately. We firstly segment RV and

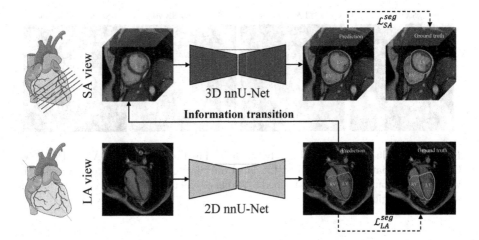

Fig. 1. The proposed RV segmentation framework for both SA and LA images. The framework includes three steps: the LA segmentation, ROI extraction from SA with assistant of LA information, and the SA segmentation. Here, the 3D cardiac image adopted from Kevil et al. [6].

LV from LA images, and then use this segmentation to localize the ventricles, which is used to guide the SA view segmentation. More specifically, we transform this information into the coordinate system of SA view, and utilize this information to crop the SA view (see Sect. 2.2). The segmentation loss functions of the framework are defined as follows,

$$\mathcal{L}_{SA}^{seg} = \mathcal{L}_{SA}^{CE} + \lambda_{SA}\mathcal{L}_{SA}^{Dice}, \tag{1}$$

$$\mathcal{L}_{LA}^{seg} = \mathcal{L}_{LA}^{CE} + \lambda_{LA}\mathcal{L}_{LA}^{Dice}, \tag{2}$$

where λ_{SA} and λ_{LA} are balancing parameters, and $\mathcal{L}^{CE}$ and $\mathcal{L}^{Dice}$ are the cross entropy (CE) loss and Dice loss, separately. Note that though our final target is to segment RV, here we also include the LV (both LV cavity and myocardium) label when minimizing the loss. We argue that the relatively stable space relationship of two ventricles can be helpful for the RV segmentation, especially in the boundary regions. Besides, we do not separate the LV cavity and myocardium to avoid overly attention on the supervision of noncritical small targets, i.e., LV myocardium.

2.2 Information Transition from the LA View

To employ the information from the LA view, we need to align the SA and LA views into a common coordinate system, as shown in Fig. 2. The transformation parameter between SA and LA views can be extracted from the header information of images. Specifically, the physical coordinates of SA and LA views can be defined as follows,

$$x'_{SA} = T_{SA}(x_{SA}), \tag{3}$$

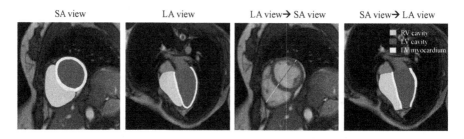

Fig. 2. Example of transformations between the label images of the SA and LA views. One can see that the transformed LA label only has one straight line traversing the ventricles. In contrast, the transformed SA label covers the whole ventricles but missing some apical regions.

$$x'_{LA} = T_{LA}(x_{LA}), \tag{4}$$

where T is the transformation matrix that converts the image coordinate x into the physical coordinate x'. We assume that the physical coordinates of SA and LA views are consist, so the transformed LA and SA views can be defined as follows,

$$x_{LA \to SA} = T_{SA}^{-1}(T_{LA}(x_{LA})), \tag{5}$$

$$x_{SA \to LA} = T_{LA}^{-1}(T_{SA}(x_{SA})). \tag{6}$$

Figure 3 presents the aligned LA and SA views in the coronal plane. One can see that in the SA view, the basal regions of the RV tends to be confused with the right atrium. In contrast, LA views can provide relatively clear boundary in the ambiguous regions. Therefore, with the assist of the LA information, one can classify the SA view as RV or non-RV regions in the coronal plane. Specifically, we employed the transformed LA segmentation (see Fig. 2) as a prior information, to extract the ROI from SA views for better segmentation. *Note that, the ROI excludes the aforementioned non-RV regions, where the SA segmentation tends to be inaccurate.*

3   Materials

3.1   Data Acquisition and Pre-processing

The dataset is from the *Multi-Disease, Multi-View & Multi-Center Right Ventricular Segmentation in Cardiac MRI (M&Ms-2)* [10] challenge event. The challenge dataset is consisted of 360 multi-center and multi-vendor subjects that are divided into three parts: 160 training data, 40 validation data, and 160 test data. It covers both healthy volunteers and patients with different pathologies in both SA and LA views, as presented in Table 1. Two pathologies (tricuspidal regurgitation and congenital arrhythmogenesis) do not appear in the training dataset, but are included in the validation and testing sets. The data setting aims to evaluate the model generalization ability to unseen pathologies.

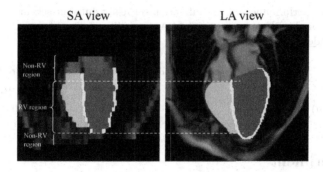

Fig. 3. The correspondence between SA and LA views in the coronal planes. There may be some inconsistencies in the apical and basal regions between the SA and LA views. Therefore, the SA plane can be marked as RV or non-RV region based on its correspondence with the LA plane. Note that here the orientation of images has been adjusted for better visualization.

Table 1. Pathology distribution among the training data, validation data, and test data. HCM: hypertrophic cardiomyopathy; CAM: congenital arrhythmogenesis; TOF: tetralogy of fallot; IC: interatrial communication; TR: tricuspidal regurgitation.

Pathology	Num. training	Num. validation	Num. test
Normal/Dilated LV/HCM	40/30/30	5/5/5	30/25/25
CAM/TOF/IC	20/20/20	5/5/5	10/10/10
Dilated RV/TR	0/0	5/5	25/25

3.2   Gold Standard and Evaluation

All the MRIs were manually delineated by experienced clinicians from the respective centers, and the label consistency between SA and LA images in basal and apical slices was confirmed. The manual segmentation includes the contours of RV, LV and LV myocardium. As this study focus on the RV segmentation, we only employ the RV manual label as the gold standard in the final evaluation.

For evaluation, Dice score (DS) and Hausdorff distance (HD) were applied. The final evaluation score is obtained by applying 0.75 and 0.25 weighting coefficients to the SA and LA segmentation accuracy, respectively.

$$\text{score} = \frac{0.75(\text{DS}_{SA} + \text{HD}_{SA}) + 0.25(\text{DS}_{LA} + \text{HD}_{LA})}{2}, \tag{7}$$

where $\text{DS}_{SA/LA} = (\text{DS}_{ED} + \text{DS}_{ES})/2$ and $\text{HD}_{SA/LA} = (\text{HD}_{ED} + \text{HD}_{ES})/2$.

3.3   Implementation

The proposed framework was implemented in PyTorch, running on a computer with a Core i7 CPU and an NVIDIA GeForce RTX 1080. To train the segmentation networks in proposed framework, λ_{SA} and λ_{LA} were set to 1 (see Eqs. 1 and 2). An Adam optimizer with an initial learning rate of 0.01 was adopted, and the networks were trained with 500 epochs.

Table 2. The performance on the validation set using different schemes to utilize LA information for the RV segmentation of SA views. The best and second results are in **bold** and underline, respectively.

Matrix	W/o-utilization	Post-utilization	Joint-utilization	Pre-utilization
DS_{SA} ↑	**0.914**	0.901	0.900	<u>0.913</u>
HD_{SA} (mm) ↓	11.2	11.3	**10.5**	<u>10.6</u>

4   Experiment

4.1   Comparison Experiment

We implemented a baseline scheme and three utilization strategies to employ LA information for the segmentation of SA views.

(1) **W/o-utilization:** one can train a nnU-Net purely on SA views without using any information from LA views. It can be considered as the baseline.
(2) **Post-utilization:** one can remove the non-RV regions of SA views via the prior segmentation of LA views.
(3) **Joint-utilization:** one can train a modified nnU-Net with an additional slice-level task at the bottom of network [14]. Here, the additional task aims to identify whether a slice includes the RV regions.
(4) **Pre-utilization:** The proposed framework. One can first perform ROI extractions on SA views via transformed LA views (see Sect. 2.2), and then train a nnU-Net to segment the RV on the ROI.

Table 2 presents the results of different strategies on validation dataset. Though w/o-utilization strategy obtained the best DS, it performed slightly worse than the joint- and pre-utilization schemes in terms of HD. The post-utilization scheme did not present any advantages compared to the baseline, and the joint-utilization strategy tended to decrease the DS. Therefore, we argue that the pre-utilization scheme is the most reliable and robust among all these strategies.

4.2   Performance on the Data with Different Pathologies

Table 3 presents the accuracy of each pathology on the test data. One can see that the best performance was obtained on the subjects with congenital arrhythmogenesis (CAM), though the most number of training data is from normal subjects. It may indicate that the accuracy of each pathology did not solely rely on the number of training data. There are two unseen pathologies in the training stage (see Table 1), i.e., dilated RV and TR. One can see that only the accuracy of dilated RV had an evident decrease for the segmentation of LA views. It can be attribute to the irregular RV shape in the dilated RV patients. Therefore, the model generalization ability on the unseen pathologies is generally promising.

Table 3. The performance on the test data for each pathology. Here, denotes the unseen pathologies in the training stage.

Pathology	DS_{SA}	HD_{SA} (mm)	DS_{LA}	HD_{LA} (mm)
Normal	0.916 ± 0.042	9.36 ± 4.14	0.931 ± 0.033	5.27 ± 3.00
Dilated LV	0.920 ± 0.069	11.2 ± 5.53	0.915 ± 0.052	6.08 ± 3.13
HCM	0.930 ± 0.052	9.11 ± 4.86	0.926 ± 0.033	5.35 ± 2.62
CAM	0.943 ± 0.025	8.52 ± 4.04	0.934 ± 0.034	5.88 ± 5.70
TOF	0.920 ± 0.035	11.8 ± 2.72	0.909 ± 0.034	7.56 ± 3.16
IC	0.915 ± 0.042	11.6 ± 4.18	0.916 ± 0.070	6.43 ± 4.49
Dilated RV	0.917 ± 0.047	11.1 ± 3.30	0.888 ± 0.136	7.80 ± 8.32
TR	0.910 ± 0.047	10.5 ± 5.52	0.915 ± 0.036	5.98 ± 3.05

Table 4. The performance on the test data for both ED and ES phases.

Phase	DS_{SA}	HD_{SA} (mm)	DS_{LA}	HD_{LA} (mm)
ED	0.933 ± 0.039	10.6 ± 4.89	0.930 ± 0.050	6.25 ± 3.73
ES	0.907 ± 0.056	10.1 ± 4.45	0.902 ± 0.080	6.10 ± 5.38
Average	0.920 ± 0.050	10.3 ± 4.67	0.916 ± 0.068	6.17 ± 4.61

4.3   Performance on the ED and ES Phase

Table 4 presents the quantitative results of the proposed method on the ED and ES phases. One can see that the performance on the ES phase was statistically significant ($p < 0.001$) worse than that on the ED phase in terms of DS, but no significant difference ($p > 0.1$) in terms of HD. As we know Dice score belongs to volumetric overlap measurement, and can be sensible to the size of target volume. Therefore, it could attribute to the larger surface of RV in the ED phase compared to that in the ES phase.

5   Conclusion

In this work, we have proposed a framework for the RV segmentation of both SA and LA views. The proposed model has been tested on 160 subjects and obtained promising results, even on the unknown pathologies. The experimental results also demonstrated the effectiveness of the proposed information transition scheme. A limitation of this work is that the SA and LA view segmentation are achieved separately, as LA segmentation is regarded as a prior for the SA segmentation. In the future, we will develop more elegant and effective information transition algorithm for the simultaneous segmentation of SA and LA views.

Acknowledgement. This work was funded by the National Natural Science Foundation of China (grant no. 61971142, 62111530195 and 62011540404) and the development fund for Shanghai talents (no. 2020015).

References

1. Ammari, A., Mahmoudi, R., Hmida, B., Saouli, R., Bedoui, M.H.: A review of approaches investigated for right ventricular segmentation using short-axis cardiac MRI. IET Image Process. **15** (2021)
2. Campello, V.M., et al.: Multi-centre, multi-vendor and multi-disease cardiac segmentation: the M&Ms challenge. IEEE Trans. Med. Imaging **40**(12), 3543–3554 (2021)
3. Chen, C., Biffi, C., Tarroni, G., Petersen, S., Bai, W., Rueckert, D.: Learning shape priors for robust cardiac MR segmentation from multi-view images. In: Shen, D., et al. (eds.) MICCAI 2019. LNCS, vol. 11765, pp. 523–531. Springer, Cham (2019). https://doi.org/10.1007/978-3-030-32245-8_58
4. Chen, J., Zhang, H., Zhang, W., Du, X., Zhang, Y., Li, S.: Correlated regression feature learning for automated right ventricle segmentation. IEEE J. Transl. Eng. Health Med. **6**, 1–10 (2018)
5. Isensee, F., Jaeger, P.F., Kohl, S.A., Petersen, J., Maier-Hein, K.H.: NNU-Net: a self-configuring method for deep learning-based biomedical image segmentation. Nat. Methods **18**(2), 203–211 (2021)
6. Kevil, C.G., Goeders, N.E., Woolard, M., Bhuiyan, M.S., Orr, A.W.: Methamphetamine use and cardiovascular disease. Arterioscler. Thromb. Vasc. Biol. **39**(9), 1739–1746 (2019)
7. Koikkalainen, J., Pollari, M., Lötjönen, J., Kivistö, S., Lauerma, K.: Segmentation of cardiac structures simultaneously from short- and long-axis MR images. In: Barillot, C., Haynor, D.R., Hellier, P. (eds.) MICCAI 2004. LNCS, vol. 3216, pp. 427–434. Springer, Heidelberg (2004). https://doi.org/10.1007/978-3-540-30135-6_52
8. Li, J., Yu, Z.L., Gu, Z., Liu, H., Li, Y.: Dilated-inception net: multi-scale feature aggregation for cardiac right ventricle segmentation. IEEE Trans. Biomed. Eng. **66**(12), 3499–3508 (2019)
9. Luo, G., An, R., Wang, K., Dong, S., Zhang, H.: A deep learning network for right ventricle segmentation in short-axis MRI. In: 2016 Computing in Cardiology Conference (CinC), pp. 485–488. IEEE (2016)
10. Martín-Isla, C., Lekadir, K.: MICCAI 2021: multi-disease, multi-view & multi-center right ventricular segmentation in cardiac MRI (2021). https://www.ub.edu/mnms-2/
11. Oghli, M.G., Mohammadzadeh, A., Kafieh, R., Kermani, S.: A hybrid graph-based approach for right ventricle segmentation in cardiac MRI by long axis information transition. Physica Med. **54**, 103–116 (2018)
12. Petitjean, C., et al.: Right ventricle segmentation from cardiac MRI: a collation study. Med. Image Anal. **19**(1), 187–202 (2015)
13. Vigneault, D.M., Xie, W., Ho, C.Y., Bluemke, D.A., Noble, J.A.: Omega-Net: fully automatic, multi-view cardiac MR detection, orientation, and segmentation with deep neural networks. Med. Image Anal. **48**, 95–106 (2018)

14. Yue, Q., Luo, X., Ye, Q., Xu, L., Zhuang, X.: Cardiac segmentation from LGE MRI using deep neural network incorporating shape and spatial priors. In: Shen, D., et al. (eds.) MICCAI 2019. LNCS, vol. 11765, pp. 559–567. Springer, Cham (2019). https://doi.org/10.1007/978-3-030-32245-8_62

15. Zhuang, X.: Challenges and methodologies of fully automatic whole heart segmentation: a review. J. Healthcare Eng. 4(3), 371–407 (2013)

16. Zhuang, X., Xu, J., Luo, X., Chen, C., Ouyang, C., Rueckert, D., Campello, V.M., Lekadir, K., Vesal, S., RaviKumar, N., et al.: Cardiac segmentation on late gadolinium enhancement mri: a benchmark study from multi-sequence cardiac MR segmentation challenge. arXiv preprint arXiv:2006.12434 (2020)

Tempera: Spatial Transformer Feature Pyramid Network for Cardiac MRI Segmentation

Christoforos Galazis[1,2(✉)], Huiyi Wu[2], Zhuoyu Li[3], Camille Petri[2],
Anil A. Bharath[4], and Marta Varela[2]

[1] Department of Computing, Imperial College London, London, UK
c.galazis20@imperial.ac.uk
[2] National Heart and Lung Institute, Imperial College London, London, UK
[3] Department of Metabolism, Digestion and Reproduction, Imperial College London,
London, UK
[4] Department of Bioengineering, Imperial College London, London, UK

Abstract. Assessing the structure and function of the right ventricle
(RV) is important in the diagnosis of several cardiac pathologies. However, it remains more challenging to segment the RV than the left ventricle (LV). In this paper, we focus on segmenting the RV in both short (SA)
and long-axis (LA) cardiac MR images simultaneously. For this task, we
propose a new multi-input/output architecture, hybrid 2D/3D geometric spatial TransformEr Multi-Pass fEature pyRAmid (Tempera). Our
feature pyramid extends current designs by allowing not only a multi-
scale feature output but multi-scale SA and LA input images as well.
Tempera transfers learned features between SA and LA images via layer
weight sharing and incorporates a geometric target transformer to map
the predicted SA segmentation to LA space. Our model achieves an average Dice score of 0.836 and 0.798 for the SA and LA, respectively, and
26.31 mm and 31.19 mm Hausdorff distances. This opens up the potential
for the incorporation of RV segmentation models into clinical workflows.

Keywords: Cardiac MRI · Right Ventricle · Segmentation ·
Multi-view · 2D/3D network · Spatial transformer · Feature pyramid

1 Introduction

Cardiac Magnetic Resonance Imaging (MRI) is the most widely used imaging
technique to quantify the structure and function of the heart [1]. As such, it
can be used to assess the right ventricle (RV) and contribute to the diagnosis
and monitoring of cardiac pathologies such as coronary heart disease, pulmonary
hypertension, dysplasia and cardiomyopathies [5,6].

Despite the RV's importance, the left ventricle (LV) has traditionally been
analysed in greater detail due to its pivotal role in a wider range of pathologies
[17]. The RV is more challenging to accurately segment for both clinicians and

© Springer Nature Switzerland AG 2022
E. Puyol Antón et al. (Eds.): STACOM 2021, LNCS 13131, pp. 268–276, 2022.
https://doi.org/10.1007/978-3-030-93722-5_29

(semi-)automated algorithms [3, 17]. This is due to the RV's more complex crescent shape, thinner ventricular wall and heavier presence of wall trabeculations compared to the LV [8, 17].

In this paper, we focus on RV automatic segmentation as part of the *"Multi-Disease, Multi-View & Multi-Center Right Ventricular Segmentation in Cardiac MRI"* (M&Ms2) challenge [16]. We propose a novel hybrid 2D/3D deep neural network that takes both short-axis (SA) and long-axis (LA) images as inputs. It includes a novel multi-input/output feature pyramid that facilitates weight sharing across SA in-plane slices and between the SA and LA views. Additionally, we include a geometric target spatial transformer that utilizes the known spatial relationship between the different cardiac views.

2   Methods

Data. The end-diastolic (ED) and end-systolic (ES) cardiac MRI image phases used for the experiments come from a 360-subject bSSFP CINE MRI dataset publicly available through the M&Ms-2 challenge [16]. The dataset is obtained across different sites, different vendors and has pathologies in the test set not included in the training. The available ground truth has manual segmentations of the LV and RV blood pool and the LV myocardium. From this set, 160 cases are dedicated for training. The remaining 40 and 160 cases are the development and test sets used for their respective phases in the competition. We further split the 160-case training set to 150 for model training and 10 for validation.

Preprocessing. As the images were acquired across different centers and vendors, they need to be standardized before being passed to the model. The resolution of both SA and LA images was resampled to 1.25 × 1.25 mm in-plane using b-spline interpolation. The through-plane spacing was left unchanged at 10 mm. We first automatically identify a region of interest (ROI) containing the RV and LV by applying Canny edge detection on the ED and ES image difference, then using the circular Hough transform to identify the heart. This allows us to identify the regions in the image that have the largest movement across the cardiac cycle, which we assume to be the heart. The image size was standardized to 192 × 192 × 17 through center ROI cropping. Finally, we standardize the cropped image such that it has a mean of zero and unit variance.

Additionally, we identify the affine transformation between the SA and LA images, which will subsequently be used in the segmentation network. The LA is treated as a 3D image (by adding a depth axis of size 1), thus allowing for a 3D/3D registration to take place. We pre-align the images based on the available file metadata of the images. For the registration, we follow a coarse-to-fine blurring approach to initially align with global features before finer ones. The registration is optimized to maximize the mutual information score. This is done using the SimpleITK package.

Model: Architecture. To simultaneously segment the SA and LA images, we propose a new hybrid 2D/3D geometric spatial TransformEr Multi-Pass fEature pyRAmid (Tempera) network, as shown in detail in Fig. 1. Our architecture

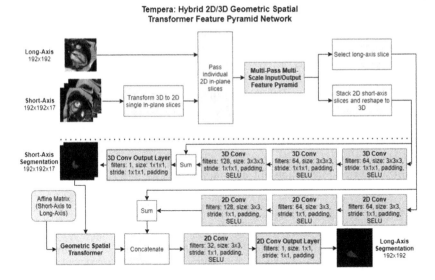

Fig. 1. An overview of the architecture of the hybrid 2D/3D Tempera network. It receives the individual 2D in-plane slices for both the SA and LA as inputs to the feature pyramid component. Then, the short-axis slices are used to reconstruct the 3D representation and each image view is passed to their respective 2D/3D convolutional branch. Finally, the predicted SA segmentation is transformed to the LA space to improve its prediction.

consists of 2D SA/LA hard-weight sharing layers, independent 3D (SA) and 2D (LA) branches and finally a geometric spatial transformer.

For the shared layers, we extend the Multi-scale Feature Pyramid network [14] to an architecture that we call Multi-Pass Feature Pyramid (MPFP). In the MPFP, the weight sharing, and thus internal (3×3) feature transfer, is achieved across the SA and LA in-plane slices and between the SA slices themselves. We are thus able to share features across all slices that relate to the RV shape, texture and contrast relative to the surroundings. The (partial) scale-invariant input/output of MPFP is designed to help Tempera generalize to unseen pathologies, such as dilated RV. An example is shown in Fig. 3 in which we base our assumption. The specific implementation details of our pyramid can be viewed in Fig. 2.

Our second component is the Geometric target Spatial Transformer (GST), depicted in Fig. 4. This is a differentiable non-trainable component, that we have built upon the spatial transformer from [2]. It applies transformations between different domain and target spaces, such as 3D to 2D and vice versa. The GST takes as input the pre-computed affine matrix to perform relevant coordinate transformations from the SA to LA. Specifically, we utilise it on the predicted SA segmentation to localize the RV in the LA.

Model: Optimizer. We use the Adam optimizer [11] with: exponential decay rates for first moment estimates $\beta_1 = 0.9$, second moment estimates $\beta_2 = 0.999$ and $\epsilon = 10^{-8}$ to avoid division with 0. Additionally, we use an empirically-chosen

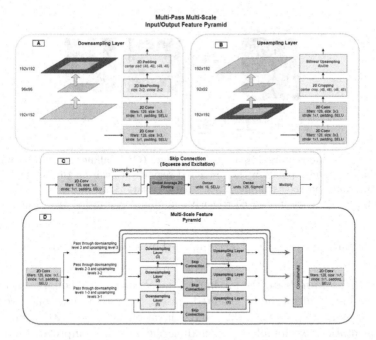

Fig. 2. Architecture of our proposed Multi-Pass and multi-scale input/output Feature Pyramid (MPFP). The downsampling component in block A consists of two 2D convolutional layers, followed by a pooling and padding layer. Similarly, the upsampling layer, block B, has two 2D convolutional layers, a cropping operation and finally bilinear upsampling. The skip connection in C contains a 2D convolutional layer followed by a summing operation and a squeeze-and-excitation component [10] to scale features. Finally, D illustrates how the components interact to form the MPFP. 2D convolutional layers are used to generate consistent feature sizes for the inputs and also to merge the outputs. After every layer we use a Scaled Exponential Linear Unit (SELU) [12] activation function, with exception the last layer of the skip connection uses a sigmoid. The input data to the pyramid is passed n times and generates n outputs, where $n = 3$ is the number of pyramid levels. Each subsequent pass goes through one less downsampling and upsampling layer.

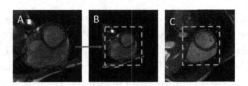

Fig. 3. Illustrative examples of how the feature pyramid may allow a hypothetical feature transfer to an unknown pathology for the network (dilated RV). Image A depicts SA basal slice of a patient with dilated RV. Image B is the downsampled version of image A by the MPFP in the first block. Finally, image C shows the SA basal slice of a healthy subject. The RV in images B and C are roughly of equal size. This allows feature transfer to untrained dilated RV cases.

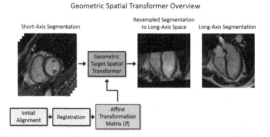

Fig. 4. The Geometric target Spatial Transformer (GST) affine transforms and resamples using linear interpolation the predicted segmentation from the SA to LA space, given an affine matrix as input. Their spatial relationship is stored in the image's metadata, in which we further correct with an affine registration algorithm.

initial learning rate of 5×10^{-4}, accompanied by a learning rate decay of 0.9 every 30 epochs.

Model: Loss Function. We minimized the combined [18] equal weighted Dice and focal loss (multiplicative weighted cross-entropy) [15] functions. We set the focal loss class weights $\alpha = 0.25$ and relaxation term $\gamma = 2.0$. Through our experimentation, we found that including the Dice loss improves the learning during initial epochs. The focal loss contributes during the later stages by dealing with class imbalance, in which background pixels outweigh foreground ones. Additionally, it helps with difficult to classify areas regardless if they are foreground or background. We set a λ of 75 for the SA loss and 1 for the LA. The equation is formed as bellow, where p is the probability of the ground truth label:

$$L_{Dice(p_t)} = 1 - \frac{2 \times true\_positive_{p_t}}{2 \times true\_positive_{p_t} + false\_positive_{p_t} + false\_negative_{p_t}} \tag{1}$$

$$L_{Focal}(p_t) = -a_t \times (1 - p_t)^\gamma \times log(p_t) \tag{2}$$

$$\begin{aligned} L_{Total}(p_{SAt}, p_{LAt}) = & \lambda_{SA} \times (L_{Dice}(p_{SAt}) + L_{Focal}(p_{SAt}) + \\ & \lambda_{LA} \times L_{Dice}(p_{LAt} + L_{Focal}(p_{LAt}) \end{aligned} \tag{3}$$

Model: Hyperparameters. We use LeCun weight initialization [12,13] from a normal distribution and set the biases to zero. This is done to maintain the self-normalizing property when using Scaled Exponential Linear Units (SELUs) [12]. Also, we set the batch size to 1 due to memory limits and train for 300 epochs, which took nearly 16 h to complete. The source code is available at https://github.com/cgalaz01/mnms2_challenge.

Model: Data Augmentation. To prevent the model from overfitting and to expand the dataset, we perform data augmentations. These include: in-plane rotations, in-plane anisotropic scalings, blurring, Gaussian noise addition, mean intensity shift and removing square segments of the image (in-painting). We use randomly selected parameters at each data iteration for each epoch.

Model: Data Postprocessing. We found that our model will occasionally identify surrounding tissues as the RV. To circumvent this, we added a postprocessing step, in which only the largest connected region will be considered as a valid prediction. Additionally, we apply a median filter to correct some of the interpolation errors caused by transforming back to the domain space using nearest neighbor interpolation.

3   Results

Tempera is able to accurately segment both the SA and LA views. On the test set we obtain 0.836 ± 0.23 Dice score and 26.31 ± 52.68 mm Hausdorff distance for the SA. The LA has a lower performance with 0.798 ± 0.28 Dice score and 31.19 ± 31.19 mm Hausdorff distance. From the predicted tests cases, there were a total of 11 SA and 20 LA that most likely failed during the ROI detection process, thus returning empty segmentations. Full details of the results are shown in Table 1.

A representative segmentation of the SA and LA images can be seen in Fig. 5. Tempera's segmentations closely match the ground truth labels, except in the LA basal slices, where Tempera occasionally under-segments the RV.

Table 1. Summary of Tempera's performance on the (competition) validation and test sets using Dice score and Hausdorff distance. The test set's evaluation was repeated (no failures), in which we removed cases where the preprocessing step failed.

Dataset	Dice score		Hausdorff distance (mm)	
	SA	LA	SA	LA
Validation	$0.895 \pm$ n/a	$0.829 \pm$ n/a	$11.8 \pm$ n/a	$15.118 \pm$ n/a
Testing	0.836 ± 0.23	0.798 ± 0.28	26.31 ± 52.68	31.19 ± 71.29
Testing (no failures)	0.896 ± 0.05	0.899 ± 0.06	18.384 ± 24.99	14.186 ± 26.86

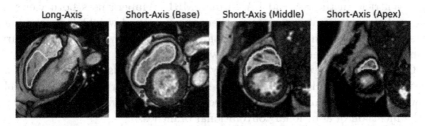

| Long-Axis | Short-Axis (Base) | Short-Axis (Middle) | Short-Axis (Apex) |

Fig. 5. A representative example from the development set of RV segmentation on the LA and SA images, showing how Tempera is able to delineate the RV successfully.

4   Discussion

Tempera network can perform RV segmentation well on a challenging dataset. By designing a model that interchanges between 2D and 3D space, we can reap several benefits. Using the SA as 2D slices rather than just a 3D volume, we are able to effectively increase the number of available training data. On the other hand, when using 3D operations the network has the option to utilize structural information between slices to improve its performance. Finally, we exploit these geometric redundancies to improve the LA segmentations by using the GST.

The preprocessing ROI detection and standardization is an important step when dealing with data variation from different scan vendors. By detecting a small ROI which contains the RV, we can minimize erroneous segmentations, minimize the network learning irrelevant background features and allow for a smaller model. However, the detection may fail in the presence of pathology, artefacted or low SNR images and, in future work, we plan to improve this heart detection step. Short-term it can benefit from restricting the detection within a central region of the image. Additionally, we can take advantage of the spatial alignment of the LA and SA to project the centres from one image space to the other. In the long term, it can be improved by using a neural network-based detector.

There has been limited work done on incorporating the multiple views of cardiac MRI to improve the segmentation of cardiac chambers. [7] *et al.* use the long axis views as shape priors for their short-axis cardiac segmentation. However, they restricted their investigations to the left myocardium, whose shape is much more regular than the RV's. Thus, their approach cannot be directly applied to the RV. Furthermore, the authors of [9] focus on preprocessing and data augmentation techniques to solve the data domain shift presented between different scan vendors and clinics, as part of the first M&Ms challenge [4]. For the RV segmentation, they obtained an average Dice score of 0.88.

5   Conclusions

For the M&Ms2 challenge, we propose the Tempera network to segment the right ventricle in the SA and LA cardiac MRI. Tempera uses both views and shares the learned features between them through hard-weight layer sharing in 2D space. A non-trainable spatial transformer, GST, allows for efficient sharing of information between the cardiac views.

Our Tempera network can seamlessly be extended to also segment the LV and the myocardium. Additionally, it can be updated to incorporate other chamber views, such as 2-Chamber and 3-Chamber. The only modification needed for this is the inclusion of additional convolutional branches.

Furthermore, we can improve the robustness of the network by utilising the temporal information (ED and ES). We can compute the deformation matrix across the cardiac cycle. Then, we can pass the deformation matrix to the GST to associate the temporal information between the ED and ES phases.

Acknowledgements. This work was supported by the UKRI CDT in AI for Healthcare http://ai4health.io (Grant No. EP/S023283/1) and the British Heart Foundation Centre of Research Excellence at Imperial College London (RE/18/4/34215).

References

1. Attili, A., Schuster, A., Nagel, E., et al.: Quantification in cardiac MRI: advances in image acquisition and processing. Int. J. Cardiovasc. Imaging **26**(Suppl. 1), 27–40 (2010). https://doi.org/10.1007/s10554-009-9571-x
2. Balakrishnan, G., Zhao, A., Sabuncu, M.R., et al.: Voxelmorph: a learning framework for deformable medical image registration. IEEE Trans. Med. Imaging **38**(8), 1788–1800 (2019)
3. Bonnemains, L., Mandry, D., Marie, P., et al.: Assessment of right ventricle volumes and function by cardiac MRI: quantification of the regional and global interobserver variability. Magn. Reson. Med. **67**, 1740–1746 (2012)
4. Campello, V.M., Gkontra, P., Izquierdo, C., et al.: Multi-centre, multi-vendor and multi-disease cardiac segmentation: the M&Ms challenge. IEEE Trans. Med. Imaging **40**(12), 3543–3554 (2021). https://doi.org/10.1109/TMI.2021.3090082
5. Caudron, J., Fares, J., Vivier, P., et al.: Diagnostic accuracy and variability of three semi-quantitative methods for assessing right ventricular systolic function from cardiac mri in patients with acquired heart disease. Eur. Radiol. **21**, 2111–2120 (2011)
6. Caudron, J., Fares, J., Lefebvre, V., et al.: Cardiac MR assessment of right ventricular function in acquired heart disease: factors of variability. Acad Radiol. **19**(8), 991–1002 (2012)
7. Chen, C., Biffi, C., Tarroni, G., Petersen, S., Bai, W., Rueckert, D.: Learning shape priors for robust cardiac MR segmentation from multi-view images. In: Shen, D., et al. (eds.) MICCAI 2019. LNCS, vol. 11765, pp. 523–531. Springer, Cham (2019). https://doi.org/10.1007/978-3-030-32245-8_58
8. Friedberg, M., Redington, A.: Right versus left ventricular failure differences, similarities, and interactions. Circulation **129**, 1033–1044 (2014)
9. Full, P.M., Isensee, F., Jäger, P.F., Maier-Hein, K.: Studying robustness of semantic segmentation under domain shift in cardiac MRI. In: Puyol Anton, E., et al. (eds.) STACOM 2020. LNCS, vol. 12592, pp. 238–249. Springer, Cham (2021). https://doi.org/10.1007/978-3-030-68107-4_24
10. Hu, J., Shen, L., Sun, G.: Squeeze-and-excitation networks. In: 2018 IEEE/CVF Conference on Computer Vision and Pattern Recognition, pp. 7132–7141 (2018) https://doi.org/10.1109/CVPR.2018.00745
11. Kingma, D.P., Ba, J.: Adam: a method for stochastic optimization (2017)
12. Klambauer, G., Unterthiner, T., Mayr, A., Hochreiter, S.: Self-normalizing neural networks. In: 31st International Conference on Neural Information Processing Systems, pp. 972–981. NIPS 2017. Curran Associates Inc. (2017)
13. LeCun, Y.A., Bottou, L., Orr, G.B., Müller, K.-R.: Efficient BackProp. In: Montavon, G., Orr, G.B., Müller, K.-R. (eds.) Neural Networks: Tricks of the Trade. LNCS, vol. 7700, pp. 9–48. Springer, Heidelberg (2012). https://doi.org/10.1007/978-3-642-35289-8_3
14. Lin, T.Y., Dollár, P., Girshick, R., et al.: Feature pyramid networks for object detection. In: 2017 IEEE Conference on Computer Vision and Pattern Recognition (CVPR), pp. 936–944 (2017). https://doi.org/10.1109/CVPR.2017.106

15. Lin, T.Y., Goyal, P., Girshick, R., et al.: Focal loss for dense object detection (2018)
16. Martín-Isla, C., Palomares, J.F.R., Guala, A., et al.: Multi-disease, multi-view & multi-center right ventricular segmentation in cardiac MRI (M&Ms-2), March 2021. https://doi.org/10.5281/zenodo.4573984
17. Petitjean, C., Zuluaga, M.A., Bai, W., et al.: Right ventricle segmentation from cardiac MRI: a collation study. Med. Image Anal. **19**(1), 187–202 (2015)
18. Zhu, W., Huang, Y., Zeng, L., et al.: AnatomyNet: deep learning for fast and fully automated whole-volume segmentation of head and neck anatomy. Med. Phys. **46**(2), 576–589 (2019)

Multi-view SA-LA Net: A Framework for Simultaneous Segmentation of RV on Multi-view Cardiac MR Images

Sana Jabbar, Syed Talha Bukhari, and Hassan Mohy-ud-Din[✉]

Department of Electrical Engineering, Syed Babar Ali School of Science and Engineering,
LUMS, Lahore 54792, Pakistan
hassan.mohyuddin@lums.edu.pk

Abstract. We proposed a multi-view SA-LA model for simultaneous segmentation of RV on the short-axis (SA) and long-axis (LA) cardiac MR images. The multi-view SA-LA model is a multi-encoder, multi-decoder U-Net architecture based on the U-Net model. One encoder-decoder pair segments the RV on SA images and the other pair on LA images. Multi-view SA-LA model assembles an extremely rich set of synergistic features, at the root of the encoder branch, by combining feature maps learned from matched SA and LA cardiac MR images. Segmentation performance is further enhanced by: (1) incorporating spatial context of LV as a prior and (2) performing deep supervision in the last three layers of the decoder branch. Multi-view SA-LA model was extensively evaluated on the MICCAI 2021 Multi- Disease, Multi-View, and Multi- Centre RV Segmentation Challenge dataset (M&Ms-2021). M&Ms-2021 dataset consists of multi-phase, multi-view cardiac MR images of 360 subjects acquired at four clinical centers with three different vendors. On the challenge cohort (160 subjects), the proposed multi-view SA-LA model achieved a Dice Score of 91% and Hausdorff distance of 11.2 mm on short-axis images and a Dice Score of 89.6% and Hausdorff distance of 8.1 mm on long-axis images. Moreover, multi-view SA-LA model exhibited strong generalization to unseen RV related pathologies including Dilated Right Ventricle (DSC: SA 91.41%, LA 89.63%) and Tricuspidal Regurgitation (DSC: SA 91.40%, LA 90.40%) with low variance (σ_{DSC}: SA < 5%, LA < 6%).

Keywords: Cardiac imaging · Magnetic resonance imaging · Right Ventricle · Segmentation · Short-axis sequence · Long-axis sequence · Deep neural network · Convolutional Neural Network · U-Net

1 Introduction

A report from World Health Organization (WHO) [1] predicts that, by 2023, approximately 23 million people will die due to cardiovascular diseases (CVDs). This will be

This work was supported by a grant from the Higher Education Commission of Pakistan as part of the National Center of Big Data and Cloud Computing and the Clinical and Translational Imaging Lab at LUMS.

© Springer Nature Switzerland AG 2022
E. Puyol Antón et al. (Eds.): STACOM 2021, LNCS 13131, pp. 277–286, 2022.
https://doi.org/10.1007/978-3-030-93722-5_30

a staggering increase of ~24% since 2019 which reported 18.6 million deaths worldwide due to CVDs [2]. These alarming statistics have sparked extensive research in the detection, staging, and treatment planning strategies of CVDs.

Cardiac MRI is the preferred imaging modality for diagnosis of CVDs. It provides high quality volumetric images of the heart noninvasively and without employing ionizing radiation. A routine cardiac MRI scan provides 3D scans of the end-diastolic (ED) and end-systolic (ES) cardiac phases. Clinical parameters of interest, including ES volume, ED volume, ejection fraction, and myocardial mass, are estimated from multiphase cardiac MRI scans (i.e., 3D ED and ES images) and greatly aid in the diagnosis and prognostication of CVDs [3].

A pivotal step towards quantification of clinical parameters of interest is segmentation of the LV blood pool, RV blood pool, and LV Myocardium on multiphase cardiac MRI scans. The gold standard approach of multi-region segmentation of 3D cardiac MRI scans is manual segmentation by expert cardiologist. Manual segmentation is an extremely cumbersome process with high inter-observer and intra-observer variability [4]. To overcome limitations of manual segmentation, it is essential to propose fully automatic (or semi-automatic), robust, and accurate algorithms for multiregional segmentation of 3D cardiac MRI scans.

In the past decade, numerous cardiac image segmentation challenges have been launched from the platform of MICCAI including LVQuan18 [5], LVQuan19 [6], EMIDEC [7], and ACDC [8]. Most of these challenges sought fully automatic or semi-automatic methods that provide robust segmentation of multiple regions on cardiac MRI scans. Two important things are noteworthy. Firstly, Convolutional Neural Networks (CNNs) have demonstrated state-of-the-art performance in multiregional segmentation of 3D multi-phase cardiac MRI scans. Secondly, the principal focus of these challenges was accurate segmentation of the epicardial and endocardial boundaries (i.e., segmentation of the LV blood pool and LV myocardium) on cardiac MRI scans. LV segmentation has been the prime focus because noninvasive quantification of LV function is essential in the diagnosis of myocardial infarction, hypertrophy, cardiomyopathy, and dilated cardiomyopathy [9].

The most recent challenge hosted from MICCAI platform was the Multi-Centre, Multi-Vendor, and Multi-Disease Cardiac Image Segmentation Challenge (M&Ms-2020). M&Ms-2020 provided a large cohort of 375 subjects representing 6 different centers, 4 different vendors, and multiple cardiac pathologies [10]. The main aim of M&Ms-2020 was to come up with a fully automatic (or semi-automatic) segmentation method that generalizes well to novel datasets acquired at multiple centers, with multiple vendors, and representing diverse cardiac pathologies. M&Ms-2020 challenge concluded that deep learning solutions yielded state-of-the-art performance on cardiac MRI segmentation task with U-Net based architectures occupying top positions [11, 12].

In 2021, a new challenge was hosted from the platform of MICCAI namely Multi-Disease, Multi-View, and Multi-Center Right Ventricular Segmentation in Cardiac MRI scans (M&Ms-2021) [13]. M&Ms-2021 specifically focused on the segmentation of RV blood pool which is essential to the study of RV related pathologies such as Dilated Right Ventricle, Tricuspid Insufficiency, Arrhythmogenesis, Tetralogy of Fallot and Interatrial Communication. Compared to LV, segmentation of RV blood pool is more challenging due to its highly complex and variable shape and ill-defined region boundaries on cardiac MRI scans. A novel aspect of M&Ms-2021 challenge is that, for each subject, it provides multi-view scans of the heart in the form of 3D short-axis images (SA) and 2D long-axis images (LA). 2D LA cardiac scans facilitate automatic definition of the basal plane of the RV which is often confused with the right atrium.

In this study, we proposed a multi-view deep neural network architecture for simultaneous segmentation of RV blood pool on SA and LA cardiac scans. The proposed architecture is called *multi-view SA-LA model*. The SA-LA model is a novel multi-encoder, multi-decoder architecture based on the U-Net model. One encoder-decoder pair segments the RV blood on SA images and the other pair on LA images. The main hypothesis of the study is that an extremely rich set of synergistic features can be assembled at the root of the encoder branch that combines feature maps extracted from matched SA and LA cardiac images. With extensive experiments on M&Ms-2021 dataset, we demonstrated that the proposed SA-LA model achieved state-of-the-art performance in the segmentation of RV blood pool on SA and LA cardiac images. Moreover, SA-LA model generalized to novel datasets, acquired at multiple centers with multiple vendors, and novel RV pathologies.

2 Material and Method

2.1 Cardiac MRI Dataset

The proposed SA-LA model was extensively evaluated on a publicly available dataset, namely M&Ms-2021 dataset, which forms the RV segmentation challenge in MICCAI 2021. M&Ms-2021 is a multi-disease, multi-view, and multi-center dataset for RV segmentation on cardiac MRI scans. M&Ms-2021 dataset is composed of 360 subjects acquired at 3 centers, with 3 vendors, and represents diverse RV related pathologies. M&Ms-2021 dataset is split into training (160 subjects), validation (40 subjects), and challenge (160 subjects) cohorts. The training and validation cohorts were publicly released and included 3D MRI scans of the ED and ES cardiac phases in the short-axis view and 2D MRI scans of the ED and ES cardiac phases in the long-axis view. Moreover, for the training cohort, manual segmentation of the LV blood pool (label = 1), LV myocardium (label = 2), and RV blood pool (label = 3) on the short-axis and long-axis views were also provided. The challenge cohort was not publicly released by the organizers. To study the generalizability of the proposed approach, the validation and challenge cohorts included novel pathologies which were not present in the training cohort. The details of M&Ms-2021 dataset is summarized in Table 1.

2.2 Data Preprocessing

The 3D multiphase cardiac MRI scans were resampled to an in-plane resolution of 1.25×1.25 mm^2 and cropped to an in-plane matrix dimension of 256×256 voxels. Furthermore, each 3D MRI volume was normalized to zero mean and unit-variance. For each subject, the 2D LA cardiac image slice (and the associated 2D manual segmentation image) was replicated to create a 3D volume of the same dimension as the corresponding

3D SA cardiac scan. This simple modification helped create a (subject-wise) matched SA-LA dataset of the same dimension which was subsequently used for training a 2D (slice-based) SA-LA model.

For accurate segmentation of RV, it is essential that the learning model captures the contextual information about RV which is implicit in cardiac MRI scans i.e., its spatial closeness to the LV. Hence, in the supervised learning framework, we also utilized voxel-wise labels of the LV which were also provided in the training cohort.

Table 1. A summary of Multi-disease, Multi-view, and Multi-center RV segmentation in cardiac MRI (M&Ms-2021) dataset. Acronyms are: end-diastolic phase (ED), end-systolic phase (ES), short-axis view (SA), and long-axis view (LA).

Dataset	Multi-disease, multi-view, and multi-center cardiac MRI scans		
Centers	Three clinical centers from Spain		
Scanners	Siemens, Philips, and General Electric		
Cohorts (subjects)	Training (160), Validation (40), and Challenge (160)		
Imaging data	3D, SA, Cardiac MRI scans of the ED and ES cardiac phases 2D, LA, Cardiac MRI scans of the ED and ES cardiac phases		
Segmentation data *Only available for training cohort*	3D segmentation maps on ED and ES cardiac phases in SA view 2D segmentation on ED and ES cardiac phases in LA view		
Segmentation labels	LV blood pool, RV blood pool, and LV Myocardium		
Pathology	*Training cohort*	*Validation cohort*	*Challenge cohort*
Normal subjects	40	5	30
Dilated left ventricle	30	5	25
Hypertrophic cardiomyopathy	30	5	25
Congenital arrhythmogenesis	20	5	10
Tetralogy of fallot	20	5	10
Interatrial communication	20	5	10
Dilated right ventricle[a]	–	5	25
Tricuspidal regurgitation[a]	–	5	25

[a]Tricuspidal Regurgitation (30 subjects) and Congenital Arrhythmogenesis (30 subjects) are only present in the validation and testing cohorts to evaluate generalization to unseen pathologies.

We make one interesting modification. We relabel the manual segmentation maps by merging the LV blood pool and LV myocardium, now assigned label = 1, and assigned label = 2 to the RV. Since the focus of M&Ms-2021 challenge is accurate segmentation of RV, this modification utilized the prior information of spatial closeness of LV and RV without the need of learning to subdivide LV into LV blood pool and LV myocardium.

2.3 Multi-view SA-LA Model

The proposed multi-view SA-LA model is based on a 2D U-Net architecture. It is formed with two U-Net architectures with asymmetrically large encoding-decoding pathways. One encoder-decoder pair segments RV on the short-axis view and the other pair on the long-axis view. The two U-Net architectures communicate encoded feature maps, extracted from matched SA and LA cardiac MR images, at the root of the encoders i.e., the highest down sampling level in the encoder branches. The synergistic information collected at the root of the encoder facilitates more accurate segmentation of RV on the SA and LA views. Figure 1 shows a schematic of the proposed multi-view SA-LA architecture.

Each 2D U-Net comprised an encoding and a decoding path, with five down/up-sampling levels each. Feature maps at each level were processed by two 3×3 convolution layers, each followed by Batch Normalization and ReLU activation. Max-pooling and Transposed Convolutional layers served as down-sampling and up-sampling layers, respectively. Feature maps from the encoder of each U-Net were forwarded to the corresponding decoder, except for the highest down-sampling level (level 5) where each decoder received a concatenation of features maps from both encoders. To improve gradient propagation and feature learning in earlier layers, we employed deep supervision on the last three layers of the decoding pathways.

2.4 Training

The matched SA-LA dataset, elaborated in Sect. 2.2, formed the training cohort in our experiments. The proposed SA-LA model was trained with five-fold cross-validation to reduce data selection biases. Random batches of 10 2D slices were sampled from the matched SA-LA dataset followed by data augmentation procedure comprising the following operations: random horizontal and vertical flipping with probability of 0.5, random rotation selected from $[-30°, +30°]$, and zooming with randomly sampled scale factor $\alpha \in [0.7, 1.4]$. The multi-view SA-LA model was initialized with He-normal weight, with a fixed seed value (46), and no regularization was used. The model was trained for 150 epochs with an unweighted sum of Soft-Dice Loss and Cross-Entropy, Adam optimizer (with $\beta_1 = 0.9$ and $\beta_2 = 0.999$), and a constant learning rate of 1×10^{-4}. One complete training session (five-fold cross validation) took 25 h GPU time.

2.5 Inference

Inference on the validation cohort was performed using the models learned with five-fold cross-validation on the training cohort. For each training fold, the optimal model for inference was selected based on the minimum validation loss[1]. For each subject in the validation cohort, the optimal model from each training fold was used to generate

---

[1] Loss on local validation dataset (20% of training dataset).

multi-class segmentation probability maps for LV and RV blood pool. The obtained probability maps from each fold were averaged to generate the final segmentation map for the SA and LA views. In the post-processing step, cluster thresholding was applied to remove isolated over-segmentations.

Inference over the challenge cohort was performed by the challenge organizers. Participants, in the M&Ms-2021 challenge, were asked to provide the inference code and trained models in a docker environment which were ultimately used to generate segmentation maps on the challenge cohort.

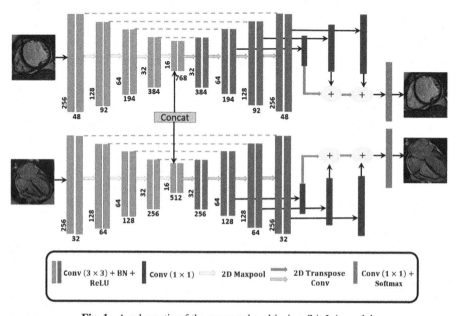

Fig. 1. A schematic of the proposed multi-view SA-LA model.

2.6 Evaluation Metrics

Segmentation performance was quantified based on the Dice Similarity Coefficient (DSC in %) and Hausdorff Distance (HD-95 in mm). We computed DSC and HD for predicted segmentation maps in the ED and ES cardiac phases in SA and LA views. Firstly, for SA and LA views, DSC and HD scores were averaged across the ED and ES cardiac phases as follows:

$$\text{DSC}_{\text{SA}} = \frac{1}{2}\left(\text{DSC}_{\text{SA}}^{\text{ED}} + \text{DSC}_{\text{SA}}^{\text{ES}}\right) \text{ and } \text{HD}_{\text{SA}} = \frac{1}{2}\left(\text{HD}_{\text{SA}}^{\text{ED}} + \text{HD}_{\text{SA}}^{\text{ES}}\right)$$

$$\text{DSC}_{\text{LA}} = \frac{1}{2}\left(\text{DSC}_{\text{LA}}^{\text{ED}} + \text{DSC}_{\text{LA}}^{\text{ES}}\right) \text{ and } \text{HD}_{\text{LA}} = \frac{1}{2}\left(\text{HD}_{\text{LA}}^{\text{ED}} + \text{HD}_{\text{LA}}^{\text{ES}}\right) \qquad (1)$$

Secondly, a unified score was computed by using the following formula provided by the M&Ms-2021 challenge:

$$\text{score} = \frac{0.75\left(\text{DSC}_{\text{SA}} + \widehat{\text{HD}}_{\text{SA}}\right) + 0.25\left(\text{DSC}_{\text{LA}} + \widehat{\text{HD}}_{\text{LA}}\right)}{2} \tag{2}$$

where $\widehat{\text{HD}}_{\text{SA}}$ and $\widehat{\text{HD}}_{\text{LA}}$ are the normalized HD scores across participants in the challenge. This unified score was used to rank various methods proposed for segmentation of RV on cardiac MRI scans.

Table 2. Quantitative comparison of a non-holistic 2D U-Net model, multi-view SA-LA model without LV context, and (the proposed) multi-view SA-LA model (with LV context) on training, validation, and challenge cohorts. Score is computed using Eq. (2). Inferences are performed with an ensemble of five models learned with five-fold cross-validation on training cohort.

Cohort (# subj)	Model	Parameters (millions)	DSC (%)			HD-95 (mm)			Score (%)
			Overall	SA	LA	Overall	SA	LA	
Training cohort (160)	Non-holistic 2D U-Net model	~39	90.14	**89.99**	90.59	5.84	6.69	3.28	–
	Multi-view SA-LA model w/o LV context	~27.2	89.62	89.42	90.25	4.96	5.82	2.36	–
	Proposed Multi-view SA-LA model (with LV context)	~27.2	**90.60**	89.59	**93.61**	**2.89**	**3.16**	**2.07**	–
Validation cohort (40)	Non-holistic 2D U-Net model	~39	90.00	89.65	**91.08**	10.19	11.52	6.21	89.65
	Multi-view SA-LA model w/o LV context	~27.2	90.00	89.92	90.26	10.20	11.42	6.60	89.92
	Proposed Multi-view SA-LA model (with LV context)	~27.2	**90.46**	**90.28**	91.00	**9.69**	**10.91**	**6.01**	**90.28**
Challenge cohort (160)	Proposed Multi-view SA-LA model (with LV context)	~27.2	90.66	91.00	89.63	10.40	11.16	8.13	91.00

3   Results

All experiments were performed on a system with 64 GB RAM and an NVIDIA RTX 2080Ti 11 GB GPU using open-source packages including Keras, Nibabel, Opencv, Numpy, Scikit-image, Pandas, and Matplotlib.

3.1   Quantitative Analysis

Table 2 provides an extensive summary of quantitative results from various experiments conducted on the M&Ms-2021 dataset. We compared our proposed multi-view SA-LA model with a non-holistic 2D U-net model and a multi-view SA-LA model without LV context. The proposed multi-view SA-LA model had considerably fewer trainable parameters (~27 million) compared to a non-holistic 2D U-net model (~39 million) which trained two independent 2D U-net architectures for segmentation of RV in SA and LA views.

On the training and validation cohorts we found that the proposed multi-view SA-LA model outperformed other approaches in terms of DSC and HD metrics. More specifically, on the validation cohort (which included novel RV related pathologies), the proposed multi-view SA-LA model yielded the highest score (90.3%). On the validation cohort, we found that the multi-view SA-LA model (with LV context) increased the DSC by 0.46% and decreased the HD by 0.51 mm.

3.2   Qualitative Analysis

Figure 2 shows predicted segmentation maps of the RV for subject 153 in the training cohort. The non-holistic 2D U-Net model under-segments the basal, midventricular, and

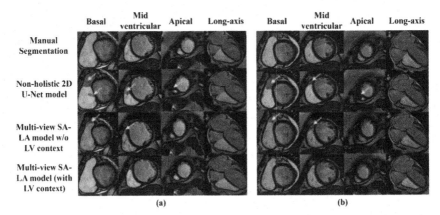

Fig. 2. Predicted segmentation maps of the RV in the (a) end-diastolic and (b) end-systolic cardiac phases. The basal, midventricular, and apical slices of the short-axis (SA) view are shown in the first three columns of (a) and (b). A 2D slice in the long-axis (LA) view is shown in the fourth column of (a) and (b). Manual segmentation maps are shown in the first row for reference. Rows 2 to 4, in (a) and (b), show predicted segmentation maps obtained with a non-holistic 2D U-Net model, a multi-view SA-LA model without LV context, and a multi-view SA-LA model (with LV context). Red arrows show under-segmented regions in the predicted segmentation maps.

apical slices in the SA view. A multi-view SA-LA model without LV context substantially improved the segmentation of RV in the SA view. The proposed multi-view SA-LA model (with LV context) predicts segmentation maps closer to manual segmentations as quantified by DSC and HD metrics.

3.3 Generalizability Study on the Challenge Cohort

The proposed multi-view SA-LA model (with LV context) performed exceedingly well (score 91%) on the challenge cohort composed of cardiac imaging scans acquired at different centers with diverse cardiac pathologies. Table 3 summarizes quantitative performance of the multi-view SA-LA model (with LV context) across eight cardiac pathologies in the challenge cohort. We find that, across pathologies, the DSC is consistently higher in the SA view ($\geq$90%) and the HD-95 is consistently lower in the LA view ($\leq$9mm). Our proposed segmentation approach also achieved superior segmentation performance on unseen RV related pathologies including Dilated Right Ventricle (DSC: SA 91.41, LA 89.63) and Tricuspidal Regurgitation (DSC: SA 91.40, LA 90.40) with low variance (σ_{DSC}: SA < 5%, LA < 6%).

Table 3. Quantitative performance of the proposed multi-view SA-LA model (with LV context) on eight different cardiac pathologies represented in the challenge cohort. 50 subjects with Dilated Right Ventricle and Tricuspidal Regurgitation help quantify generalizability of the segmentation algorithm as they are not represented in the training cohort. Acronyms are: DSC in short-axis view (DSC-SA), DSC in long-axis view (DSC-LA), HD-95 in short-axis view (HD-SA), and HD-95 in long-axis view (HD-LA).

Pathology	No of subjects	DSC-SA	DSC-LA	HD-SA	HD-LA
Normal subjects	30	90.52 ± 5.21	91.40 ± 572	9.95 ± 5.32	6.22 ± 5.05
Dilated left ventricle	25	91.73 ± 7.63	86.80 ± 17.12	12.03 ± 5.97	7.10 ± 5.60
Hypertrophic cardiomyopathy	25	89.99 ± 8.39	90.06 ± 7.44	11.80 ± 7.59	5.97 ± 2.95
Congenital arrhythmogenesis	10	91.92 ± 6.07	90.81 ± 7.22	11.60 ± 8.83	7.15 ± 5.98
Tetralogy of fallot	10	91.65 ± 3.72	90.74 ± 3.30	11.56 ± 3.65	7.81 ± 3.65
Interatrial communication	10	89.55 ± 6.87	86.17 ± 17.34	12.11 ± 4.43	9.44 ± 9.55
Dilated right ventricle	25	91.41 ± 4.64	89.63 ± 5.31	11.59 ± 4.73	8.89 ± 5.59
Tricuspidal regurgitation	25	91.40 ± 4.10	90.40 ± 4.52	9.97 ± 3.83	6.80 ± 4.31

4 Conclusion

In this study, we proposed a multi-view SA-LA model (with LV context) for simultaneous segmentation of RV on short-axis and long-axis views. SA-LA model has a multi-encoder, multi-decoder U-Net shaped structure where one encoder-decoder pair segments the RV blood on SA images and the other pair on LA images. We demonstrated, with extensive experiments on M&Ms-2021 dataset, that the proposed SA-LA model achieves state-of-the-art performance in the segmentation of RV on multi-view and multi-phase cardiac MRI scans. Our experiments also revealed that adding LV contextual information boosted the segmentation performance on training and validation cohorts. Moreover, SA-LA model (with LV context) shows high generalizability on novel RV related pathologies not represented in the training cohort.

References

1. Heart Disease and Stroke Statistics–At-a-Glance. https://www.heart.org/-/media/phd-files-2/science-news/2/2021-heart-and-strokestatpdate/2021_heart_disease_and_stroke_statistics_update_fact_sheet_at_a_glance.pdf
2. Roth, G.A., et al.: Global burden of cardiovascular diseases and risk factors, 1990–2019: update from the GBD 2019 study. J. Am. College Cardiol. **76**(25), 2982–3021 (2020)
3. Lebenberg, J., et al.: Improved estimation of cardiac function parameters using a combination of independent automated segmentation results in cardiovascular magnetic resonance imaging. PLoS ONE **10**(8), e0135715 (2015)
4. Luijnenburg, S.E., et al.: Intra-observer and inter-observer variability of biventricular function, volumes and mass in patients with congenital heart disease measured by CMR imaging. Int. J. Cardiovasc. Imaging **26**(1), 57–64 (2009)
5. Left Ventricle Full Quantification Challenge MICCAI (2018). lvquan18.github.io/
6. Left Ventricle Full Quantification Challenge MICCAI (2019). lvquan19.github.io/
7. Lalande, A., et al.: EMIDEC: a database usable for the automatic evaluation of myocardial infarction from delayed-enhancement cardiac MRI. Data **5**(4), 89 (2020)
8. Bernard, O., et al.: Deep learning techniques for automatic MRI cardiac multi-structures segmentation and diagnosis: is the problem solved? IEEE Trans. Med. Imaging **37**(11), 2514–2525 (2018)
9. Tao, Q., et al.: Deep learning–based method for fully automatic quantification of left ventricle function from cine MR images: a multivendor, multicenter study. Radiology **290**(1), 81–88 (2019)
10. Campello, V.M., et al.: Multi-center, Multi-vendor, and Multi-disease cardiac segmentation: the M&Ms challenge. IEEE Trans. Med. Imaging **40**(12), 3543–3554 (2020). https://doi.org/10.1109/TMI.2021.3090082
11. Full, P.M., Isensee, F., Jäger, P.F., Maier-Hein, K.: Studying robustness of semantic segmentation under domain shift in cardiac MRI. In: Puyol Anton, E., et al. (eds.) STACOM 2020. LNCS, vol. 12592, pp. 238–249. Springer, Cham (2021). https://doi.org/10.1007/978-3-030-68107-4_24
12. Zhang, Y., et al.: Semi-supervised cardiac image segmentation via label propagation and style transfer. In: Puyol Anton, E., et al. (eds.) STACOM 2020. LNCS, vol. 12592, pp. 219–227. Springer, Cham (2021). https://doi.org/10.1007/978-3-030-68107-4_22
13. Multi-Disease, Multi-View, and Multi-Center Right Ventricle Segmentation in Cardiac MRI (M&Ms-2). https://www.ub.edu/mnms-2/

Right Ventricular Segmentation in Multi-view Cardiac MRI Using a Unified U-net Model

Sandro Queirós[1,2]([✉]) [ID]

[1] Life and Health Sciences Research Institute (ICVS), School of Medicine,
University of Minho, Braga, Portugal
sandroqueiros@med.uminho.pt
[2] ICVS/3B's - PT Government Associate Laboratory, Braga/Guimarães, Portugal

Abstract. Accurate ventricle segmentation is an essential step towards cardiac function quantification from magnetic resonance imaging (MRI). Despite the vast efforts made over the last few years to automate this step, few works have specifically targeted the right ventricle (RV) assessment, and even fewer have attempted to integrate the information provided by both short- and long-axis cine MRI views. In this work, a novel automatic method for multi-view RV segmentation in cardiac MRI is proposed, which contemplates: (1) a preprocessing stage that re-orients the standard views according to their DICOM-provided 3D pose; (2) a novel multi-view augmentation module that augments on the fly the input images while conserving the 3D relationship between them; and (3) a unified cross-view model based on the U-net architecture (named xUnet) that simultaneously processes both views. The proposed method was evaluated on the M&Ms-2 challenge and achieved a combined average Dice score and Hausdorff distance of 0.918 and 9.25 mm, and 0.916 and 9.78 mm, on the validation and test sets, respectively. These results demonstrate the potential of the proposed unified model (and associated training scheme) towards accurate RV segmentation in cardiac MRI.

Keywords: M&Ms-2 challenge · Cardiac magnetic resonance imaging · Deep learning · U-net · Multi-view segmentation · Right ventricle

1 Introduction

Cardiovascular diseases (CVDs) are the leading cause of death worldwide [14]. This dismal fact highlights the need for accurate and efficient techniques for diagnosis and treatment follow-up of patients with CVDs. Assessment of ventricular morphology and function quantification using non-invasive cardiac imaging play a vital role towards this goal. Cardiac magnetic resonance imaging (MRI) is considered the gold-standard modality for such assessment, requiring the delineation of the heart ventricles from cine MRI images. This is a tedious and labour-intensive clinical task, prone to subjective biases and lack of reproducibility [1].

Over the past decades, countless efforts have been made to fully automate this task [6–10], although few have successfully found their way towards clinical

© Springer Nature Switzerland AG 2022
E. Puyol Antón et al. (Eds.): STACOM 2021, LNCS 13131, pp. 287–295, 2022.
https://doi.org/10.1007/978-3-030-93722-5_31

practice. More recently, deep learning (DL) techniques, and particularly convolutional neural networks (CNNs), have brought forth a newly revived interest on this topic, demonstrating an outstanding potential [1,2]. However, two major remarks can still be made: (1) right ventricle (RV) functional assessment has received considerably less attention from the research community, with most works targeting it as a way to improve left ventricle (LV) assessment only [7]; and (2) the vast majority of the literature has targeted the segmentation of short-axis (SAx) image stacks only [1,2], disregarding the potential interest of analyzing them together with long-axis (LAx) image views.

To address it, this work proposes a novel automatic method for multi-view RV segmentation in cardiac MRI, which incorporates and takes advantage of the 3D spatial context provided by the complementary SAx and LAx image views. The proposed method was evaluated in the context of the Multi-Disease, Multi-View and Multi-Center Right Ventricular Segmentation (M&Ms-2) challenge, showing a competitive performance against the remaining challenge submissions.

2   Methods

2.1   Overview

The proposed method rests on three core modules: (1) a preprocessing routine that re-orients the standard views (SAx and 4-chamber LAx) according to their DICOM-provided 3D pose (Sect. 2.2); (2) a unified cross-view model, henceforth named xUnet, that simultaneous segments the 3D SAx stack and the 2D LAx image (Sect. 2.3); and (3) a training scheme that includes a novel multi-view augmentation module, allowing on-the-fly augmentation of both input image views, while conserving the 3D relationship between them (Sect. 2.4). During inference, a postprocessing routine is also employed (Sect. 2.5).

For the remaining of the document, the term image(s) is interchangeably used to refer to the full SAx stack and/or the LAx image.

2.2   Data Preprocessing

By exploiting the DICOM-provided pose information, one re-orients the standard views to a reference pose (Fig. 1). To this end, both image views are centered on the average position between their original centroid positions (see orange dot). In addition, one rotates both images to match their axes, *i.e.* the LAx image is rotated so that its Y-axis matches the SAx stack's Z-axis, and the SAx stack is rotated on the XY plane so that its X-axis matches the rotated LAx image's X-axis (X'_{LA}). Simultaneously, the applied transformation matrices also resample both images to a pre-defined XY resolution (1.125×1.1986 mm^2 and 1.1986×1.1986 mm^2 for LAx and SAx views, respectively). The remaining preprocessing steps are applied on the fly during training (Sect. 2.4).

2.3 Network Architecture

In this work, the potential of a unified multi-view segmentation model, xUnet, is investigated. The proposed model is based on the U-net architecture [12], but integrates two parallel streams consisting in one 3D U-net and one 2D U-net that process, respectively, the SAx stack and the LAx image (Fig. 2).

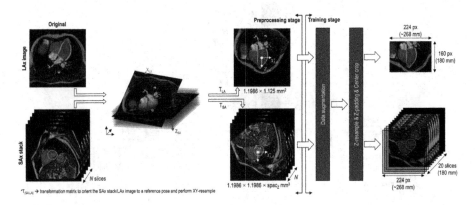

Fig. 1. Illustration of the proposed preprocessing and data augmentation routines.

To exploit the complementary spatial context provided by both views, novel cross-view modules (Fig. 3) are employed at the end of the three lowest levels of each U-net contraction path (dark blue and dark green blocks in Fig. 2). In short, each cross-view module is responsible for concatenating the features learnt at a given contraction level with the respective features from the complementary stream, and compute a new set of spatially-relevant features through a 1×1 convolution layer. Note that, since the input images have been transformed to a reference (known) pose, it is possible to keep the spatial coherence during feature concatenation by applying a carefully ordered set of resize, transpose, pooling and/or repeat operations. Overall, the proposed modules act as a spatial attention mechanism, where relevant spatial information provided by the complementary view is embedded before proceeding with the expansion path.

Each U-net stream is largely based on the configurations proposed in [3]. However, following some initial experiments, one replaced the leaky ReLUs with the swish activation function [11]. In addition, each stream's input size was set according to the normalized input spacing defined in Sect. 2.2, therefore guaranteeing that the images' matched axes cover the same physical space (Fig. 1).

2.4 Model Training

The xUnet model is trained with a batch size of 2 for a fixed number of epochs (variable across the experiments), with one epoch being defined as an iteration over the entire training set. The ADAM optimizer [4] is used, with the initial

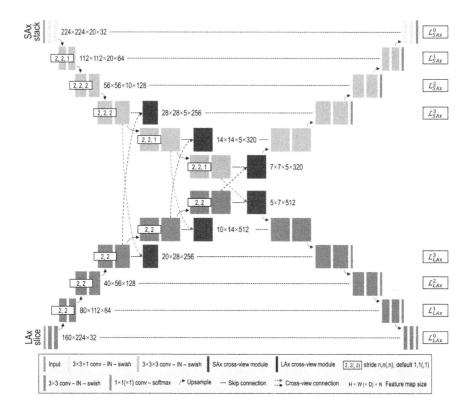

Fig. 2. Proposed cross-view U-net (xUnet) architecture.

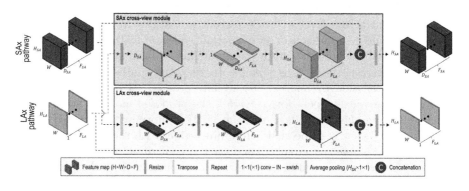

Fig. 3. Proposed SAx and LAx cross-view modules.

learning rate set to 1×10^{-3} and updated using a 'cosine decay' schedule [5]. Similar to [3], the training objective is given by the sum of both cross-entropy and Dice losses over the four highest resolution levels of both U-net streams ($\mathcal{L}_j^i$, with $i \in \{0, 1, 2, 3\}$ and $j \in \{SAx, LAx\}$; Fig. 2), with the weights halved with each decrease in resolution ($w_i = 2^{-i}$). No weight regularization is applied.

During training, a variety of data augmentation techniques are applied on the fly, including spatial (mirroring, rotation, scaling, and translation) and intensity (Gaussian noise, brightness, contrast and gamma correction) transformations. To preserve the 3D relationship between images, a custom augmentation module was implemented using SOLT [13]. In short, one randomly determines the set of augmentations to apply for each image pair on the batch, with the respective transformation being applied to both image views coherently. For example, if the LAx image is mirrored on the Y-axis, the respective SAx stack is mirrored on the Z-axis. A similar principle applies for scaling and translation (in the latter, one must consider each image pixel spacing). Given the anisotropic spacing of the SAx stack, rotations were applied on the XY plane only. However, to account for possible rotational misalignments between images, small rotations were also randomly applied on the LAx image. For the remaining spatial transformations, any of the three axes may be randomly augmented. Of note, the exact set of transformations used varies between experiments (details in Sect. 3).

Upon augmenting each image pair, the final steps of preprocessing are applied, consisting on resampling and padding operations along the Z-axis of the SAx stack, plus center cropping of both images (Fig. 1). Note that the Z-resample operation is implemented as to preserve the augmentation-based scaling factor.

2.5 Inference and Postprocessing

Distinct inference routines are investigated in Sect. 3. This includes test-time augmentation (TTA) by mirroring along the X and Y axes (4 variations in total), plus (optionally) ensembling of xUnet models. All predictions are combined through averaging of the softmax probabilities. Then, connected component-based postprocessing and hole filling operations are applied to eliminate spurious false positive and false negative pixel predictions, respectively.

2.6 Cardiac MRI Dataset

The experiments in this study were performed on the cardiac cine MRI data of the M&Ms-2 challenge[1], whose cohort is composed of 360 patients with different RV and LV pathologies, as well as healthy subjects, scanned in 4 clinical centers with distinct scanners. The training set comprises 160 annotated cases, including contours of LV and RV blood pools, plus LV myocardium, on both end-diastolic (ES) and end-systolic (ES) phases. Another 40 patients (10 with pathologies not found on the training set), without annotations, are also publicly shared, serving as validation set. The testing set includes another 160 cases. In the latter, participants only submit the pre-trained prediction model(s), and associated code, and the inference is run on the organizers' docker environment. Despite providing LV delineations, the challenge's evaluation focuses on the RV only.

---

[1] https://www.ub.edu/mnms-2/.

3 Results

Two measures were used to assess the RV segmentation performance on each image view: the Dice coefficient and the Hausdorff distance (HD; in mm).

The first set of experiments focused on investigating the added value of the core modules of the proposed method on the validation set (Table 1). Each model was trained for 1000 epochs, with random rigid spatial transformations only.

Table 1. Ablation study on the core modules proposed

Preprocessing	Augmentation	Model	SAx		LAx	
			Dice	HD	Dice	HD
Independent	Independent	Independent	0.891	12.88	0.902	6.94
Fused	Independent	Independent	0.898	12.39	0.912	6.25
Fused	Independent	Combined	0.892	12.77	0.908	6.37
Fused	Fused	Combined	0.904	11.46	0.911	6.28
Fused	Fused	Fused	**0.909**	**11.34**	**0.914**	**6.11**

Independent means that each image is preprocessed/augmented/processed without influence from the other view. Combined means that both SAx and LAx networks are trained together, but without cross-view connections. Fused stands for the proposed approaches. No TTA or ensembling is used.

Table 2. Influence of intensity-based augmentation, no. of epochs, TTA and ensembling

Intensity-based augmentation	# Epochs	TTA	Ensemble	SAx		LAx	
				Dice	HD	Dice	HD
No	1000	No	1	0.909	11.34	0.914	6.11
No	1000	Yes	1	0.912	10.75	0.915	5.97
Yes	1000	Yes	1	0.911	10.77	**0.916**	**5.80**
Yes	2000	Yes	1	0.916	10.36	0.913	6.05
Yes	1500/2000	Yes	2	**0.918**	10.38	0.914	6.08
Yes	1500/2000	Yes	4	**0.918**	**10.33**	**0.916**	5.99

Intensity-based augmentation includes Gaussian noise, brightness, contrast and gamma correction. 'Ensemble' indicates the number of xUnet models ensembled.

Next, one investigated the influence of adding intensity-based augmentations, increase the number of epochs or use different inference routines, including the use of TTA or an ensemble (using models trained with slight changes in training epochs or augmentation frequency). In all tests, the proposed core modules were employed. Table 2 summarizes the obtained results. The best combined score was obtained by the 4-way ensemble (with TTA and intensity-based augmentations), with a combined average Dice and HD of 0.918 and 9.25 mm, respectively.

Given the above results, the ensemble of 4 xUnet models was evaluated on the test set. The proposed model obtained an average Dice score of 0.916 and 0.915 for the SAx and LAx views (combined score of 0.916), and an average HD of 10.91 mm and 6.41 mm (combined average of 9.78 mm), respectively.

4 Discussion

In this work, a novel unified U-net model, xUnet, was proposed and evaluated for automatic multi-view RV segmentation on cardiac MRI images. For its adequate implementation and training, the proposed architecture was integrated with a preprocessing module aimed at re-orienting both cine views to a reference pose, as well as with a novel data augmentation routine that simultaneously augments both image views without compromising their 3D spatial relationship.

As demonstrated by the ablation study performed (Table 1), these three core contributions led to a significant accuracy improvement over the baseline approach (>1.5% Dice increase), resulting from incremental gains given by each module. Indeed, both metrics improved with the use of the proposed preprocessing module alone (2nd vs 1st row), which may be linked to the reduced variability on ventricles' initial pose. A similar impact was observed when further activating the 'fused' augmentation (4th vs 3rd row), even if separately (data not show) or jointly ('Combined') training both U-net models. Note, however, that simply training both models together results in a lower performance (3rd vs 2nd row), most probably because the optimizer must now minimize (and balance) both models' losses. The introduction of the xUnet model, with its cross-view connections, further improves the segmentation (5th vs 4th row). When qualitatively analyzing the results, this accuracy boost seems linked to the 3D context provided by the complementary view. This is particularly noticeable in the SAx stack, with less spurious RV predictions on slices above the basal plane. It is nevertheless important to highlight that the number of cross-view connections is key, as both extremes are detrimental to the model's performance (data not shown). Decreasing it to 1 (or 2) linked levels closes in on the 'Combined' scenario, as the 3D spatial context is only shared on extremely low (and uninformative) resolutions. In opposition, increasing it to 6 (i.e. all levels) means the assumption of a perfect match between images' spatial coverage, which is a flawed premise given the misalignment frequently found among SAx slices or between image views.

Using this novel model as baseline, a second set of experiments (Table 2) evaluated, and verified, the added value of employing intensity-based augmentations (especially noticeable in longer training schemes; data not shown), a higher number of training epochs, TTA and ensemble modeling, as described in the literature [2,3]. However, these experiments also revealed a slight overfit on the LAx U-net stream with the increased number of training epochs (4th vs 3rd row), which was not observed for the SAx stream within the same range (a more complex task). This behaviour advocates for future refinements of the xUnet model, namely by considering stream-specific weight regularization schemes. Alternatively, one may also consider modifying the frequency or even adding new augmentation transforms. In this context, one evaluated the use of elastic deformations (implemented as to compute a 3D in-plane displacement field and apply it coherently to the image pair), but no overall accuracy improvement was found (data not shown) while significantly increasing the training time.

On the test set, the proposed approach corroborated its accuracy, obtaining a similar combined score over the 160 unseen test cases. Remarkably, the obtained segmentation scores compare favourably to the related literature [1,2,7], even though a direct comparison is deceiving. When dividing the test set per pathology (data not shown), no significant differences (in an one-way ANOVA) were observed on the average Dice score (and similarly for HD, except for a slightly lower score for LAx images on patients with interatrial communication). Given that two pathologies had not been included in the training set, this result ultimately demonstrates the generalisation capacity of the proposed model.

Acknowledgements. This work has been funded by national funds, through the Foundation for Science and Technology (FCT, Portugal) in the scope of the projects UIDB/50026/2020, UIDP/50026/2020 and PTDC/EMD-EMD/1140/2020. Financial support to S. Queirós from FCT (CEECIND/03064/2018) is also gratefully acknowledged.

References

1. Bernard, O., et al.: Deep learning techniques for automatic MRI cardiac multi-structures segmentation and diagnosis: is the problem solved? IEEE Trans. Med. Imaging **37**(11), 2514–2525 (2018)
2. Campello, V.M., et al.: Multi-centre, multi-vendor and multi-disease cardiac segmentation: the M&Ms challenge. IEEE Trans. Med. Imaging **40**, 3543-3554 (2021)
3. Isensee, F., Jaeger, P.F., Kohl, S.A., Petersen, J., Maier-Hein, K.H.: nnU-Net: a self-configuring method for deep learning-based biomedical image segmentation. Nature Methods **18**(2), 203–211 (2021)
4. Kingma, D.P., Ba, J.: Adam: a method for stochastic optimization. arXiv preprint arXiv:1412.6980 (2014)
5. Loshchilov, I., Hutter, F.: SGDR: stochastic gradient descent with warm restarts. arXiv preprint arXiv:1608.03983 (2016)
6. Petitjean, C., Dacher, J.N.: A review of segmentation methods in short axis cardiac MR images. Med. Image Anal. **15**(2), 169–184 (2011)
7. Petitjean, C., et al.: Right ventricle segmentation from cardiac MRI: a collation study. Med. Image Anal. **19**(1), 187–202 (2015)
8. Queirós, S., et al.: Multi-centre validation of an automatic algorithm for fast 4D myocardial segmentation in cine CMR datasets. Eur. Heart J. Cardiovasc. Imaging **17**(10), 1118–1127 (2016)
9. Queirós, S., et al.: Fast automatic myocardial segmentation in 4D cine CMR datasets. Med. Image Anal. **18**(7), 1115–1131 (2014)
10. Queirós, S., Vilaça, J.L., Morais, P., Fonseca, J.C., D'hooge, J., Barbosa, D.: Fast left ventricle tracking using localized anatomical affine optical flow. Int. J. Numer. Methods Biomed. Eng. **33**(11), e2871 (2017)
11. Ramachandran, P., Zoph, B., Le, Q.V.: Searching for activation functions. arXiv preprint arXiv:1710.05941 (2017)

12. Ronneberger, O., Fischer, P., Brox, T.: U-net: convolutional networks for biomedical image segmentation. In: Navab, N., Hornegger, J., Wells, W.M., Frangi, A.F. (eds.) MICCAI 2015. LNCS, vol. 9351, pp. 234–241. Springer, Cham (2015). https://doi.org/10.1007/978-3-319-24574-4_28
13. Tiulpin, A.: SOLT: streaming over lightweight transformations, July 2019. https://doi.org/10.5281/zenodo.3702819
14. World Health Organization: World health statistics 2018 (2018)

Deformable Bayesian Convolutional Networks for Disease-Robust Cardiac MRI Segmentation

Mitchell J. Fulton[✉], Christoffer R. Heckman, and Mark E. Rentschler

University of Colorado Boulder, Boulder, CO 80309, USA
{mitchell.fulton,christoffer.heckman,mark.rentschler}@colorado.edu

Abstract. Automating cardiovascular disease diagnosis via medical image segmentation could save many lives across the world by bringing more cost-efficient technologies to developing countries. While neural networks currently perform well on many tasks, they can often fail when deployed to other domains, such as patients with different diseases. Therefore the M&Ms-2 challenge was created to help aid this effort to develop a neural network robust to domain shift in cardiac imaging. We use deformable Bayesian convolutional networks (DBCNs) [5] inserted into the nnUNet [6] framework to approach this problem and provide a solution to this challenge. We then explore the effects of both training time and network size on generalizability for both 2D long-axis and 3D short axis cardiac MRIs and find that the optimum training and network configuration is dependent both on the dataset size and task. We then enter the final test set of the competition to achieve competitive results.

Keywords: Machine learning · Domain generalization · Heart diseases · Neural networks · Medical imaging

1 Introduction

Cardiovascular disease (CVD) is the leading cause of death worldwide, accounting for 32% of all fatalities [10]. Diagnosing some CVDs can require an MRI, CT, or ultrasound scan and manual segmentation of regions of the heart during different stages of the pumping cycle. Automating this process would save physician time, reducing costs for both patients and hospitals. However, despite recent attempts toward automating this segmentation, it is still an open challenge. Even though several datasets for segmentation exist and many different techniques have been developed, the small datasets have trouble representing the diverse patient populations and imaging hardware present throughout the world. Furthermore, there are not uniform labeling practices across these datasets, preventing accurate tests of generalization [2]. With CVD mostly affecting less wealthy nations, it is important that automated detection technology is robust to differences found throughout the world, including different medical hardware and patient populations.

© Springer Nature Switzerland AG 2022
E. Puyol Antón et al. (Eds.): STACOM 2021, LNCS 13131, pp. 296–305, 2022.
https://doi.org/10.1007/978-3-030-93722-5_32

In recent years neural networks have become the standard for new medical imaging tasks. However, in many areas of machine learning it has been shown that neural networks struggle to generalize well across domains such as patient population and imaging hardware [3]. There are three main approaches to handle this problem of domain shift: transfer learning, domain adaptation, and domain generalization. Transfer learning starts the training process using a pretrained model, then fine-tunes the weights using labeled data from the target domain. Domain adaptation trains a network using data from a source domain, then adapts using unlabeled data from the target or test domain. Domain generalization (DG) assumes that no data will be available from the target domain and learns robust features from only the source domain. As new medical data is often infeasible to procure, DG is needed to enable the use of automated medical imaging software for all patients.

While there is a significant body of research regarding domain generalization, few studies have been applied to the medical realm. This is due in part to a lack of medical domain generalization datasets, with the few current attempts in the medical realm using makeshift datasets which do not necessarily have consistent labeling procedures [2,12]. To help address this dataset scarcity, the M&Ms dataset [1] was created by collecting data from different patient populations, different medical centers, and different scanner types from across the world. The first competition using this dataset tested a network's ability to generalize both across medical centers and different scanner types. This ability is of paramount importance when deploying a medical technology and the competition yielded many impressive results, mostly centered around data augmentation (DA). In fact, the top three performing teams each used the identical networks with variations on DA to achieve their results including: hand-engineered DA [4], label propagation and style transfer [11], and histogram matching [7]. Despite these results, DA-based methods still rely on the augmentations' ability to accurately represent the shifted domain, which is not always possible. Therefore a new approach is needed for domain generalization within the medical realm that is not dependent on manual tuning of DA parameters.

The second competition using the M&Ms dataset, the M&Ms-2 competition, is testing generalizability of right ventricle (RV) segmentations across different heart diseases and utilizes both 3D short-axis (SA) scans and 2D long-axis (LA) scans. With a previous single-center RV segmentation dataset, the RV's variable size and shape as well as its ill-defined borders proved difficult for segmentation models [8]. The M&Ms-2 competition adds the difficulty of testing RV segmentation generalizability, an important task for deployment and real-world use in the medical realm. In this paper we take a new approach to domain generalization for the M&Ms-2 competition by using a novel structural method rather than solely data augmentation. We utilize a new technique known as deformable Bayesian convolutional networks (DBCNs) [5] and use these within the nnUNet framework [6]. By replacing a single convolutional layer in the nnUNet structure, DBCNs allow for a drop-in convolution replacement for generalization without the manual tuning of hand-engineered data augmentations and yield impressive results.

2   Methods

2.1   Competition Description

The M&Ms-2 competition dataset is aimed toward testing domain generalizability and robustness across different kinds of diseases that are present in the heart. The competition dataset has both 3D short-axis (SA) and 2D long-axis (LA) scans for each patient segmented at both the end-diastole (ED) and end-systole (ES) phases. As seen in Table 1, the data was separated into train, validation and test sets of 160, 40, and 160 subjects in each set, respectively. To test generalizability across different ailments, two of the eight represented diseases were not included in the train set. To train the networks, segmentations for the left ventricle, myocardium, and right ventricle were provided for the train set, with only the right ventricle (RV) segmentation output counting toward the final ranking score.

Table 1. The subjects are separated by disease for the train, validation, and test sets. Note that two of the diseases are left out of the train set to test generalizability.

Disease	# Train	# Validation	# Test
Normal subjects	40	5	30
Dilated left ventricle	30	5	25
Hypertrophic cardiomyopathy	30	5	25
Congenital arrhythmogenesis	20	5	10
Tetralogy of fallot	20	5	10
Interatrial communication	20	5	10
Dilated right ventricle	0	5	25
Tricuspidal regurgitation	0	5	25
Totals	**160**	**40**	**160**

The rankings of the competition are based off of a weighted combination of the dice coefficient (DC) and Hausdorff distance (HD) of the RV segmentations. The SA scans will have a weighting coefficient of 0.75 and the LA scans will have a weighting coefficient of 0.25. To combine these, the averaged ED and ES HD of the SA or LA will be normalized to a range of [0,1] then added to the averaged DC and multiplied by the weighting coefficient. Then the weighted SA and LA scores will then be averaged, or in equation form,

$$score = \frac{0.75(DC_{SA} + \widehat{HD}_{SA}) + 0.25(DC_{LA} + \widehat{HD}_{LA})}{2} \tag{1}$$

where DC is defined by

$$DC = \frac{DC_{ED} + DC_{ES}}{2} \tag{2}$$

and HD is defined by

$$HD = \frac{HD_{ED} + HD_{ES}}{2}. \tag{3}$$

2.2   Deformable Bayesian Convolutional Networks

Deformable Bayesian convolutional networks (DBCNs) were recently developed to approach the domain generalization problem on small medical datasets. They combine the generalizability and quick training of Bayesian convolutions with the increased accuracy and receptive field size of deformable convolutions [5]. As opposed to other methods that require significant structural differences DBCNs use a single drop in replacement for a convolutional layer to achieve significant generalizability gains, even without any data augmentation in the training.

To generalize well and learn quickly, Bayesian convolutions rethink a standard convolution by learning a weight distribution rather than a single weight value. This distribution is formed by having a mean and variance convolution describe a normal distribution to represent a weight. The operation used to achieve this is described by

$$b = A * \mu + \epsilon \odot \sqrt{A^2 * (\alpha \odot \mu^2)} \tag{4}$$

where b is an activation, $A * \mu$ is the mean convolution, and $\epsilon \odot \sqrt{A^2 * (\alpha \odot \mu^2)}$ is a sampling of $\epsilon \sim \mathcal{N}(0,1)$ from the standard deviation [9]. Having a weight distribution allows for each subject during training to be thought of as a sample from a posterior distribution, where during training the Kullback-Leibler (KL) divergence is to be minimized. These new weights not only speed up the training of the network, but help with domain generalization.

Deformable convolutions change standard convolutions by learning a sampling offset in addition to the standard convolution weights. This convolution operation can be described by

$$y(p) = \sum_{k=1}^{K} w_k \cdot x(p + p_k + \Delta p_k) \cdot \Delta m_k \tag{5}$$

where $x(p+p_k)$ is a standard convolution, Δp_k and Δm_k are the learnable offsets and a modulation scalar, w_k is the convolution weights, and $y(p)$ is the final activation [13]. The deformable convolution has shown increases in performance as a drop-in replacement for standard convolutions. It also allows for a larger and flexible receptive field without the need for extra convolutional layers, especially useful when much of the image space is not relevant to the desired target.

To create DBCNs and get benefits from both Bayesian and deformable convolutions, the mean and variance convolutions of Eq. 4 are replaced by the deformable convolutions of Eq. 5, sharing the learned deformable offsets. These operations can then directly replace standard convolutions, with the weight distributions being trained by adding a KL-divergence term into the network's loss function.

2.3   Implementation

To apply our DBCN to the M&Ms-2 challenge we modified the nnUNet [6] framework for biomedical image segmentation and replaced the lowest, or bottleneck, layer with a single deformable Bayesian convolution. We then trained two separate networks, one for LA images and one for SA images. Each network consisted of five downsampling layers, a DBCN bottleneck layer, and five upsampling layers with skip connections between the down- and upsampling layers. Each of the layers consisted of either two or three Convolution-InstanceNorm-LeakyReLU (Conv-IN-LReLU) blocks, with the first block having a stride of two for downsampling. The instance norm was chosen as it yielded better results than the batch norm, likely due to both the batch norm and the deformable Bayesian convolutions providing a smoothing effect to oversmooth the final output.

To train the network we used the default training procedure of nnUNet, or SGD with momentum and a combination dice and cross-entropy loss with an initial learning rate of 0.01. To train the Bayesian distributions, we added the KL-divergence of the Bayesian weights to the loss function and multiplied it by a weight coefficient, β. We chose β to be $1e - 7$, or a value such that $0.5 \leq \beta * KL \leq 1$ at the end of the first epoch. Even though DBCNs have been shown to perform well under no data augmentation, they have slight performance gains if augmentations are added [5]. We chose to use identical data augmentation to [4], which uses many operations (added noise, brightness, affine deformations, etc.) tuned to perform well in the first M&Ms challenge.

For both the SA and LA networks we initially trained on both all three labels (LV, MYO, RV) and on only the tested RV labels and compared the results. The LA network performed around thirty points better when trained on the LV, MYO, and RV even though only the RV was the only tested label. We believe that this produced the best results because it helped prevent overfitting in the small dataset. For the SA network we found that only training on the RV produced similar training performance as training on all three labels, but yielded a roughly 1.5 point increase on the validation set. We believe this was due to the variation across the depth of the 3D images preventing the overfitting seen in the LA data while allowing the network to focus on one task only. We then ran a five fold cross-validation over the LA or SA training data for each respective network to produce a final model to test.

3   Experiments and Results

Since deformable Bayesian convolutional networks (DBCNs) are a new technique, prior to our final submission we experimented with various training procedures and network structures. We first tested the effects of different training lengths and two different network sizes, then used our best configurations for the final test set.

3.1   Training Length

To test the effects of training length on generalization performance we trained the first fold of our five fold cross-validation for every 50 epochs up to 400 epochs, then every 100 epochs up to a maximum number of 1000 epochs. An epoch was defined as training on 250 batches with the batch size being 32 2D slices for the SA network and 16 2D images for the LA network. We then chose to run the full five fold cross-validation for the networks the four highest training scores, or 250, 300, 600, and 1000 epochs.

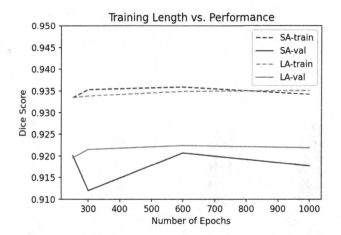

Fig. 1. Of the four training lengths (250, 300, 600, 1000) that performed best on the first fold, all had varied validation set performance after training all five folds. However, when the networks were trained longer the training performance was a better predictor of validation set performance.

As can be see in Fig. 1, there is a significant drop in validation performance versus training performance for all four training lengths, which is to be expected. It can also be seen that at lower training lengths, the training performance is a less reliable indicator of the validation performance. While it is slightly more reliable for the LA scans, this is to be expected as the variation between images in the 2D LA dataset is less than the volumetric slices in the SA scans. In addition, the LA scans continue to improve over longer training times and we believe this is due to being trained on three labels rather than one, so the fine-tuning of features takes longer. For a good tradeoff between performance and training time, we decided to train our networks for 600 epochs.

3.2   Network Size

After deciding on a training procedure, we then tested the effects of two different network sizes on the generalization performance. The difference between the two networks is described by two parameters: the maximum number of features in the network and the number of Conv-IN-LReLU blocks per layer. In the smaller network, the initial layer has 30 channels with each layer doubling the number of channels until the maximum number of 480 channels is reached. Each layer in this network consists of two Conv-IN-LReLU blocks, other than the bottleneck layer and the transpose convolutions in the decoder. In the larger network, the initial number of features is 32, which doubles until the maximum number of 640 channels is reached. In each of the layers for the larger network, the number of Conv-IN-LReLU blocks was increased to three.

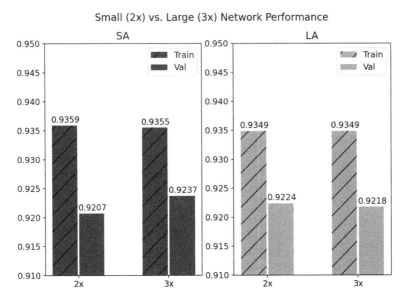

Fig. 2. Between the training and validation sets the smaller network with 2 Conv-IN-LReLU blocks per layer (2x) has a 22% smaller drop in performance in the LA vs the SA data. The large network with 3 blocks per layer (3x) is the opposite with a 5% smaller drop in validation performance for the SA than the LA images.

The performance for both the smaller (2x) and larger (3x) networks is shown in Fig. 2. For both SA and LA data the training performance was nearly identical between the 2x and 3x networks. However, the validation performance for the 3x yielded better results for the SA images while the validation performance for the 2x yielded better results for the LA images.

3.3   Final Test Results

For the final testing phase two submissions were allowed, so we submitted both the 2x and 3x networks that had been trained for 600 epochs. The results of the test submissions are shown in Table 2 alongside the validation results and the percent change for each network. The percent differences are much larger in the HD score which may be due to it not being part of the loss function during training. It is interesting to note that between the two network sizes, the one producing the better score changes in two of the categories (DC-LA and HD-SA) when evaluated on the test set. We believe the shift in validation versus test

Table 2. The 2x and 3x network both show large differences between the validation and test set. In two of the four scoring categories the best performing network changes. This shift may be due to patient disease distribution differences, as seen in Table 1. The larger performance drop in HD is likely due to only DC being part of the loss function. Note that a higher HD is worse and shown as a negative percentage.

Phase	Network	DC-SA	DC-LA	HD-SA	HD-LA
Validation	2x	0.9207	0.9224	9.0382	5.5909
	3x	0.9237	0.9218	8.8528	5.4494
Test	2x	0.9219	0.9174	9.8208	6.5264
	3x	0.9223	0.9192	10.1451	6.0810
Change (%)	2x	+0.1	−0.5	−8.7	−16.7
	3x	−0.2	−0.3	−14.6	−11.6

Table 3. Our final test set submission scored separated by performance shows the strengths and weaknessed of the 2x network used for SA and the 3x used for LA. The 2x can overgeneralize and have trouble on seen categories (e.g. Interatrial Communication DC or Tetralogy of Fallot HD) and the 3x can cause some overfitting to the training categories. *These diseases only present in test set*

Disease	DC-SA	DC-LA	HD-SA	HD-LA
Normal subjects	0.9192	0.9305	8.8741	5.3060
Dilated left ventricle	0.9236	0.9149	10.3452	5.8327
Hypertrophic cardiomyopathy	0.9319	0.9300	9.7751	5.0517
Congenital arrhythmogenesis	0.9383	0.9307	8.4113	5.9288
Tetralogy of fallot	0.9265	0.9140	11.1666	7.3339
Interatrial communication	0.8888	0.9229	11.5157	6.5408
Dilated right ventricle*	0.9224	0.8984	10.3090	7.2910
Tricuspidal regurgitation*	0.9180	0.9162	9.3379	6.4543
Totals	**0.9219**	**0.9192**	**9.8208**	**6.0810**

performance is due to the different makeup of the test set with respect to the proportion of subjects with each disease, as can be seen in Table 1.

The per-disease scores for the test set highlight both the strengths and weaknesses of the 2x and 3x networks. These results are shown in Table 3 for the final submission using a 2x SA network and 3x LA network. The 2x network excelled in its performance for the unseen disease categories but may have overgeneralized and underperformed in the seen categories, like the DC-SA of the Interatrial Communication or the HD-SA of the Tetralogy of Fallot. Conversely, the 3x network performed well on all of the seen categories but underperformed on the unseen categories. We hypothesize that the addition of extra standard convolutional blocks in the 3x network caused the DBCN to lose some of its generalization power, causing it to overfit slightly to the training set.

4 Conclusion

In this paper we have used deformable Bayesian convolutional networks (DBCNs) to provide a solution to the M&Ms-2 challenge. Using the DBCN operation as a simple drop-in replacement for a standard convolution yielded competitive results in the generalization challenge of the final test set. We also explored the effects of training length on generalization performance as well as network size. Using our final submission's training procedure and configurations we also tested the 2D nnUNet with only standard convolutions. Even with heavy data augmentation it achieved scores of 0.004 DC and 106.7 HD for SA and 0.7476 DC and 17.83 HD for LA. This highlights the generalization power of DBCNs regardless of well-tuned data augmentation. While further tuning of the data augmentation and network configuration would likely achieve better results in the standard nnUNet network, for this competition we chose to only pursue further tuning of the DBCNs. In the future we plan to calculate confidence inherently present in DBCNs to aid in training and produce better estimates. We also plan to use DBCNs in other network configurations and test on other datasets to further show theie generalization power. We hope that in the future DBCNs will provide an easily implemented tool to help new medical technologies generalize to all populations without the need to gather extra training data.

Acknowledgements. This research was supported in part by the National Science Foundation (NSF) (Grant No. 1849357). This work also used the Extreme Science and Engineering Discovery Environment (XSEDE), supported by National Science Foundation grant number ACI-1548562.

References

1. Campello, V.M., et al.: Multi-centre, multi-vendor and multi-disease cardiac segmentation: the m&ms challenge. IEEE Trans. Med. Imaging **40**, 1 (2021). https://doi.org/10.1109/TMI.2021.3090082
2. Chen, C., et al.: Deep learning for cardiac image segmentation: a review. Front. Cardiovasc. Med. **7**, 25 (2020)
3. D'Amour, A., et al.: Underspecification presents challenges for credibility in modern machine learning (2020)
4. Full, P.M., Isensee, F., Jäger, P.F., Maier-Hein, K.: Studying robustness of semantic segmentation under domain shift in cardiac MRI (2020)
5. Fulton, M., Heckman, C., Rentschler, M.: Deformable bayesian convolutional networks enable domain shift robustness in small training sets (2021). in review
6. Isensee, F., Jaeger, P.F., Kohl, S.A., Petersen, J., Maier-Hein, K.H.: NNU-NET: a self-configuring method for deep learning-based biomedical image segmentation. Nature Methods **18**(2), 203–211 (2021)
7. Ma, J.: Histogram matching augmentation for domain adaptation with application to multi-centre, multi-vendor and multi-disease cardiac image segmentation (2020)
8. Petitjean, C., et al.: Right ventricle segmentation from cardiac MRI: a collation study. Med. Image Anal. **19**(1), 187–202 (2015)
9. Shridhar, K., Laumann, F., Liwicki, M.: A comprehensive guide to bayesian convolutional neural network with variational inference (2019)

10. WHO: Cardiovascular diseases (CVD). https://www.who.int/news-room/fact-sheets/detail/cardiovascular-diseases-(cvds). Accessed 26 July 2021
11. Zhang, Y., et al.: Semi-supervised cardiac image segmentation via label propagation and style transfer (2020)
12. Zhou, K., Liu, Z., Qiao, Y., Xiang, T., Loy, C.C.: Domain generalization: a survey. arXiv preprint arXiv:2103.02503 (2021)
13. Zhu, X., Hu, H., Lin, S., Dai, J.: Deformable convnets v2: more deformable, better results (2018)

Consistency Based Co-segmentation for Multi-view Cardiac MRI Using Vision Transformer

Zheyao Gao and Xiahai Zhuang[✉]

School of Data Science, Fudan University, Shanghai, China
zxh@fudan.edu.cn

Abstract. Segmentation of cardiac structures in magnetic resonance imaging is essential to the diagnosis of many cardiovascular diseases. However, sometimes it is challenging to accurately define the right ventricle (RV) structure due to the complex texture. It requires the collaboration of both short-axis (SA) images and long-axis (LA) images. Current deep learning methods trained with single-view data neglect the spatial relations between SA and LA images and could fail at the basal and apex plane of the RV. In order to properly handle the geometrical relations, we proposed a consistency based co-training method that involves an extra penalty for confusing images at the basal and apex plane. At test phase, the anatomical relations are also used for post-processing which further improves the robustness of segmentation. Besides, we incorporate the transformer network in U-Net architecture to enhance its ability to model long-range dependency. We evaluated the proposed method in the M&Ms-2 challenge and obtained promising performance for the segmentation of both SA and LA images.

Keywords: Co-training · Transformer · Semantic segmentation · Multi-view MRI

1 Introduction

Cardiovascular disease is one of the main causes of death and disability worldwide. Clinically, diagnosis of cardiac disease is commonly based on cardiac magnetic resonance (CMR) images. Accurate segmentation of CMR images is important to the assessment of many major cardiovascular diseases. The process of manual segmentation requires the clinician to accurately and consistently annotate the cardiac boundaries across all image slices and cardiac phases [1]. Such annotation is time-consuming and suffers from interrater-variability. Therefore, computer-aided automatic segmentation approaches have been studied for a long time.

Recently, a plenty of deep learning methods have been proposed for the segmentation of cardiac structure. In the past few years, convolutional neural networks (CNN) have been dominant in medical image segmentation. Especially, U-Net [8], which consists of an encoder-decoder structure with skip connections

© Springer Nature Switzerland AG 2022
E. Puyol Antón et al. (Eds.): STACOM 2021, LNCS 13131, pp. 306–314, 2022.
https://doi.org/10.1007/978-3-030-93722-5_33

and its variants have achieved tremendous success in many medical applications including CMR image segmentation [10].

Despite the efficiency and strong representation power of CNN in image segmentation, there still exhibits some deficiencies. The local receptive field limits the modeling of long-range dependency, which is important in semantic segmentation. Besides, the weights of convolution kernels are independent of input, which is not flexible and may restrict its ability to model heterogeneous distributions [7]. In order to handle the above limitations, some studies propose to use self-attention mechanisms on CNN features [9]. To further improve the representation ability, transformers, designed for machine translation, have been applied to image segmentation [12]. The transformer architecture is a fully attention network that is aimed at modeling global representations. Most recently, several works have adopted the transformer structure in medical image segmentation and outperforms the state-of-the-art CNN networks in many applications [4]. However, the optimal structure of transformer-based segmentation network is still an open question.

Apart from network structure, exploiting the geometrical prior knowledge is also important in CMR image segmentation. Experienced clinicians usually assess cardiac structure from multiple standard views. Inspired by this, many studies have been proposed to focus on multi-view CMR image segmentation. One of the solution is to ensemble a 2D and a 3D network [6] and thus the multi-view relations are captured by the 3D network. However, training a 3D network is not computationally efficient. Another approach is to incorporate the geometric prior knowledge in the 2D segmentation network through a learned representation of LA images [3]. However, all the above methods model the spatial relations of multi-view images in an implicit way, which is highly relied on the representation ability of the network. Therefore, we propose an co-training method that explicitly utilizes the anatomical priors to improve the segmentation accuracy of SA images.

The main contributions of this paper are the following: a) We proposed a consistency based co-segmentation method that explicitly utilizes the geometric relations for the segmentation of multi-view CMR images. b) We re-examined the transformer architecture for semantic segmentation and proposed a transformer enhanced network in U-Net structure. c) We evaluated the proposed method in M&Ms-2 challenge, demonstrating that the proposed method produced more accurate results than other state-of-the-art networks in the same experiment setting.

2   Method

The proposed method consists of two novel mechanisms: 1) A consistency based co-segmentation strategy that explicitly incorporates the anatomic prior knowledge of multi-view CMR images. 2) A transformer enhanced U-Net architecture that establishes self-attention mechanisms on convolutional features to model global representations.

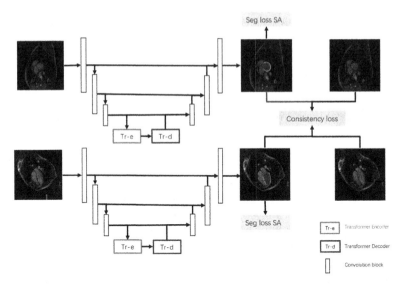

Fig. 1. Overview of the co-training framework. A pair of SA and LA image are input to two transformer enhanced U-Net of same structure and the segmentation loss is calculated based on the label. Beside the supervised loss, given corresponding masks derived from the input images, a consistency loss is also calculated between two predictions

2.1 Co-segmentation

Before the co-training process, masks indicating the intersection line between paired SA and LA images (shown in Fig. 1) should be calculated through coordinates transformation. In this work, we assume the paired SA and LA images have already been aligned such that the scanner coordinates of both images are matched. In this case, given a point (i, j, k) in voxel coordinates, the scanner RAS+ coordinates (x, y, z) corresponding to the voxel coordinates can be derived by:

$$\begin{bmatrix} x \\ y \\ z \\ 1 \end{bmatrix} = M \begin{bmatrix} i \\ j \\ k \\ 1 \end{bmatrix}, \tag{1}$$

where $M \in \mathbb{R}^{4 \times 4}$ is the affine matrix that has been adjusted and saved during registration. Thus, given a voxel coordinate $A \in \mathbb{R}^4$ in SA image, the corresponding voxel coordinate $B \in \mathbb{R}^4$ in LA image can be calculated by:

$$B = M_2^{-1} M_1 A, \tag{2}$$

where M_1 and M_2 are affine matrices of SA image and LA image. In this way, we can calculate all cross points given a pair of SA and LA images and generate a mask for each image.

Multi-view Co-training. The input data of the co-training framework is a tuple denoted by $\{X_{SA}, X_{LA}, Y_{SA}, Y_{LA}, S_{SA}, S_{LA}, g\}$, where $\{X_{SA}, X_{LA}\}$ represents the paired 2D slices, $\{Y_{SA}, Y_{LA}\}$ are segmentation labels and $\{S_{SA}, S_{LA}\}$ denotes the set of corresponding voxel coordinates on the intersection line of SA and LA slices. Apart from images, labels and coordinates, we also generate the gate value denoted by $g \in \{0, 1\}$, which is derived from the spatial position of SA slices. More specifically, the gate value is set to 1 if the slice is on the apex or basal plane and 0 elsewhere.

As shown in Fig. 1, the image pairs are simultaneously input to two U-Net like networks F_1, F_2 of same architecture. The overall loss function of the whole framework is defined as follows:

$$\mathcal{L} = \mathcal{L}_{SAseg} + \mathcal{L}_{LAseg} + g \cdot w\mathcal{L}_{consistency}, \tag{3}$$

where the w is the weight parameter that controls the importance of consistency loss. The segmentation loss $\mathcal{L}_{SAseg}, \mathcal{L}_{LAseg}$ used for SA network and LA network is cross entropy loss. And the consistency loss $\mathcal{L}_{consistency}$ is defined by:

$$\mathcal{L}_{consistency} = |F_1(X_{SA})[S_{SA}] - F_2(X_{LA})[S_{LA}]|, \tag{4}$$

where $F_1(X_{SA})[S_{SA}]$ denotes the network output at the intersection line of the input SA slice.

According to the above formulation, we incorporate extra penalty at the segmentation of basal and apex plane which is difficult to assess using SA or LA image alone. The consistency loss forces the prediction to be the same at the intersection line of paired SA and LA slice through L1 distance. In this way, the network explicitly learns extra knowledge from multi-view images and could perform better at "hard to define" regions.

Multi-view Post-processing. At test phase, to make the segmentation results of SA slice more robust to noise and artifacts, we propose to use the prediction of LA slice for amendment. Since it is easier to assess the basal and apex region on LA images, we use the prediction of LA network at the cross line to confine that of SA network. If we define the class indicator as c, the post-processing can be formulated as:

$$F'_{1c}(X_{SA}) = \begin{cases} 0, & \text{if } \forall s \in S_{LA}, \arg\max_{c'} F_{2c'}(X_{LA})[s] \neq c \\ F_{1c}(X_{SA}), & \text{if } \exists s \in S_{LA}, \arg\max_{c'} F_{2c'}(X_{LA})[s] = c \end{cases}, \tag{5}$$

where $F'_{1c}(X_{SA})$ denotes the output probability for class c after post-processing. According to Eq. (5), the prediction of class c is set to 0 if there is no prediction for class c by LA network at the intersection line. Because the SA network may confuse RV with the right atrium at basal plane which are more distinguishable in LA view, we can eliminate the inaccurate prediction by using the output of LA network.

2.2   Transformer Enhanced U-Net

CNN Encoder and Decoder. Due to the high computation complexity of transformer architecture, the proposed network uses convolutional encoder and decoder and the global representation is learned based on high-level features produced by CNN encoder. Given an input 2D slice of shape $1 \times H \times W$, we use the first three stages of ResNet18 as encoder to progressively downsample the input to $C \times H/8 \times W/8$. The decoder contains three upsample blocks using transpose convolution to resize the transformer output to original size. Each upsample block in decoder contains two convolution layers. Skip connections are applied to keep high resolution features for detailed segmentation.

Transformer Architecture. Different from TransUNet proposed in [4], we use both transformer encoder and decoder for feature representation. The detailed structure is similar to [2]. Since transformer is a sequence-to-sequence mode, we first transform the CNN encoder output to a sequence of shape $\frac{H}{8} \cdot \frac{W}{8} \times C$, where $\frac{H}{8} \cdot \frac{W}{8}$ denotes the sequence length and C is the embedding dimension of each entity in the sequence. In this way, global attentions are calculated between each embedding of 8×8 patch in original images.

Different from [2], we applied fixed positional encoding in our framework. Because unlike natural image analysis, the training data for medical image segmentation is extremely limited, it is difficult to learn an accurate positional encoding in low data scenario. And we have found in experiments that inaccurate positional encoding may cause severe shift of the segmentation results. To be more specific, the positional encoding we use is calculated by the sine and cosine functions with different frequencies [5], shown as follow:

$$\begin{cases} PE_d(pos, 2i) = sin(pos \cdot v) \\ PE_d(pos, 2i + 1) = cos(pos \cdot v) \end{cases}, \tag{6}$$

where pos is the spatial position, i is the dimension, $d \in \{h, w\}$ indicates hight and width dimensions and the frequency $v = 1/10000^{2i/\frac{C}{3}}$. PE_h and PE_w are then concatenated as the 2D positional encoding and combined with the input sequence by element-wise summation.

3   Experiments

3.1   Dataset

The dataset from M&Ms-2 challenge is composed of 360 patients with different right ventricle and left ventricle pathologies as well as healthy subjects. All subjects were scanned in three clinical centres from Spain using three different magnetic resonance scanner vendors (Siemens, General Electric and Philips).

The training set contains 200 annotated images from four different centres. The CMR images have been segmented by experienced clinicians from the respective institutions, contours for the left (LV) and right ventricle blood pools, as well as for the left ventricular myocardium (MYO). Among the training set, 40 cases are kept as validation data whose annotations are not available locally during validation and test phase. The test set includes 160 cases with same pathology distribution as validation data. Two pathologies (Tricuspidal Regurgitation and Congenital Arrhythmogenesis) is not present in the training set but in the validation and testing sets to evaluate generalisation to unseen pathologies.

3.2 Implementation Details

Preprocess and Data Augmentation. Before training, all input slices were resampled to 1.25 mm/px along each direction. And we applied intensity and geometric augmentation to both SA and LA images, which includes flipping, rotation, Gaussian blurring, additive Gaussian noise, contrast adjustment and brightness adjustment.

Training and Post-processing. For training, we used the Adam optimizer to update the network parameters (weight decay = 1e-4, momentum = 0.9). The initial learning rate was set to 1e-3 and the learning rate was updated using the poly policy as implemented in [11] where current learning rate equals to the base one multiplying $(1 - \frac{iter}{max\_iter})^{power}$. The power in our experiments was set to 0.9. The weight for consistency loss was set to a constant value 1 during training. In our experiment, *we only employed the consistency loss to update SA network* since we found that LA network could properly segment the basal and apex region without extra loss. For post-processing, Besides the multi-view constraint as described in Sect. 2.1, we also removed inaccurate prediction caused by unexpected disturbation in the test data by keeping the largest connected component.

3.3 Results

We first evaluated our method on the training set that consists of 160 cases, where 128 cases were used for training and the rest for validation. We employed dice similarity for evaluation.

Table 1. Results on training set.

Method	SA			LA		
	LV	MYO	RV	LV	MYO	RV
Res-UNet	0.8605	0.7915	0.8469	0.9104	0.8128	0.8704
U-Net++ [13]	0.9043	0.8123	0.8812	0.9047	0.8098	0.8877
TU-Net	0.936	0.8552	0.9013	0.9344	0.8593	0.8948
Proposed	0.9412	0.8475	0.9257	0.9408	0.8645	0.9058

As shown in Table 1, we compared the proposed transformer enhanced U-Net (TU-Net) with other state-of-the-art network architectures. Res-UNet in the table

SA prediction raw   after process   LA prediction and mask

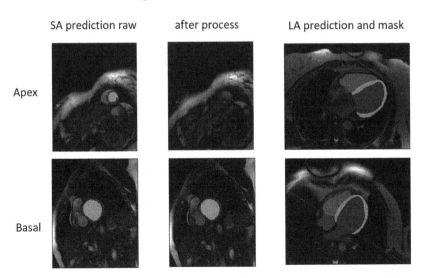

Fig. 2. The segmentation of SA slice at basal or apex plane may be confusing, while the segmentation of LA slice is almost accurate. After post-process via the relations with LA prediction, SA predictions are recalibrated

has the same structure in convolutional blocks but without transformer block. The third method means that proposed network is trained without consistency loss and the post-processing method is not applied during evaluation neither. It is observed that the incorporation of transformer network could obviously improve the performance of fully convolutional network. And the co-segmentation method proposed in this work further elaborates the result.

Table 2. Ablation study on validation set.

Method	SA		LA	
	Dice	Hausdorff	Dice	Hausdorff
Res-UNet	0.8519	16.28	0.8596	9.559
TU-Net	0.8585	15.05	0.8698	8.055
TU-Net w/co-tr	0.8697	14.93	0.8844	7.983
TU-Net w/post	0.8797	13.81	0.882	8.055
Proposed	0.8812	13.66	0.8844	7.983

We then performed an ablation study for our method on the validation set, whose annotations were held out by the organizers. The output from our prediction pipeline were submitted to the organizers to get the results, including average dice similarity and Hausdorff distance (mm). Table 2 shows that the transformer network slightly improve the results for both SA and LA images. And the co-segmentation strategy obviously enhance the robustness of SA segmentation on validation set. We have shown some predictions on validation set which are refined by the post-process strategy (see Fig. 2).

Table 3. Results on test set.

Method	SA		LA	
	Dice	Hausdorff	Dice	Hausdorff
TU-Net w/post	0.8264	16.75	0.8754	9.049
Proposed	0.8329	16.45	0.8689	9.689

At test phase, we submitted our code for online test twice with different model. The reported results only included the dice similarity and Hausdorff distance of RV prediction for both SA and LA images. As shown in Table 3, the first model was not trained with consistency loss but was post-processed. The results further proved that even on unseen test data, the model trained with consistency loss could still achieve more accurate results on SA images.

4 Conclusion

In this work, we designed a co-segmentation strategy that employed the geometric prior knowledge to derive consistency penalty for the segmentation of obscure and complex slices of SA images. Different form previous work, our method explicitly models the relations and does not rely on the implicit representation learning of LA images. The results showed that models co-trained with the proposed consistency loss could perform more robust segmentation and the post-processing that applied multi-view relations also refined the prediction for SA images. Besides, we developed a transformer enhanced U-Net with fixed position encoding that was able to achieve superior performance in all of our experiments.

References

1. Campello, V.M., et al.: Multi-centre, multi-vendor and multi-disease cardiac segmentation: the m&ms challenge. IEEE Trans. Med. Imaging **40**, 1 (2021). https://doi.org/10.1109/TMI.2021.3090082
2. Carion, N., Massa, F., Synnaeve, G., Usunier, N., Kirillov, A., Zagoruyko, S.: End-to-end object detection with transformers (2020)
3. Chen, C., Biffi, C., Tarroni, G., Petersen, S., Bai, W., Rueckert, D.: Learning shape priors for robust cardiac MR segmentation from multi-view images. Med. Image Comput. Comput. Assist. Interv. MICCAI **2019**, 523–531 (2019)
4. Chen, J., et al.: Transunet: transformers make strong encoders for medical image segmentation (2021)
5. Gehring, J., Auli, M., Grangier, D., Yarats, D., Dauphin, Y.N.: Convolutional sequence to sequence learning (2017)

6. Pop, M., et al. (eds.): STACOM 2017. LNCS, vol. 10663. Springer, Cham (2018). https://doi.org/10.1007/978-3-319-75541-0

7. Khan, S., Naseer, M., Hayat, M., Zamir, S.W., Khan, F.S., Shah, M.: Transformers in vision: a survey (2021)

8. Ronneberger, O., Fischer, P., Brox, T.: U-net: convolutional networks for biomedical image segmentation (2015)

9. Schlemper, J., et al.: Attention gated networks: learning to leverage salient regions in medical images (2019)

10. Yu, Q., Xie, L., Wang, Y., Zhou, Y., Fishman, E.K., Yuille, A.L.: Recurrent saliency transformation network: incorporating multi-stage visual cues for small organ segmentation (2018)

11. Zhao, H., Shi, J., Qi, X., Wang, X., Jia, J.: Pyramid scene parsing network (2017)

12. Zheng, S., et al.: Rethinking semantic segmentation from a sequence-to-sequence perspective with transformers (2021)

13. Zhou, Z., Siddiquee, M.M.R., Tajbakhsh, N., Liang, J.: Unet++: a nested u-net architecture for medical image segmentation (2018)

Refined Deep Layer Aggregation for Multi-Disease, Multi-View & Multi-Center Cardiac MR Segmentation

Di Liu[1]($\boxtimes$), Zhennan Yan[2], Qi Chang[1], Leon Axel[3], and Dimitris N. Metaxas[1]

[1] Department of Computer Science, Rutgers University, Piscataway, NJ 08854, USA
`di.liu@rutgers.edu`
[2] SenseBrain Technology, Princeton, NJ 08540, USA
[3] Department of Radiology, New York University Grossman School of Medicine, New York, NY 10016, USA

Abstract. Efficient and accurate segmentation of the heart is important for analysis of cardiac magnetic resonance imaging (MRI). Although many convolutional neural networks (CNNs) have been proposed to address cardiac segmentation in cine MRI, the task is still an open and challenging problem due to highly complex and variable cardiac shape in various pathologies and the ill-defined borders, particularly near the base and apex of the heart. Existing methods typically only segment the heart on either short-axis images or long-axis images without employing any complementary information from multi-view images for better segmentation of the ventricles. This can be problematic in the basal region, where the ventricles can be easily confused with the atria. In this paper, we propose a novel framework to jointly segment the ventricles in both short- and long-axis images. The method has two stages: 1) segment the two views independently and then 2) refine their segmentations by fusing the complementary information from the other views. The proposed method was evaluated on the MICCAI 2021 Multi-Disease, Multi-View & Multi-Center Cardiac MR Segmentation Challenge (M&Ms-2) dataset. The result shows improvement on the segmentation of the left and right ventricular cavities and the myocardium for both short-axis and long-axis cardiac MR images compared to the conventional state-of-the-art segmentation methods.

Keywords: Cardiac MRI · Segmentation · Deep learning

1 Introduction

Cardiac magnetic resonance (MR) imaging is widely used in clinical diagnosis and offers a non-invasive way for the evaluation of cardiac function and internal structure [1,6,13]. Cardiac MR imaging is often considered as the gold standard for the quantification of cardiac function through its high-resolution imaging of multiple internal structures such as the left ventricle (LV), right ventricle (RV), and myocardium (Myo). However, disease diagnosis and assessment of

© Springer Nature Switzerland AG 2022
E. Puyol Antón et al. (Eds.): STACOM 2021, LNCS 13131, pp. 315–322, 2022.
https://doi.org/10.1007/978-3-030-93722-5_34

cardiac function require careful examination and delineation of the related cardiac structures in multiple phases, including end systole (ES) and end diastole (ED), which is labor-intensive and time-consuming. Recently, a number of approaches have been proposed to address the automatic segmentation task. The traditional approaches include deformable models [11] and level set methods [4]; however, these classic methods are limited in their robustness for external testing cases [14]. Recently, a variety of learning-based methods have been introduced for cardiac MR segmentation [15,17,20]. A temporal regression network using convolutional neural networks (CNNs) and a recurrent neural network was initially proposed by [8] to detect the ED and ES frames from the cardiac cycle. Another method, combining deep learning and level sets was introduced by [12] for LV segmentation. Avendi [5] integrated cardiac segmentation and pathology evaluation into an end-to-end pipeline, while Khened [7] introduced an ensemble approach for the automatic diagnosis of cardiac disease. These learning-based methods have shown promising results in assessing different diseases from cardiac MR images by different scanner vendors. More recently, a network based on Deep Layer Aggregation (DLA) [19] has been proposed to merge features from shallow layers to deep layers iteratively, to better fuse information across layers. In further development, Li et al. [9] successfully employed this DLA network to address the LV segmentation and quantification and won the LVQuan18 MICCAI challenge. Li et al. [10] also adapted and improved it by integrating channel attention block (CAB) [18], obtaining promising results on the ACDC 2017 challenge dataset [1].

These previously developed deep networks have shown advantages in cardiac MR segmentation. However, most methods are primarily 2D, because short-axis sequences are acquired with relatively large between-slice distance, which leads to inevitable misalignment between slices [16] due to the gaps (as well as to potential breath-hold inconsistency), and thus the performance of 3D networks is inferior compared to 2D networks [5]. Although 2D methods have achieved good results for the middle region of the ventricles, the apical and basal regions are still challenging, because the apical region is relatively small in the image and the basal region has unclear boundaries between the ventricles and the atria.

In this paper, we propose a novel cardiac MR segmentation method for both short- and long-axis images. It has a 2D segmentation stage followed by a refinement stage. This refinement fuses cross-view information and can adjust minor errors of the delineation in the single view. The results of the proposed method on the MICCAI 2021 Multi-Disease, Multi-View & Multi-Center Right Ventricular Segmentation in Cardiac MRI (M&Ms-2) challenge [2] show that the proposed method contributes to the improvement of LV, RV, and Myo segmentation.

2   Methods

2.1   Overview of the Framework

The overview of the proposed framework which consists of two stages, is presented in Fig. 1. The short-axis and long-axis images are segmented independently first, and then are refined jointly to adjust minor errors, which could be hard to tackle in only a single view. There are four networks in total, including two segmentation networks in the first stage for SA and LA, respectively, and two refinement networks in the second stage. These networks share the same architecture, except that the refinement networks take more channels as input, including the original image, the corresponding 2D segmentation, and the aligned segmentation from the other view. In this work, the four networks were trained independently.

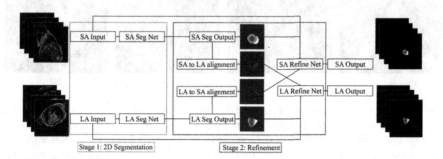

Fig. 1. Overall framework of the proposed refined deep layer aggregation for cardiac MR segmentation.

2.2 Modified Deep Layer Aggregation for Cardiac MR Segmentation

The backbone of the networks in our framework is similar to the modified Deep Layer aggregation (DLA) used in [10], which is able to extract discriminative and representative features from the cardiac MR image slices. Specifically, a backbone of DLA-34 stride-2 network [19] is implemented, followed by a bottleneck layer with 32 1×1 convolutional kernels and an upsample layer with 4 de-convolutional kernels of size 4, stride 2, and padding size 1 in order to produce the same-size output as the input images.

2.3 Refinement Through Cross-View Alignment

Since views of both the short- and long- axis are provided and require segmentation, the standard segmentation network is followed by a refinement stage to fuse the complementary features from the other view. We first compute the intersections of the SA slices and the LA slice (only a 4-chamber view is included in the M&Ms-2 dataset) based on their world coordinates, as illustrated by the blue lines in Fig. 2(a) and (b). Taking the short-axis image (Fig. 2a) as an example, through the intersection lines, we can extract segments of the masks from the LA slice and map them to the SA image space, as shown in Fig. 2(c). Then, we concatenate the SA image, its corresponding SA prediction (3 labels for LV, RV and Myo), and the aligned LA prediction (the same 3 labels) together to form a 7-channel input to train the refinement network for SA. Similarly, we can align the masks from SA image spaces into the LA image space, as shown in Fig. 2(d) and train another refinement network for LA. In this way, the first-stage prediction results of the cardiac structures can be refined to produce the final predictions.

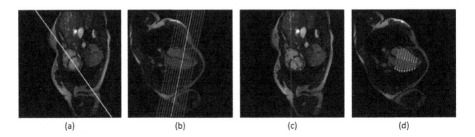

Fig. 2. Cross-view alignment of SA and LA images, as well as corresponding masks. The blue lines in (a) and (b) are the intersections of the corresponding SA and LA images. The green lines in (b) are the intersections of other SA slices. The line segments in (c) and (d) are aligned masks extracted from the other view (green: LV, magenta: Myo, cyan: RV). (Color figure online)

3   Experiments

3.1   Dataset

We validated our method on the Multi-center, Multi-vendor, and Multi-disease Cardiac Image Segmentation (M&Ms-2) challenge dataset [2]. This challenge cohort contains a training set of 160 data sets with ground truth, and an online validation set of 40 data sets without ground truth. The datasets were collected from subjects with different RV and LV pathologies in three clinical centers from Spain using three different magnetic resonance scanner vendors (Siemens, General Electric and Philips). The acquired cine MR images were delineated by experienced clinical doctors, including the left ventricle (LV), right ventricle (RV) and the left ventricular myocardium (Myo). The whole cine MR sequences with varied number of frames are provided, but only ED and ES frames are labeled in both short-axis and long-axis views. Although the task of the challenge only focuses on the RV segmentation, all the three labels could be used for training. There is also a offline testing set of 160 data sets for the final ranking, including the 40 online validation examples. The training set has data from subjects of Normal, Dilated Left Ventricle (DLV), Hypertrophic Cardiomyopathy (HC), Congenital Arrhythmogenesis (CA), Tetralogy of Fallot (TF) and Interatrial Comunication (IC). The online validation set and final testing set include two new pathologies Dilated Right Ventricle (DRV) and Tricuspid Regurgitation (TR).

3.2 Implementing Details and Results

The proposed method was trained on the 160 training image sets with manual labels, following a strategy of 5-fold cross validation. We randomly split the 160 training image sets into five groups for the cross validation. Specifically, 128 image sets are selected from the 160 image sets and used for training, while the rest 32 image sets for evaluation. We used all the three labels for training. The models were trained by a combination of Dice loss and focal loss, with Adam optimization in 300 epochs. The learning rate was initially 0.001 and decreased by 99% per epoch. All the experiments were implemented with PyTorch on a Linux system with eight Quadro RTX 8000 GPUs. The data was resampled to 1.25 mm resolution in preprocessing. During training, data augmentation strategies were also utilized to avoid overfitting, including random histogram matching, rotation, shifting, scaling, elastic deformation and mirroring.

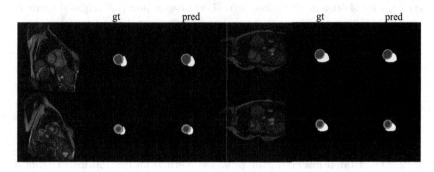

Fig. 3. Two examples for the short-axis segmentation. GT is the ground truth mask, pred is the output from the proposed segmentation model.

Example results of the proposed method are randomly selected from the 32 validation data of fold 2 and are presented in Fig. 3 for the short-axis segmentation and Fig. 4 for the long-axis segmentation. We also present a comparison of the proposed method with different state-of-the-art cardiac MR segmentation methods, including the 2D U-net [5], the modified DLA [10], and a semi-supervised method [3] which can utilize the unlabeled frames in the cine MR sequence. Here we report the results for the online validation set with 40 images used in the public leaderboard. The details are shown in Table 1. We can see that the proposed method outperforms the other methods in terms of both Dice and the Hausdorff coefficients for the RV segmentation task. Although the modified DLA has shown promising performance on the ACDC 2017 dataset in [10] as well as the M&Ms-2 challenge dataset in this work, our proposed refinement method can further improve the accuracy, by fusing the complementary information from different views.

Table 2 illustrates the performance of the proposed method for the final testing set of 160 data sets, obtained offline by submitted code and trained models.

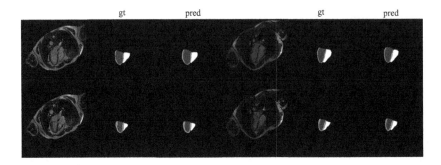

Fig. 4. Two examples for the long-axis segmentation. GT is the ground truth mask, pred is the output from the proposed segmentation model.

Table 1. Quantitative comparison of related methods on the 40 online validation data sets, in terms of Dice and Hausdorff distance of RV. The bold text indicates the best performance for short-axis (SA), long-axis (LA) images, and the weighted average.

Methods	Dice			Hausdorff (mm)		
	SA	LA	Average	SA	LA	Average
2D U-net [5]	0.8938	0.8935	0.8938	12.80	9.18	11.89
Semi-supervised [3]	0.8891	**0.9064**	0.8934	12.84	6.18	11.18
Modified DLA [10]	0.9065	**0.9064**	0.9065	11.23	6.18	9.97
Proposed	**0.9088**	0.9052	**0.9079**	**10.65**	**6.02**	**9.49**

Table 2. Quantitative results of the proposed method on the 160 final testing data sets, in terms of Dice and Hausdorff distance of RV.

Methods	Dice (%)			Hausdorff (mm)		
	SA	LA	Average	SA	LA	Average
Proposed	0.9170	0.9122	0.9158	10.69	6.69	9.69
Normal	0.9199 ± 0.06	0.9347 ± 0.03	–	9.03 ± 4.3	5.23 ± 2.8	–
DLV	0.9159 ± 0.07	0.901 ± 0.07	–	12.6 ± 8.3	7.09 ± 4.4	–
TF	0.9143 ± 0.04	0.9204 ± 0.04	–	11.99 ± 3.5	7.35 ± 3.8	–
IC	0.9013 ± 0.06	0.8733 ± 0.2	–	11.99 ± 3.8	9.52 ± 11	–
CA	0.942 ± 0.03	0.9336 ± 0.03	–	10.24 ± 5.7	5.03 ± 1.9	–
HC	0.9184 ± 0.08	0.9269 ± 0.05	–	10.56 ± 6.0	5.63 ± 2.6	–
TR	0.9133 ± 0.04	0.9078 ± 0.05	–	9.71 ± 3.4	6.51 ± 6.8	–
DRV	0.9149 ± 0.04	0.8902 ± 0.07	–	11.07 ± 6.8	8.59 ± 7.2	–

The first row is the overall results and the following rows are accuracies per pathology. The results indicate that the generalization of the proposed method is promising, even for unseen pathologies.

4 Conclusion

Accurate segmentation of cine MR images is an open and challenging task due to highly complex and variable heart wall shape in various pathologies and ill-defined borders, particularly around the base and apex of the heart. Current deep networks have shown advantages in cardiac MR segmentation. However, most methods are mainly 2D, because short-axis sequences are acquired with relatively large between-slice distance, which leads to misalignment between slices. In this paper, we propose a novel framework to jointly segment the ventricles in both short- and long-axis images. The method has two stages to segment the two views independently and then refine their segmentations by fusing the complementary information from the other view. The results of the proposed method on the MICCAI 2021 Multi-Disease, Multi-View & Multi-Center Right Ventricular Segmentation in Cardiac MRI (M&Ms-2) challenge [2] show that the proposed method contributes to the improvement of LV, RV, and Myo segmentation. In the future, the proposed method could be further improved, by correcting misalignment due to respiratory motion, or not only fusing first-step predictions but also combining image features from different views.

Acknowledgement. The authors declare that the segmentation method implemented for participation in the M&Ms-2 challenge has not used any pre-trained models nor additional MRI datasets other than those provided by the organizers.

References

1. Bernard, O., et al.: Deep learning techniques for automatic MRI cardiac multi-structures segmentation and diagnosis: Is the problem solved? IEEE Trans. Med. Imaging **37**(11), 2514–2525 (2018)
2. Campello, V.M., et al.: Multi-centre, multi-vendor and multi-disease cardiac segmentation: the M&Ms challenge. IEEE Trans. Med. Imaging **40**, 3543–3554 (2021)
3. Chang, Q., Yan, Z., Lou, Y., Axel, L., Metaxas, D.N.: Soft-label guided semi-supervised learning for bi-ventricle segmentation in cardiac cine MRI. In: 2020 IEEE 17th International Symposium on Biomedical Imaging (ISBI), pp. 1752–1755. IEEE (2020)
4. Han, X., Xu, C., Prince, J.L.: A topology preserving deformable model using level sets. In: Proceedings of the 2001 IEEE Computer Society Conference on Computer Vision and Pattern Recognition. CVPR 2001, vol. 2, pp. 2. IEEE (2001)
5. Isensee, F., Jaeger, P.F., Full, P.M., Wolf, I., Engelhardt, S., Maier-Hein, K.H.: Automatic cardiac disease assessment on cine-MRI via time-series segmentation and domain specific features. In: Pop, M., et al. (eds.) STACOM 2017. LNCS, vol. 10663, pp. 120–129. Springer, Cham (2018). https://doi.org/10.1007/978-3-319-75541-0_13
6. Karamitsos, T.D., Francis, J.M., Myerson, S., Selvanayagam, J.B., Neubauer, S.: The role of cardiovascular magnetic resonance imaging in heart failure. J. Am. Coll. Cardiol. **54**(15), 1407–1424 (2009)
7. Khened, M., Kollerathu, V.A., Krishnamurthi, G.: Fully convolutional multi-scale residual densenets for cardiac segmentation and automated cardiac diagnosis using ensemble of classifiers. Med. Image Anal. **51**, 21–45 (2019)

8. Kong, B., Zhan, Y., Shin, M., Denny, T., Zhang, S.: Recognizing end-diastole and end-systole frames via deep temporal regression network. In: Ourselin, S., Joskowicz, L., Sabuncu, M.R., Unal, G., Wells, W. (eds.) MICCAI 2016. LNCS, vol. 9902, pp. 264–272. Springer, Cham (2016). https://doi.org/10.1007/978-3-319-46726-9_31

9. Li, J., Hu, Z.: Left ventricle full quantification using deep layer aggregation based multitask relationship learning. In: Pop, M., et al. (eds.) STACOM 2018. LNCS, vol. 11395, pp. 381–388. Springer, Cham (2019). https://doi.org/10.1007/978-3-030-12029-0_41

10. Li, Z., et al.: Fully automatic segmentation of short-axis cardiac MRI using modified deep layer aggregation. In: 2019 IEEE 16th International Symposium on Biomedical Imaging (ISBI 2019), pp. 793–797. IEEE (2019)

11. Metaxas, D.N.: Physics-based Deformable Models: Applications to Computer Vision, Graphics and Medical Imaging, vol. 389. Springer Science & Business Media, Boston (2012)

12. Ngo, T.A., Lu, Z., Carneiro, G.: Combining deep learning and level set for the automated segmentation of the left ventricle of the heart from cardiac cine magnetic resonance. Med. Image Anal. **35**, 159–171 (2017)

13. Peng, P., Lekadir, K., Gooya, A., Shao, L., Petersen, S.E., Frangi, A.F.: A review of heart chamber segmentation for structural and functional analysis using cardiac magnetic resonance imaging. Mag. Reson. Mater. Phys. Biol. Med. **29**(2), 155–195 (2016). https://doi.org/10.1007/s10334-015-0521-4

14. Petitjean, C., Dacher, J.N.: A review of segmentation methods in short axis cardiac MR images. Med. Image Anal. **15**(2), 169–184 (2011)

15. Poudel, R.P.K., Lamata, P., Montana, G.: Recurrent fully convolutional neural networks for multi-slice MRI cardiac segmentation. In: Zuluaga, M.A., Bhatia, K., Kainz, B., Moghari, M.H., Pace, D.F. (eds.) RAMBO/HVSMR -2016. LNCS, vol. 10129, pp. 83–94. Springer, Cham (2017). https://doi.org/10.1007/978-3-319-52280-7_8

16. Queirós, S., et al.: Fast automatic myocardial segmentation in 4d cine CMR datasets. Med. Image Anal. **18**(7), 1115–1131 (2014)

17. Tan, L.K., Liew, Y.M., Lim, E., McLaughlin, R.A.: Convolutional neural network regression for short-axis left ventricle segmentation in cardiac cine MR sequences. Med. Image Anal. **39**, 78–86 (2017)

18. Yu, C., Wang, J., Peng, C., Gao, C., Yu, G., Sang, N.: Learning a discriminative feature network for semantic segmentation. In: Proceedings of the IEEE Conference on Computer Vision and Pattern Recognition, pp. 1857–1866 (2018)

19. Yu, F., Wang, D., Shelhamer, E., Darrell, T.: Deep layer aggregation. In: Proceedings of the IEEE Conference on Computer Vision and Pattern Recognition, pp. 2403–2412 (2018)

20. Zhuang, X., Shen, J.: Multi-scale patch and multi-modality atlases for whole heart segmentation of MRI. Med. Image Anal. **31**, 77–87 (2016)

A Multi-view Crossover Attention U-Net Cascade with Fourier Domain Adaptation for Multi-domain Cardiac MRI Segmentation

Marcel Beetz[(✉)], Jorge Corral Acero, and Vicente Grau

Institute of Biomedical Engineering, Department of Engineering Science,
University of Oxford, Oxford OX3 7DQ, UK
`marcel.beetz@eng.ox.ac.uk`

Abstract. Cardiac image segmentation is a crucial step in clinical practice as it allows for the assessment of cardiac morphology and the quantification of image-based biomarkers. While deep learning methods have recently achieved near human-level performance on large, single-domain cine MRI datasets, their accuracy decreases considerably in more complex multi-domain settings, limiting their clinical applicability. To this end, we propose a novel multi-view crossover cascade approach combined with both shape and appearance augmentations for effective multi-domain cardiac image segmentation. Our cascade consists of two Attention U-Net paths that share information across different views and an intermediate heart location crop to reduce variance and improve label balance. In addition to multiple shape augmentations (scaling, elastic deformations, grid distortions, etc.) and histogram matching, we introduce multi-scale Fourier Domain Adaptation to cardiac image analysis. We evaluate both the crossover cascade and the augmentations on the cine MRI dataset of the M&Ms-2 challenge and outperform a U-Net benchmark by respective Dice score increases of ∼0.02 and ∼0.03.

Keywords: Multi-view cascade · Crossover cascade · Fourier domain adaptation · Attention U-net · Multi-domain cardiac image segmentation · Histogram matching · Deep learning · Cardiac MRI

1 Introduction

Cardiac Magnetic Resonance (CMR) imaging is the gold standard for the accurate non-invasive quantification of cardiac function and structure [20]. Anatomical segmentation is a standard pre-requisite for the calculation of both 2D and 3D image-based biomarkers [1,3,4,18] with proven prognostic value in the management of cardiac diseases [12]. Facilitated by the expansion of big data in the cardiovascular medicine field, deep learning models have marked a watershed moment towards the automation of the analysis [8,19]. Nevertheless, CMR analysis mostly remains manual in clinical practice, with the incurred time burden and associated costs

© Springer Nature Switzerland AG 2022
E. Puyol Antón et al. (Eds.): STACOM 2021, LNCS 13131, pp. 323–334, 2022.
https://doi.org/10.1007/978-3-030-93722-5_35

[19]. The main challenges limiting the automation include: (a) the intrinsic complexity of cardiac dynamics and the anatomical variability of the heart; (b) the heterogeneity in imaging acquisitions, introduced by different scanners, centers, and protocols; and (c) the limited availability of clinical data, constrained by financial, technical, ethical and confidentiality issues [13,19,21]. In order to address these challenges, data augmentation has become standard practice in fully automatic method designs to enlarge datasets and expose the network to higher degrees of variability. This includes simple, routinely-used approaches (cropping, translation, rotation, and flipping) [5,7], deformation-based techniques [11], as well as augmentations addressing imaging heterogeneity [5,23]. The above-mentioned methods aim to overcome the segmentation challenges by exploiting the available training data. Alternatively, a number of architectural variations have been introduced to improve segmentation performance, such as Attention U-Nets [16], cascaded approaches [10,22], domain adaptation methods [5,9], multi-view techniques [6], and ensembles [2,5]. In this work, we explore both data-driven methods and architectural enhancements in the context of the Multi-Disease, Multi-View & Multi-Center RV Segmentation in Cardiac Magnetic Resonance Imaging (MRI) Challenge (M&Ms-2) [15]. Its main goal is the segmentation of the right ventricle (RV), which is especially challenging given its highly complex and variable shape and its sometimes ill-defined borders. Thus, we deploy a novel multi-view crossover cascade pipeline, based on Attention U-Nets, that integrates the information from short and long axis. In addition, we train our approach with augmentations to populate both spaces of anatomical variability and imaging heterogeneity and thereby introduce a multi-scale version of Fourier Domain Adaptation [23] to cardiac image analysis as a new cross-domain data augmentation approach.

2   Methods

We first briefly describe the dataset of the M&Ms-2 challenge before giving an explanation of our proposed approaches.

2.1   Dataset

The M&Ms-2 training dataset consists of both short-axis (SA) and long-axis (LA) cine MRI acquisitions of 40 healthy subjects and 120 patients. It exhibits considerable variation in terms of scanner type, acquisition center and protocol, and contains multiple diseases affecting both ventricles. The pixel sizes vary from ∼0.68 to ∼1.63 mm for the SA data and from ∼0.68 to ∼1.72 mm for the LA data while the image resolution ranges from 192 to 512 pixels for SA and from 208 to 512 pixels for LA images. The left ventricular (LV) bloodpool, the LV myocardium, and the right ventricular (RV) bloodpool were manually segmented by clinical experts and serve as the ground truth for evaluation. We randomly split the dataset into train, validation, and test sets of 120, 8, and 32 cases respectively. In addition to the challenge training dataset available for method development, a test dataset with 160 cases is withheld from all challenge participants by the organizers and used only for the final evaluation of all challenge submissions.

2.2 Preprocessing

As preprocessing steps, we first adjust the pixel sizes of all SA images to ~1.33 mm and all LA images to ~1.43 mm using linear interpolation for the MR images and nearest neighbor interpolation for the segmentation masks. We then either crop or pad the resulting images to the same resolution of 256 × 256 pixels for both SA and LA data. Finally, we apply min-max normalization and intensity clipping (10th and 96th percentiles) to each SA and LA image.

2.3 Multi-view Crossover Cascade

A core feature of the M&Ms-2 dataset is the availability of paired SA and LA images. This allows information of the respective other view to be used in the segmentation task which is especially beneficial for determining the often difficult-to-pinpoint RV boundary or when segmenting the basal slices to avoid confusion between the right atrium and ventricles. In order to enable this information sharing between the SA and LA views in a deep learning approach, we propose a novel multi-view crossover cascade pipeline of Attention U-Nets (Fig. 1).

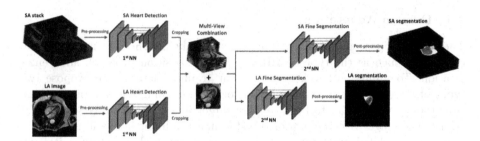

Fig. 1. Overview of the proposed multi-view crossover cascade pipeline. Two separate neural networks first locate the heart in the long-axis and short-axis input images respectively. Their predictions are then used to crop out a rectangular patch around the heart from the input images. The cropped images from both views are then combined into a new crossover volume that is fed into two additional separate neural networks which output the final predictions of the segmentation masks corresponding to the SA and LA input images respectively.

The pipeline consists of four main steps. First, two Attention U-Nets locate the heart in the LA and SA images respectively. Second, the heart location predictions are used to apply a crop of size 128 × 128 pixels centered at the heart's center of mass to both SA and LA images. Third, the information of both views is combined into a single volume. For LA segmentation, we concatenate the cropped LA image and the three cropped mid-cavity SA slices, as their segmentation quality is usually very high and it thus provides reliable information about the LV and RV boundaries and shapes to the LA segmentation task. The choice of three SA slices reduces the negative effects of potential segmentation errors on individual SA slices while still maintaining a high probability that ventricular anatomy is present in the given slices. For SA segmentation, we concatenate the respective cropped SA slice with the cropped LA image enabling access of the SA slices to further anatomical information in the basal and apical heart areas. Finally, each of the two volumes is passed through one of two additional Attention U-Nets to produce the final segmentation masks.

2.4 Shape Augmentations

Our first proposed set of augmentations is aimed at representing the considerable spatial variability of anatomical shapes and sizes, both of the heart and its surrounding tissue. In addition to commonly used affine transforms (flipping, rotating, and translating), we therefore use scaling, elastic deformations, grid distortions, and optical distortions as additional augmentation methods. We hypothesize these to be crucial for our highly multi-domain dataset and effective in capturing different disease phenotypes, such as the size variations of the dilated left and right ventricle condition. This is especially important considering the unknown diseases in the external test dataset of the M&Ms-2 challenge.

2.5 Domain Adaptation

The general appearance of cine MR images is heavily influenced by its acquisition conditions, such as the type of MRI scanner, acquisition protocol, and center location. To mimic such differences in our training dataset, we propose two types of appearance augmentations: histogram matching and Fourier Domain Adaptation (FDA) [23]. In both cases, a target image is randomly selected for each input image with the aim of transforming the input image in a way that matches the global appearance of the target image without altering its core structural content. Figure 2 shows the effect of each on both a SA (top row) and LA sample (bottom row) image.

Histogram matching achieves this similarity between images by first determining the histogram of both the input and target images, then deriving the respective cumulative probability distributions from each histogram, and finally creating a mapping function from each graylevel value in the input image to the target image. This mapping function can then be applied to the input image to obtain the transformed image. In Fourier Domain Adaptation, both the input and target images are first converted into Fourier space using a standard 2D discrete Fourier transform. A rectangular patch centered at the Fourier image center is then exchanged between the input and target Fourier images before both are converted back to the spatial domain using the inverse Fourier transform. Since the central locations in Fourier space represent low-frequency signals, only global image information is transferred from the target to the input image while its structural integrity remains mostly unchanged. The size of the patch is controlled by a parameter β that is crucial for maintaining a good trade-off between excessive and barely noticeable appearance changes. In this work, we

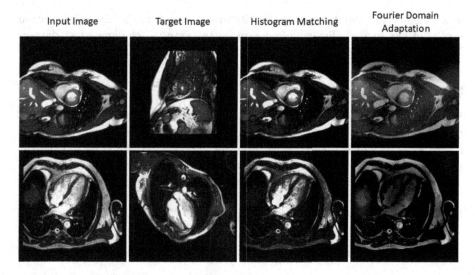

Fig. 2. Effect of histogram matching and Fourier Domain Adaptation on one sample short-axis (top row) and one sample long-axis image (bottom row).

randomly sample the β values from a range between 0.001 and 0.01, thereby extending the original method to a multi-scale FDA setting.

2.6 Postprocessing

As postprocessing steps, we apply hole closing and select only the largest connected RV component in the prediction mask. We also completely remove the right ventricle from slices which do not contain any predicted left ventricle.

2.7 Implementation

We conduct all experiments and method development on a GeForce RTX 2070 Graphics Card with 8 GB memory. All networks are implemented using the PyTorch deep learning framework [17] and trained using the Adam optimizer [14] with a batch size of 4. We stop the training process once network performance on the validation dataset has not improved for 50 epochs and select the best checkpoint as our final model.

3   Experiments

In order to evaluate both the architectural (Sec. 3.1) and data-driven (Sec. 3.2) approaches we select a baseline method as a benchmark for each of the two and compare it to the performance of the proposed changes using the Dice coefficients of three anatomical structures (LV bloodpool, LV myocardium, RV bloodpool) in a 5-fold cross-validation setup. Furthermore, we report the RV segmentation results of our final challenge submission on the external challenge test dataset in terms of both Dice scores and Hausdorff distances (Sec. 3.3).

3.1 Architectural Methods

We choose a standard U-Net as a benchmark to assess the effect of three architectural changes, an Attention U-Net [16] and a cascaded Attention U-Net approach, each of which trained separately for SA and LA data, as well as the proposed multi-view crossover Attention U-Net cascade (Sec. 2.3). All methods are trained using the same baseline augmentations (flipping, rotating, translating) to enable a fair comparison. The results on the test dataset for both SA and LA data are depicted in Table 1.

Table 1. Segmentation results of proposed architectures.

Image type	Method	Dice score		
		LV bloodpool	LV myocardium	RV bloodpool
Short-axis	Baseline U-Net	0.88 ($\pm$ 0.1)	**0.80** ($\pm$ 0.08)	0.86 ($\pm$ 0.08)
	Attention U-Net	**0.89** ($\pm$ 0.08)	**0.80** ($\pm$ 0.07)	**0.87** ($\pm$ 0.08)
	Cascade	0.88 ($\pm$0.11)	**0.80** ($\pm$ 0.12)	0.84 ($\pm$ 0.15)
	Crossover cascade	**0.89** ($\pm$ 0.14)	**0.80** ($\pm$ 0.13)	0.85 ($\pm$ 0.17)
Long-axis	Baseline U-Net	0.89 ($\pm$ 0.15)	0.77 ($\pm$ 0.14)	0.83 ($\pm$ 0.21)
	Attention U-Net	**0.92** ($\pm$ 0.09)	**0.80** ($\pm$ 0.12)	0.87 ($\pm$ 0.15)
	Cascade	0.91 ($\pm$ 0.11)	0.77 ($\pm$ 0.13)	0.87 ($\pm$ 0.14)
	Crossover cascade	**0.92** ($\pm$ 0.13)	**0.80** ($\pm$ 0.13)	**0.88** ($\pm$ 0.15)

All values represent mean ($\pm$ standard deviation).

We find similar results across all four architectures in the SA dataset, while the standalone Attention U-Net and the crossover cascade achieve the highest scores for the LA data.

When assessing the methods' performance on individual cases and slices, we observe that prediction quality is relatively similar for mid-cavity SA images, but varies to a much larger extent for basal and apical slices. Figure 3 depicts the qualitative prediction results for three sample cases.

For example, the first row in Fig. 3 shows a basal SA slice for which the crossover pipeline was able to correctly detect the absence of the heart while the other approaches erroneously predicted a left ventricular structure.

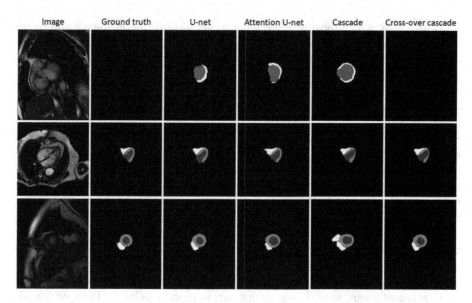

Fig. 3. Qualitative segmentation results of all analyzed architectures on three sample images.

3.2 Data-Driven Methods

In order to evaluate the effects of our proposed data-driven changes, we choose a U-Net architecture with basic augmentations (flipping, rotation, translation) as our baseline. We then compare it to both a U-Net with basic and shape augmentations (Sec. 2.4) and a U-Net with basic and appearance augmentations (Sec. 2.5) for both SA and LA data (Table 2).

Table 2. Segmentation results of baseline, shape, and appearance augmentations.

Image type	Method	Dice score		
		LV bloodpool	LV myocardium	RV bloodpool
Short-axis	Baseline	0.88 (± 0.10)	0.80 (± 0.08)	0.86 (± 0.08)
	+ Shape	**0.90** (± 0.08)	0.81 (± 0.09)	**0.88** (± 0.07)
	+ Appearance	**0.90** (± 0.06)	**0.82** (± 0.06)	**0.88** (± 0.06)
Long-axis	Baseline	0.89 (± 0.15)	0.77 (± 0.14)	0.83 (± 0.21)
	+ Shape	0.90 (± 0.11)	0.80 (± 0.12)	0.86 (± 0.16)
	+ Appearance	**0.93** (± 0.08)	**0.82** (± 0.09)	**0.88** (± 0.13)

All values represent mean (± standard deviation).

Both shape and appearance augmentations improve performance compared to the baseline in both views with differences between methods larger for LA data. Appearance augmentations achieve the highest Dice scores for each cardiac substructure and view.

3.3 Ensemble

Following the experimental results described in Sec. 3.1 and Sec. 3.2, we combine the architectural and data-driven changes that showed improvements over their respective baselines in an ensemble approach as our final submission to the M&Ms-2 challenge. To this end, we train both the crossover cascade and the normal cascade with Attention U-Net architectures and both augmentation types (baseline plus shape and appearance) in a 5-fold cross-validation setting on the training dataset of the challenge and average the outputs of each network to obtain the ensemble predictions. The results of this approach on the external challenge test dataset are depicted in Table 3 as reported by the challenge organizers. Both Dice scores and Hausdorff distances between predicted and ground truth segmentations were chosen as evaluation metrics using a 2D-based formulation for LA images and a 3D-based one for SA data.

Table 3. Results of our ensemble method on the external test dataset of the M&Ms-2 challenge as provided by the challenge organizers. Row 1 and row 2 show mean values for each image type. Row 3 represents the weighted average values across both image types. The respective weight coefficients were selected by the challenge organizers and set to 0.75 for SA data and 0.25 for LA images [15].

Image type	Right ventricle	
	Dice score	Hausdorff distance
Short-axis	0.83	17.62
Long-axis	0.85	10.95
Average	0.84	15.95

We observe an overall better performance of our method on long-axis images than on short-axis images in terms of both metrics. Dice scores for both image types are lower compared to the ones obtained in our experiments on the challenge training dataset.

The challenge organizers also provide our method's segmentation results separated by the pathologies present in the external challenge test dataset (Table 4).

Overall, our method achieves both its best and worst results on pathological imaging data while intermediate scores are recorded for healthy subjects. The highest Dice scores and lowest Hausdorff distances for long-axis images are obtained for tetralogy of fallot and hypertrophic cardiomyopathy patients respectively. Furthermore, we observe the best performance on short-axis images for congenital arrhythmogenesis patients in terms of both evaluation metrics.

Table 4. Results per pathology of our ensemble method on the external test dataset of the M&Ms-2 challenge as provided by the challenge organizers.

Image type	Pathology	Right ventricle	
		Dice	Hausdorff
Short-axis	Tricuspid Regurgitation	0.82 (± 0.13)	19.46 (± 9.99)
	Dilated Right Ventricle	0.86 (± 0.13)	16.90 (± 7.57)
	Interatrial Communication	0.77 (± 0.26)	19.11 (± 10.67)
	Dilated Left Ventricle	0.81 (± 0.18)	19.21 (± 11.07)
	Congenital Arrhythmogenesis	**0.88** (± 0.05)	**13.49** (± 4.49)
	Hypertrophic Cardiomyopathy	0.81 (± 0.20)	18.78 (± 11.37)
	Tetralogy of Fallot	0.85 (± 0.09)	16.75 (± 8.56)
	Normal	0.85 (± 0.11)	15.57 (± 8.95)
Long-axis	Tricuspid Regurgitation	0.84 (± 0.24)	9.61 (± 12.61)
	Dilated Right Ventricle	0.82 (± 0.20)	13.51 (± 14.46)
	Interatrial Communication	0.81 (± 0.28)	11.81 (± 12.15)
	Dilated Left Ventricle	0.88 (± 0.12)	8.02 (± 7.25)
	Congenital Arrhythmogenesis	0.86 (± 0.21)	8.99 (± 9.96)
	Hypertrophic Cardiomyopathy	0.89 (± 0.14)	**7.44** (± 8.79)
	Tetralogy of Fallot	**0.90** (± 0.04)	8.45 (± 4.08)
	Normal	0.84 (± 0.18)	16.52 (± 22.07)

All values represent mean (± standard deviation).

4   Discussion and Conclusion

In this work, we have presented a novel multi-view crossover cascade pipeline with both shape and appearance augmentations capable of improving performance for challenging multi-domain cardiac image segmentation.

In our experiments, the crossover cascade slightly outperforms a standard cascade with separate view processing, showing the utility of information sharing across different views. This is especially effective in the LA segmentation task, indicating that the highly reliable mid-cavity SA information is helpful in delineating anatomical LA boundaries. Since the stand-alone Attention U-Net achieved similar results as the crossover Attention U-Net cascade, we conclude that the usage of attention blocks is equally effective for accurately focusing on the correct heart location compared to a cascaded approach in the given dataset. However, we also find noticeable variations in performance between different slices, which shows that cascaded approaches can learn correct mappings in situations where the Attention U-Net does not. In general, we observe a large variability in Dice scores across the 5-fold cross-validation experiments, which could be a reflection of the highly variable dataset (regarding pixel size, image resolution, disease type, etc.) and indicates that our method would benefit from further experiments.

Regarding the data-driven changes, we find positive effects of both shape and appearance augmentations. This demonstrates the special importance of augmentation methods in highly variable multi-domain datasets since they expose the network to a larger variety of inputs during training which in turn strengthens its generalization ability. Furthermore, we observe that interchanging domain information between images as part of the appearance augmentations leads to greater performance increases and smaller standard deviations than the analyzed shape variations which were constrained to individual images in this work. The increased robustness could be explained by the small amounts of additional noise introduced into the images to a varying degree by the FDA approach, forcing the network to be less susceptible to image artifacts.

Our ensemble method performs better on the LA images than on the SA images of the external challenge test dataset in terms of both evaluation metrics, which is in line with our own experimental results. The considerably larger Hausdorff distance indicates that more extreme outliers are present in the RV predictions of the short-axis data. We hypothesize, that this might be caused by the 3D-based metrics used to evaluate the performance on the SA stack where distances between wrong and correct predictions are captured along all three spatial dimensions instead of the 2D-based calculations used for the long-axis data. Since the right ventricle is often only present in some of the slices of the SA stack, small erroneous predictions in slices far away from the true anatomy can lead to large Hausdorff distances despite negligible effects on 3D Dice scores. In addition, the results indicate that the cross-view information exchange in our proposed cascade architecture is more important for LA data since it only consists of one 2D image as opposed to the multiple slices available in the SA view which facilitate the network's segmentation task.

When analyzing the pathology-specific results, we find that our ensemble approach achieves similar levels of performance for the previously unseen disease cases (tricuspid regurgitation and dilated right ventricle) and for the healthy subjects. On the one hand, this shows that the Dice score decrease between the results observed in our own experiments and the ones on the external challenge dataset was likely due to other domain changes unrelated to disease phenotypes. On the other hand, it demonstrates that our method was able to generalize well to new pathologies. In this regard, the access to multi-view information in the crossover pipeline might have played an important role, for example when determining the boundary of the dilated RV. Besides this, the good generalization results could also be attributed to the proposed augmentation strategies since, for example, the scaling and deformation-based shape changes might have mimicked the dilation of the RV while the domain adaption techniques could have exposed the network to a wider variety of acquisition conditions during training.

Acknowledgments. The authors declare that the segmentation methods they developed for participating in the M&Ms-2 challenge did not use any additional datasets besides the ones provided by the organizers. The work of M. Beetz was supported by the Stiftung der Deutschen Wirtschaft (Foundation of German Business). The work of J. Corral-Acero was supported by the EU's Horizon 2020 research and innovation program under the Marie Sklodowska Curie (g.a. 764738 to JCA). The work of V. Grau was supported by the CompBioMed 2 Centre of Excellence in Computational Biomedicine (European Commission Horizon 2020 research and innovation programme, grant agreement No. 823712).

References

1. Alfakih, K., Plein, S., Thiele, H., Jones, T., Ridgway, J.P., Sivananthan, M.U.: Normal human left and right ventricular dimensions for MRI as assessed by turbo gradient echo and steady-state free precession imaging sequences. J. Magn. Reson. Imaging Official J. Int. Soc. Magn. Reson. Med. **17**(3), 323–329 (2003)
2. Audelan, B., Hamzaoui, D., Montagne, S., Renard-Penna, R., Delingette, H.: Robust fusion of probability maps. In: Martel, A.L., et al. (eds.) MICCAI 2020. LNCS, vol. 12264, pp. 259–268. Springer, Cham (2020). https://doi.org/10.1007/978-3-030-59719-1_26
3. Bai, W., et al.: A bi-ventricular cardiac atlas built from 1000+ high resolution mr images of healthy subjects and an analysis of shape and motion. Med. Image Anal. **26**(1), 133–145 (2015)
4. Beetz, M., Banerjee, A., Grau, V.: Biventricular surface reconstruction from cine MRI contours using point completion networks. In: 2021 IEEE 18th International Symposium on Biomedical Imaging (ISBI), pp. 105–109 (2021)
5. Campello, V.M., et al.: Multi-centre, multi-vendor and multi-disease cardiac segmentation: The M&Ms challenge. IEEE Trans. Med. Imaging **40**, 3543–3554 (2021)
6. Chen, C., Biffi, C., Tarroni, G., Petersen, S., Bai, W., Rueckert, D.: Learning shape priors for robust cardiac mr segmentation from multi-view images. In: Shen, D., et al. (eds.) MICCAI 2019. LNCS, vol. 11765, pp. 523–531. Springer, Cham (2019). https://doi.org/10.1007/978-3-030-32245-8_58
7. Chen, C., et al.: Deep learning for cardiac image segmentation: a review. Front. Cardiovasc. Med. **7**, 25 (2020)
8. Corral-Acero, J., et al.: The digital twin to enable the vision of precision cardiology. Eur. Heart J. **41**(48), 4556–4564 (2020)
9. Corral Acero, J., Sundaresan, V., Dinsdale, N., Grau, V., Jenkinson, M.: A 2-step deep learning method with domain adaptation for multi-centre, multi-vendor and multi-disease cardiac magnetic resonance segmentation. In: Puyol Anton, E., et al. (eds.) STACOM 2020. LNCS, vol. 12592, pp. 196–207. Springer, Cham (2021). https://doi.org/10.1007/978-3-030-68107-4_20
10. Corral Acero, J., et al.: Left ventricle quantification with cardiac MRI: deep learning meets statistical models of deformation. In: Pop, M., et al. (eds.) STACOM 2019. LNCS, vol. 12009, pp. 384–394. Springer, Cham (2020). https://doi.org/10.1007/978-3-030-39074-7_40
11. Corral Acero, J., et al.: SMOD - data augmentation based on statistical models of deformation to enhance segmentation in 2d cine cardiac MRI. In: Coudière, Y., Ozenne, V., Vigmond, E., Zemzemi, N. (eds.) FIMH 2019. LNCS, vol. 11504, pp. 361–369. Springer, Cham (2019). https://doi.org/10.1007/978-3-030-21949-9_39
12. Dall'Armellina, E.: From recognized to novel quantitative CMR biomarkers of lv recovery: a paradigm shift in acute myocardial infarction imaging (2017)
13. Dey, D., et al.: Artificial intelligence in cardiovascular imaging: jacc state-of-the-art review. J. Am. Coll. Cardiol. **73**(11), 1317–1335 (2019)
14. Kingma, D.P., Ba, J.: Adam: a method for stochastic optimization. arXiv preprint arXiv:1412.6980 (2014)

15. Martin-Isla, C.: Multi-disease, multi-view & multi-center right ventricular segmentation in cardiac MRI (M&Ms-2) (2021). https://www.ub.edu/mnms-2/
16. Oktay, O., et al.: Attention u-net: learning where to look for the pancreas. arXiv preprint arXiv:1804.03999 (2018)
17. Paszke, A., et al.: Pytorch: an imperative style, high-performance deep learning library. Adv. Neural Inf. Process. Syst. **32**, 8026–8037 (2019)
18. Petersen, S.E., et al.: Reference ranges for cardiac structure and function using cardiovascular magnetic resonance (cmr) in caucasians from the uk biobank population cohort. J. Cardiovasc. Mag. Reson. **19**(1), 18 (2017)
19. Shameer, K., Johnson, K.W., Glicksberg, B.S., Dudley, J.T., Sengupta, P.P.: Machine learning in cardiovascular medicine: are we there yet? Heart **104**(14), 1156–1164 (2018)
20. Stokes, M.B., Roberts-Thomson, R.: The role of cardiac imaging in clinical practice. Aust. Prescriber **40**(4), 151 (2017)
21. Tao, Q., et al.: Deep learning-based method for fully automatic quantification of left ventricle function from cine mr images: a multivendor, multicenter study. Radiology **290**(1), 81–88 (2019)
22. Vigneault, D.M., Xie, W., Ho, C.Y., Bluemke, D.A., Noble, J.A.: ω-net (omega-net): fully automatic, multi-view cardiac mr detection, orientation, and segmentation with deep neural networks. Med. Image Anal. **48**, 95–106 (2018)
23. Yang, Y., Soatto, S.: FDA: fourier domain adaptation for semantic segmentation. In: Proceedings of the IEEE/CVF Conference on Computer Vision and Pattern Recognition, pp. 4085–4095 (2020)

Multi-disease, Multi-view and Multi-center Right Ventricular Segmentation in Cardiac MRI Using Efficient Late-Ensemble Deep Learning Approach

Moona Mazher[1]([⊠]), Abdul Qayyum[2], Abdesslam Benzinou[2], Mohamed Abdel-Nasser[1,3], and Domenec Puig[1]

[1] Department of Computer Engineering and Mathematics, University Rovira I Virgili, Tarragona, Spain
moona.mazher@estudiants.urv.cat
[2] ENIB, UMR CNRS 6285 LabSTICC, 29238 Brest, France
[3] Faculty of Engineering, Department of Electrical Engineering, Aswan University, Aswan, Egypt

Abstract. In many computer vision areas, deep learning-based models achieved state-of-the-art performances and started catching the attention in the context of medical imaging. The emergence of deep learning is a cutting-edge for the state-of-the-art methods of cardiac magnetic resonance (CMR) segmentation. For generalization and better optimization of current deep learning models for CMR segmentation problems, the M&Ms-2 (Multi-Disease, Multi-View & Multi-Center Right Ventricular Segmentation in Cardiac MRI) challenge proposed data that are acquired from three clinical centers of Spain and Germany using three different magnetic resonance scanner vendors (Siemens, General Electric and Philips). To cater to the generalization issue on a multi-Disease dataset, the proposed model used lightweight convolutional layers' blocks before the proposed residual block which have been used as a skip connection with a series of several layers for boundary and structural information preservation. The residual blocks that are used after every encoder block help bridge the semantic gap between the encoder and decoder. The efficient expansion, depth-wise, and projection block (EDP) is used at each decoder block for efficiently improving the segmentation maps. The proposed 2D-based model is used to segment the right ventricle (RV) in short-axis (SA) images, as well as in long-axis (LA) cardiac MR images without any additional dataset and pretrained weights. The proposed model produced optimal dice coefficients (DC) and Hausdorff distance (HD) scores in validation and testing images and could be useful for the segmentation of the RV and LA in cardiac MRI.

Keywords: CMR segmentation · Deep learning models · Right ventricle · Post ensemble · TTA

© Springer Nature Switzerland AG 2022
E. Puyol Antón et al. (Eds.): STACOM 2021, LNCS 13131, pp. 335–343, 2022.
https://doi.org/10.1007/978-3-030-93722-5_36

1   Introduction

The assessment and diagnose of major cardiovascular diseases [1] are an important task and in clinical practice, accurate segmentation of cardiovascular magnetic resonance (CMR) images plays an important role for precise assessment of cardiovascular diseases. For a clinician, manual annotation of the cardiac boundaries for all image slices and cardiac phases is a time-consuming task, and accurate annotation requires a significant amount of time and cost. Multiple approaches such as statistical shape models [2] or cardiac atlases [3] have been proposed for the automation of such a tedious and time-consuming task. Nowadays, the availability of large and relevant datasets in medical imaging claims for the study of a machine learning approach and more particularly for a deep learning one for CMR segmentation. Indeed, in the past few years, deep learning has become a breakthrough technology in a wide range of problems such as image classification, action recognition, automatic labeling of an image, image segmentation to name a few. In many computer vision areas, deep learning-based models achieved state-of-the-art performances and started to catch the attention and effort in the context of medical imaging. However, most of these techniques have been trained and evaluated using cardiac imaging samples collected from single clinical centers using similar imaging protocols. While these works have advanced the state-of-the-art in deep learning-based cardiac image segmentation, their high performances were reported on samples with relatively homogeneous imaging characteristics. As an example, the CMR datasets from the Automated Cardiac Diagnosis Challenge (ACDC) dataset [4] have been extensively used to build and test new implementations of deep neural networks for cardiac image segmentation. In 2020, Multi-Centre, Multi-Vendor and Multi-Disease Cardiac Segmentation challenge [5] has been proposed and deep learning-based models produced an excellent performance on this diverse dataset.

The different 3D and 2D-based deep learning models have been proposed for the segmentation of the M&Ms dataset. Most deep learning models are based on encoder-decoder type architectures and these architectures also have been widely used for numerous medical image segmentation applications. In 2021, M&Ms-2 (Multi-Disease, Multi-View & Multi-Center Right Ventricular Segmentation in Cardiac MRI) challenge focuses on segmenting the RV in short-axis (SA) images, as well as in long-axis (LA) cardiac MR images. However, the segmentation of the right ventricle (RV) is a challenging task due to the highly complex and variable shape of the RV and its ill-defined borders in cardiac MR images. We need a solution based on a deep learning model to minimize the error in the presence of RV-related pathologies (e.g., Dilated Right Ventricle, Tricuspid Insufficiency, Arrhythmogenesis, Tetralogy of Fallot and Interatrial Communication).

In this paper, we proposed a single 2D deep learning model solution that has the following contribution for segmentation of the right ventricle (RV) in short-axis (SA) images, as well as in long-axis (LA) cardiac MR images without any additional dataset.

1. The simple and lightweight convolutional layers' blocks before the proposed residual block have been presented for RV segmentation.
2. The efficient expansion, depth-wise, and projection block (EDP) has been proposed and used at each decoder block for efficiently improving the segmentation maps.
3. The cross-validation-based proposed post models' predictions have been ensembled for segmentation and extensive experiments have been performed with different hyperparameters that were used for training and optimization of the proposed model

with the test-time augmentation (TTA) technique. The performance of the proposed model has been compared with basic UNet and other pretrained existing models.

2   Material and Methods

2.1   M&Ms-2 Dataset Descriptions

The challenge cohort was composed of 360 patients with different right ventricle and left ventricle pathologies as well as healthy subjects. All subjects were scanned in three clinical centers from Spain using three different magnetic resonance scanner vendors (Siemens, General Electric, and Philips). The training set contained 200 annotated images from four different centers. The CMR images have been segmented by experienced clinicians from the respective institutions, including contours for the left (LV) and right ventricle (RV) blood pools, as well as for the left ventricular myocardium (MYO). Two pathologies (Tricuspidal Regurgitation and Congenital Arrhythmogenesis) will be not present in the training set but the validation and testing sets will be used to evaluate generalization to unseen pathologies.

Table 1. The number of cases used in training and validation.

Pathology	Num. studies training	Num. studies validation
Normal subjects	40	5
Dilated left ventricle	30	5
Hypertrophic cardiomyopathy	30	5
Congenital arrhythmogenesis	20	5
Tetralogy of fallot	20	5
Interatrial communication	20	5
Dilated right ventricle	0	5
Tricuspidal regurgitation	0	5

Table 2. Number of cases used in test data.

Pathology	Num. studies test
Normal subjects	30
Dilated left ventricle	25
Hypertrophic cardiomyopathy	25
Congenital arrhythmogenesis	10
Tetralogy of fallot	10
Interatrial communication	10
Dilated right ventricle	25
Tricuspidal regurgitation	25

A detailed description of the dataset can be found [5] in Table 1 and Table 2. The training and validation dataset with pathology distribution has been represented in the Table 1 while the details about the test data is given in Table 2.

2.2 Proposed Method

A framework of the proposed HeartSeg-DeepSRU model is presented as an encoder, a decoder, and a baseline module. The proposed model receives RV and LV as input and produces a mask for each class (LV, MYO, and RV in both SA and LA view with a variety of difficult RV pathologies as well as LV pathologies). We have trained a 2D single model for each slice of the SA and LA dataset. The proposed module at the encoder side is based on expansion, depth-wise, and projection block (EDP). The complete architecture of the proposed CardiacSeg-RUNET is shown in Fig. 1.

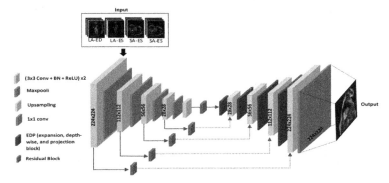

Fig. 1. The proposed model for RV segmentation. Each color represents a different number of blocks. (Color figure online)

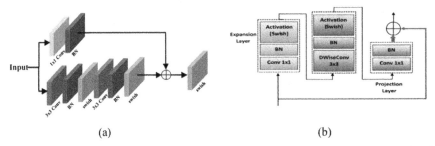

Fig. 2. (a) Proposed residual block (b) EDP (expansion, depth-wise, and projection block) decoder block.

The detail of the residual module is shown in Fig. 2(a). The Residual network (ResNets) has been widely used for deep learning classification and segmentation and this model is built on hundreds of layers [5]. In this paper, residual blocks have been used as a skip connection with series of several layers for boundary and structural information preservation in the residual block. The structural information for feature maps could be restored by the addition of the residual blocks that aimed to preserve the fine-grained structures that would be useful and play an important role in medical image segmentation.

On the decoder side, EDP (expansion, depthwise, and projection layer block) has been proposed. The expansion layer increases the number of feature maps, and in the projection, the layer decreases the feature maps with some regularization layers such as batch normalization and activation. Different depths (number of filters) using expansion and projection 1 × 1 Conv layer were used, to extract useful information with a smaller number of computational complexities. Instead of using the maxpooling layer, the input image size is downsampled after each expansion block using a depthwise convolutional layer with stride 2. The depth-wise convolutional module with BN and swish activation effectively captures the features as compared to the standard convolutional module. It is inspired by two widely used convolutional-based methods such as group convolutions and depth-wise separable convolutions [6]. The swish activation function proposed by Ramachandran et al. [7] effectively works in automatic search techniques based on reinforcement learning. The decoder block based on our proposed module is shown in Fig. 2(b).

We have used a 5-fold cross-validation method to divide the training subjects into training and validation. At each fold, 80% dataset was used for training and 20% used for validation. The five proposed model trained weights after the Test Time augmentation technique are used in post ensemble based on the max majority voting technique. Similar to what data augmentation is doing to the training set, the purpose of test time augmentation is to perform random modifications to the test images. Thus, instead of showing the regular, "clean" images, only once to the trained model, we will show the augmented images several times and then average the predictions of each corresponding image and take that as the final prediction. We used flipping and rotation with various angles (0, 90, 180, and 270) as test time augmentation in our proposed solution and received fruitful results. Figure 3 shows the complete post ensemble process to get the final prediction on test samples.

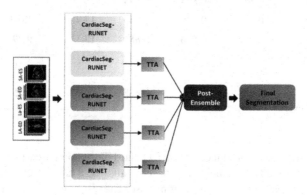

Fig. 3. Post ensemble based on majority voting method with Test Time Augmentation.

The proposed model is optimized using the Adam optimizer with a learning rate (0.0003). The binary cross-entropy function is used as a loss function between ground-truth and predication mask. The data augmentation such as resize (224, 224), HorizontalFlip, VerticalFlip, Transpose, Rotate, and GridDistortation have been used during training and resize augmentation during validation with data normalization method (zero mean and unit standard deviation). The 150 number epochs with 16 batch sizes have been used for training and optimizing the proposed model. The PyTorch Lightning has been used for training and the PyTorch library for model development.

3   Results and Discussion

3.1   Quantitative Results

Quantitative analysis for proposed and existing deep learning model on performance measurements based on Dice coefficients (DC) and Hausdorff Distance (HD) for SA and LA views is given in Table 3. The higher the DC, the better the segmentation results. Similarly, lower the HD means the higher the model performance. In Table 3 the proposed model produced overall higher DC and lower HD as compared to other existing state-of-the-art models.

Table 3. Performance analysis on validation dataset (40 cases) using proposed CardiacSeg-RUNET and existing state-of-the-art methods (The best results are given in bold) as well as performance analysis of the proposed CardiacSeg-RUNET on test dataset (160 cases).

Models	DC (SA)	DC (LA)	Average DC	HD (SA)	HD (LA)	Average HD
Validation data						
UNet [8]	0.837	0.779	0.822	18.438	18.483	18.449
UNet-DeepLabv3	0.857	0.804	0.844	18.108	14.215	17.135
UNet-Resnet18	0.856	0.828	0.849	14.453	11.010	13.592
Proposed CardiacSeg-RUNET	**0.890**	**0.856**	**0.881**	**13.112**	**9.297**	**12.158**
Test data						
Proposed CardiacSeg-RUNET	0.894	0.870	0.888	14.94	10.55	13.84

The proposed model produced better DC and HD as compared to basic UNet and other pretrained based UNet models. These results were taken from the leaderboard provided by the organizers in the validation phase. The results on the test dataset provided by the organizer during the test phase are shown in Table 3. The proposed model produced average HD (0.8885%) and HD (13.84) on the test dataset that is comparably close to the average dice score (0.890) and HD (12.158) on the validation dataset.

3.2 Qualitative Results

We are providing some qualitative results on validation samples that were used during internal validation for SA_ED and SA_ES. For example, the proposed model produced better 3D segmentation volume for Sub_2 and Sub_27. In Sub_2, the 3D segmentation volume produced by the proposed model is not good as compared to ground-truth as shown in the second row in Fig. 4.

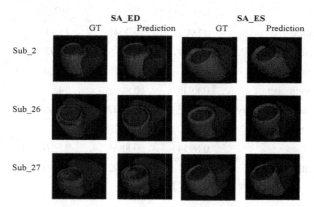

Fig. 4. 3D Segmentation Volume predicted by proposed model and ground-truth for SA-ED and SA-ES. The ITK-SNAP software is used to show a 3D segmentation map.

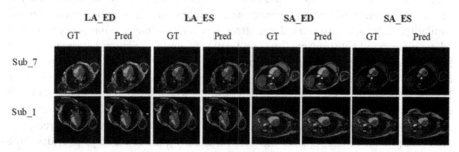

Fig. 5. 2D segmentation map for ground truth and prediction by the proposed model based on Sub_1, Sub_7. These subjects are used for validation.

The two different subjects (Sub_1, and Sub_7) were used to check the visualization segmentation map of the proposed model. In Sub_7 for LA_ED, the proposed model did not produce a better segmentation map (first row, second column). The performance of the proposed model seems optimal for other slices in Sub_7 as shown in Fig. 5. The overall Dice coefficient computed on the test dataset for different pathology is shown in Table 4. The Dice coefficient produced by the proposed model for Interatrial Communication and Dilated Left Ventricle pathology is worst as compared to other pathology in LA. Similar HD for each pathology on the test dataset is shown in Table 4. The model produced relatively worst HD on Dilated Left Ventricle pathology as compared to other pathology for SA and LA. Also, the model produced bad HD for SA in Dilated Right Ventricle and Congenital Arrhythmogenesis pathology as shown in Table 4.

Table 4. Dice coefficients (DC) and Hausdorff Distance (HD) Scores were produced by the proposed model on a test dataset for different pathology.

Pathology	DC (SA)	DC (LA)	HD (SA)	HD (LA)
Tetralogy of fallot	0.8909	0.9041	13.11	7.77
Tricuspidal regurgitation	0.9039	0.9185	11.92	6.16
Interatrial communication	0.8910	0.8255	12.20	12.47
Congenital arrhytmogenesis	0.8938	0.8932	14.60	7.17
Hypertrophic cardiomyopathy	0.8916	0.8797	13.86	7.37
Dilated right ventricle	0.9040	0.8629	15.23	10.17
Normal	0.8954	0.8865	11.26	7.60
Dilated left ventricle	0.8801	0.7992	25.10	23.73

The proposed model performance is comparatively good on Dice and a little less in overall HD for all pathology in this diverts dataset. The proposed model may produce an overestimation of the boundary of the segmentation map in diverse cases. The subtle and complex surrounding boundaries of organs, insufficient uniform region and similarity, image noise, partial volume effect, low contrast intensity inhomogeneity, and other artifacts could obstruct precise identification of abrupt variations between organs of interest. Based on observation, the various stages in the encoder and decoder-based segmentation models can recognize the fruitful features in diverse consistency appearance. The network encodes fine spatial resolution in the lower stage that has poor semantic consistency without the guidance of spatial context and its small receptive view. While it has strong semantic consistency in the higher stage because of the large receptive view that provides coarse spatial information. Globally, the lower stage provides more accurate spatial predictions, whereas the higher stage produced more accurate semantic predictions. A combination of the higher and lower stage can make the more powerful network for semantic segmentation.

4   Conclusion and Future Work

In this paper, deep learning-based models have been proposed for RV and LV segmentation tasks. The proposed model produced optimal performance and produced average DC (0.8885%) and HD (13.84) on the test dataset. The results on the test dataset are encouraging and a little worse than on the validation dataset. The proposed model could be used as the first step towards correct diagnoses and prediction of RV segmentation. In the future, other 2D and 3D deep learning models (transformer-based) will be tested to further enhance RV segmentation.

References

1. Cetin, I., et al.: A radiomics approach to computer-aided diagnosis with cardiac cine-MRI. In: Pop, M., et al. (eds.) STACOM 2017. LNCS, vol. 10663, pp. 82–90. Springer, Cham (2017). https://doi.org/10.1007/978-3-319-75541-0_9
2. Albà, X., Lekadir, K., Pereañez, M., Medrano-Gracia, P., Young, A.A., Frangi, A.F.: Automatic initialization and quality control of large-scale cardiac MRI segmentations. Med. Image Anal. **43**, 129–141 (2018)
3. Bai, W., Shi, W., Ledig, C., Rueckert, D.: Multi-atlas segmentation with augmented fea-tures for cardiac MR images. Med. Image Anal. **19**(1), 98–109 (2015)
4. Bernard, O., et al.: Deep learning techniques for automatic MRI cardiac multi-structures seg-mentation and diagnosis: Is the problem solved? IEEE Trans. Med. Imaging **37**, 2514–2525 (2018)
5. Campello, V.M., et al.: Multi-centre, multi-vendor and multi-disease cardiac segmentation: the M&Ms challenge. IEEE Trans. Med. Imaging **40**(12), 3543–3554 (2021)
6. Chollet, F.: Xception: deep learning with depthwise separable convolutions. In: Proceedings of the IEEE Conference on Computer Vision and Pattern Recognition, pp. 1251–1258 (2017)
7. Shankaranarayana, S.M., Ram, K., Mitra, K., Sivaprakasam, M.: Fully convolutional networks for monocular retinal depth estimation and optic disc-cup segmentation. IEEE J. Biomed. Health Inform. **23**(4), 1417–1426 (2019)
8. Ronneberger, O., Fischer, P., Brox, T.: U-net: convolutional networks for biomedical image segmentation. In: Navab, N., Hornegger, J., Wells, W.M., Frangi, A.F. (eds.) Medical Image Computing and Computer-Assisted Intervention – MICCAI 2015: 18th International Confer-ence, Munich, Germany, October 5-9, 2015, Proceedings, Part III, pp. 234–241. Springer, Cham (2015). https://doi.org/10.1007/978-3-319-24574-4_28

Automated Segmentation of the Right Ventricle from Magnetic Resonance Imaging Using Deep Convolutional Neural Networks

Kumaradevan Punithakumar[1,2(✉)], Adam Carscadden[1,2], and Michelle Noga[1,2]

[1] Department of Radiology and Diagnostic Imaging, University of Alberta,
Edmonton, Canada
punithak@ualberta.ca

[2] Servier Virtual Cardiac Centre, Mazankowski Alberta Heart Institute,
Edmonton, Canada

Abstract. Although the left ventricle (LV) is commonly assessed in current clinical practice, the assessment of the right ventricle (RV) also plays an important role in the diagnosis of cardiovascular disease. RV failure has numerous causes, including pulmonary hypertension, myocardial infarction, and congenital heart disease. However, assessment of the RV is more challenging than the LV due to its complex shape and thin walls. This study proposes an automated approach to delineate the RV from magnetic resonance imaging (MRI) scans using a deep convolutional neural network approach. The proposed method uses nnU-Net, a self-adapting framework based on the U-Net neural network approach for the segmentation of the RV from short and long axis MRI images at the end-systolic and end-diastolic phases of the heart. The proposed neural network models were trained using the datasets provided by Multi-Disease, Multi-View & Multi-Center Right Ventricular Segmentation in Cardiac MRI Challenge hosted by MICCAI 2021 conference. The quantitative evaluations were performed by the challenge organizers on a test set consisting of MRI scans acquired from 160 patients where the images and ground truth were blinded to the challenge participants. The proposed method yielded an overall Dice metric of 92.47% with 92.73% and 91.71% for short and long axis images, respectively. The corresponding Hausdorff distance values were 9.08 mm, 10.05 mm, and 6.16 mm, respectively.

Keywords: Cardiac magnetic resonance imaging · Deep convolutional neural networks · Right ventricle · Image segmentation

1 Introduction

The importance of analyzing right ventricular (RV) function has been emphasized in several studies [5]. Magnetic resonance imaging (MRI) offers an excellent non-invasive option to image the RV due to its high spatial resolution and superior soft-tissue contrast. However, MRI generates hundreds of images in each scan

E. Puyol Antón et al. (Eds.): STACOM 2021, LNCS 13131, pp. 344–351, 2022.
https://doi.org/10.1007/978-3-030-93722-5_37

from a patient, and automating the assessment of the RV function from MRI is of great interest to reduce the intense manual postprocessing required for producing clinical measurements. Typically, an anatomical cine MRI scan consists of multiple short and long axis images. Among different long axis sequences, the RV is typically imaged along with LV and atria in the four-chamber view. The segmentation of the RV at the end-diastolic and end-systolic phases is required to compute clinical measures such as RV ejection fraction. Automated and semi-automated methods to delineate the RV from MRI sequences have been proposed by several research groups [10]. In addition, recent public challenges on cardiac ventricular segmentation also included the RV in addition to the LV to encourage and promote automated approaches to solve the problem [3,4].

The majority of the existing automated and semi-automated ventricular segmentation algorithms from cardiac MRI focused on the LV. The RV segmentation is more challenging than LV due to its complex shape and thinner wall. In addition, the longitudinal shortening of the RV is larger than the radial shortening [6]. Over the past few decades, many traditional computer vision and image processing algorithms [8–12] were proposed for the segmentation of the RV from MRI image sequences. Recently, deep convolutional neural network based approaches have been proposed for the RV segmentation from MRI [2,13].

This study proposes a deep convolutional neural network approach based on nnU-Net [7] to delineate the RV from short and long axis MRI sequences. The nnU-Net is a self-adapting framework based on the U-Net [14] neural network approach. The nnU-Net has been shown to be the best performing segmentation algorithm in public segmentation challenges [1,4] on various organs including cardiac chambers. Separate neural networks are used for the RV segmentation from short and long axis images, and the neural networks are trained using the five-fold cross validation approach. The proposed neural networks were trained using MRI datasets acquired from 160 patients along with their corresponding manual annotations. The MRI datasets and ground truth annotations were provided by the Multi-Disease, Multi-View & Multi-Center Right Ventricular Segmentation in Cardiac MRI challenge organizers. The proposed method was evaluated quantitatively over validation and test datasets using an automated scoring system.

2   Methodology

The overall process showing the training and inference steps are given in Fig. 1. We used the nnU-Net without any modifications for the segmentation from short-axis sequences. However, for the long-axis sequences, we oriented the images and labels to be parallel to coordinate axes during the training process.

2.1   Training Data

The training data consists of short and long axis MRI images acquired from 160 patients and ground truth delineations for the end-systolic and end-diastolic phases. The short-axis scans were saved as 3D volumes, and their sizes varied from ($256 \times 208 \times 6$) to ($512 \times 512 \times 17$). The four chamber long axis scans

Training

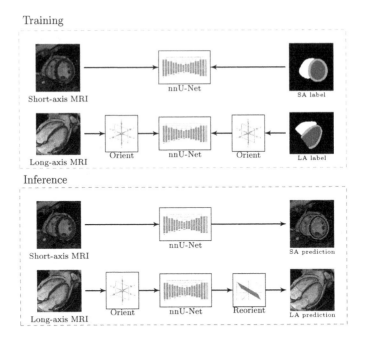

Fig. 1. The proposed neural network based solution to segment the ventricles from short and long axis MRI sequences.

were 2D images and their size varied from (208×256) to (512×512). A total of 320 3D datasets from short axis used for training the neural networks, where 160 volumes correspond to the end-diastole and another 160 volumes correspond to end-systole. Similarly a total of 320 2D images from four chamber long-axis images were used for training the neural networks. The annotated ground truth data consists of labels corresponding to the LV endocardium, LV myocardium, and RV. However, the evaluations were performed only for the RV.

2.2   Neural Network and Training

Two separate nnU-Net [7] based neural networks were used in this study, where one of them was used for predicting the results on short-axis images and the other one was used for the long-axis images. The default configuration was used for both neural networks and a combination of Dice and cross-entropy loss was used for neural network training [7]. We used a five-fold cross validation approach for training the neural networks. The nnU-Net utilizes real-time data augmentation strategy where rotations, scaling, and elastic deformations are applied randomly along with gamma correction and mirroring.

For short axis datasets, the neural network was trained using 2D U-Net and 3D U-Net with full resolution options and the final neural network model was based on an ensemble of these neural networks. For long-axis images, the nnU-Net was trained only with 2D U-Net option. All neural networks were trained for 1000 epochs. The neural network training took around three days with the short-axis data and around one day for long-axis images on NVIDIA Tesla V100 (16 GB memory) graphics processors. We relied on the nnU-Net framework to identify the best configuration for both neural networks. The neural network model for the nnU-Net was implemented using the PyTorch module.

2.3 Preprocessing of Images and Postprocessing of Labels

The short-axis datasets were used in their original 3D format for neural network training and prediction. However, for the long-axis datasets, we removed the header information consisting of the position and orientation information before applying the neural network for training and prediction. The default nnU-Net preprocessing option for MRI in normalizing the pixel values was used for both short and long axis datasets. The output labels were automatically postprocessed to retain only the largest connected component for each predicted label corresponding to the LV endocardium, LV myocardium, and RV.

3 Results

The proposed method was quantitatively evaluated over validation and test MRI datasets acquired from 40 and 160 patients, respectively. The evaluations were performed automatically through the challenge results submission system[1] and the images and ground truth for the test sets were not revealed to the challenge participants. For the validation set, only the MRI images were shared with participants without the ground truth annotations. The agreement between the segmentation predictions by the neural network and ground truth delineations was assessed in terms of the Dice metric and Hausdorff distance.

Table 1 reports the mean accuracy values for the validation and test sets for short and long axis images as well as the entire set of images. Similar performances were observed for both validation and test sets, and the algorithm yielded an average Dice metric value over 91% for all the cases. The example segmentation predictions on validation sets for the LV endocardium, LV myocardium, and RV are given in Fig. 2.

Table 2 reports the performance of the proposed algorithm for images acquired at the end-systolic and end-diastolic phases of the cardiac cycle from short and long axis sequences. The mean and standard deviation values for each evaluation metric are reported in the table. The proposed neural network approach yielded better performance for the end-diastolic phase than the end-systolic phase in terms of the Dice metric. All the average Dice metric values were above 90%. Figure 3 shows the performance of the proposed approach for short and long axis sequences at end-systolic and end-diastolic phases using box plots.

---

[1] https://competitions.codalab.org/.

Table 1. The performance of the proposed algorithm on the validation and test MRI datasets in terms of the Dice metric and Hausdorff distance. The table reports mean evaluation metrics for short and long axis images as well as the overall average value.

Evaluation	Short-axis	Long-axis	Overall
Validation set ($n = 40$)			
Dice metric (%)	92.16	92.22	92.18
Hausdorff distance (mm)	9.56	5.44	8.53
Test set ($n = 160$)			
Dice metric (%)	92.73	91.71	92.47
Hausdorff distance (mm)	10.05	6.16	9.08

Patient 161

Patient 181

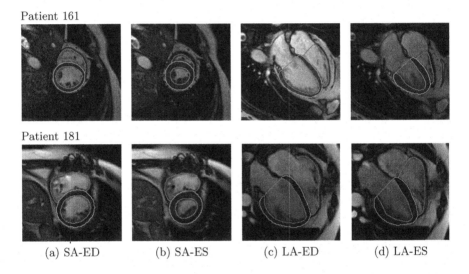

(a) SA-ED   (b) SA-ES   (c) LA-ED   (d) LA-ES

Fig. 2. Example prediction results by the proposed method on short and long axis sequences from the validation set. The green, orange, and blue contours indicate the LV endocardial, LV myocardial, and RV borders, respectively. (Color figure online)

Table 2. Overall performance of the proposed method evaluated over MRI test datasets acquired from 160 patients in terms of Dice metric and Hausdorff distance for the RV segmentation.

Evaluation metric	Short-axis end-diastole	Short-axis end-systole	Long-axis end-diastole	Long-axis end-systole
Dice (%)	94.03 ± 3.65	91.42 ± 5.94	93.05 ± 8.21	90.37 ± 6.74
Hausdorff (mm)	10.12 ± 7.55	9.99 ± 6.03	6.34 ± 4.78	5.98 ± 4.06

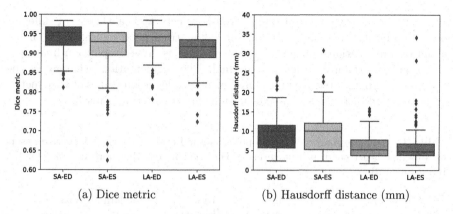

(a) Dice metric (b) Hausdorff distance (mm)

Fig. 3. The box plots showing the performance of the proposed approach over MRI test sets acquired from 160 patients for the RV segmentation. The accuracy of the proposed method was evaluated for short axis images at end-diastole (SA-ED) and end-systole (SA-ES) as well as long axis images at end-diastole (LA-ED) and end-systole (LA-ES). The evaluations were performed using the Dice metric, and Hausdorff distance.

We also evaluated the RV segmentation results for the test set in terms of a reliability measure defined as the complementary cumulative distribution function (CCDF) of the obtained Dice metric. The reliability metric, $R(a)$ indicates the number of segmentations with a Dice value higher than a:

$$R(a) = Pr(Dice > a) \tag{1}$$

The reliability curves for the RV delineations from short and long axis images at the end-diastolic and end-systolic phases are given in Fig. 4.

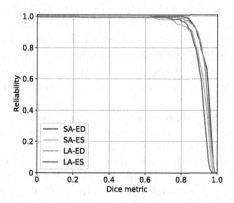

Fig. 4. Reliability curves for the short and long axis sequences at the end diastolic (SA-ED and LA-ED) and end-systolic (SA-ES and LA-ES) phases.

The test datasets were acquired from subjects with a variety of different heart conditions, including tetralogy of Fallot, tricuspidal regurgitation, interatrial communication, congenital arrhythmogenesis, hypertrophic cardiomyopathy, dilated cardiomyopathy, and normal. The segmentation accuracy of the proposed method in terms of mean and standard deviation values of Dice metric and Hausdorff distance for these conditions are reported in Table 3.

Table 3. The performance of the proposed method evaluated over MRI test datasets for different heart conditions in terms of Dice metric and Hausdorff distance for the RV segmentation.

Heart condition	Dice short-axis	Dice long-axis	Hausdorff short-axis	Hausdorff long-axis
Tetralogy of fallot	92.98 ± 3.00	91.42 ± 3.41	10.77 ± 2.31	7.35 ± 3.33
Tricuspidal regurgitation	92.13 ± 3.74	91.50 ± 3.61	9.94 ± 4.19	5.96 ± 2.83
Interatrial communication	91.15 ± 4.79	88.61 ± 20.55	11.37 ± 4.60	7.55 ± 9.79
Congenital arrhythmogenesis	93.97 ± 4.15	93.46 ± 3.45	8.71 ± 6.03	5.01 ± 2.09
Hypertrophic cardiomyopathy	93.70 ± 5.76	92.83 ± 3.36	9.33 ± 5.38	5.47 ± 3.09
Dilated right ventricle	92.34 ± 4.06	89.77 ± 10.38	10.09 ± 3.66	7.48 ± 5.12
Dilated left ventricle	93.04 ± 6.75	91.48 ± 5.83	12.72 ± 13.22	6.29 ± 4.31
Normal	92.51 ± 5.27	93.30 ± 3.28	8.27 ± 4.49	5.19 ± 3.33

4   Conclusion

In this study, we applied a deep convolutional neural network based nnU-Net approach to automatically delineate the RV from MRI images at the end-systole and end-diastole from short and long axis sequences. The proposed approach was evaluated using validation and test sets acquired from 40 and 160 patients, respectively. The evaluations were performed by the challenge organizers through an automated system. The results demonstrated a very high agreement between the automated predictions and manual delineations in terms of the Dice metric and Hausdorff distance. Future work will utilize cross-reference information from short and long axis sequences to improve segmentation accuracy.

Acknowledgment. The authors wish to thank the challenge organizers for providing train and test datasets as well as performing the algorithm evaluation. The authors of this paper declare that the segmentation method they implemented for participation in the M&Ms challenge has not used any pre-trained models nor additional MRI datasets other than those provided by the organizers. A. Carscadden was supported by an Undergraduate Student Research Award by the Natural Sciences and Engineering Research Council of Canada (NSERC). This research was enabled in part by computing support provided by Compute Canada (www.computecanada.ca) and WestGrid.

References

1. Antonelli, M., et al.: The medical segmentation decathlon. arXiv:2106.05735, June 2021
2. Avendi, M.R., Kheradvar, A., Jafarkhani, H.: Automatic segmentation of the right ventricle from cardiac MRI using a learning-based approach. Magn. Reson. Med. **78**(6), 2439–2448 (2017)
3. Bernard, O., et al.: Deep learning techniques for automatic MRI cardiac multi-structures segmentation and diagnosis: is the problem solved? IEEE Trans. Med. Imaging **37**(11), 2514–2525 (2018)
4. Campello, V.M., et al.: Multi-centre, multi-vendor and multi-disease cardiac segmentation: the M&Ms challenge. IEEE Trans. Med. Imaging **40**(12), 3543–3554 (2021). https://doi.org/10.1109/TMI.2021.3090082
5. Caudron, J., Fares, J., Lefebvre, V., Vivier, P.H., Petitjean, C., Dacher, J.N.: Cardiac MRI assessment of right ventricular function in acquired heart disease: Factors of variability. Acad. Radiol. **19**(8), 991–1002 (2012)
6. Haddad, F., Hunt, S.A., Rosenthal, D.N., Murphy, D.J.: Right ventricular function in cardiovascular disease, Part I: anatomy, physiology, aging, and functional assessment of the right ventricle. Circulation **117**(11), 1436–1448 (2008)
7. Isensee, F., et al.: nnU-Net: self-adapting framework for U-Net-based medical image segmentation. arXiv:1809.10486, September 2018
8. Mahapatra, D., Buhmann, J.M.: Automatic cardiac RV segmentation using semantic information with graph cuts. In: 2013 IEEE 10th International Symposium on Biomedical Imaging (ISBI), pp. 1106–1109. IEEE (2013)
9. Nambakhsh, C.M., Peters, T.M., Islam, A., Ayed, I.B.: Right ventricle segmentation with probability product kernel constraints. In: Medical Image Computing and Computer-Assisted Intervention (2013)
10. Petitjean, C., et al.: Right ventricle segmentation from cardiac MRI: a collation study. Med. Image Anal. **19**(1), 187–202 (2015)
11. Punithakumar, K., Boulanger, P., Noga, M.: A GPU-accelerated deformable image registration algorithm with applications to right ventricular segmentation. IEEE Access **5**, 20374–20382 (2017)
12. Punithakumar, K., Noga, M., Ayed, I.B., Boulanger, P.: Right ventricular segmentation in cardiac MRI with moving mesh correspondences. Comput. Med. Imaging Graph. **43**, 15–25 (2015)
13. Punithakumar, K., Tahmasebi, N., Boulanger, P., Noga, M.: Convolutional neural network based automated RV segmentation for hypoplastic left heart syndrome MRI. In: 8th International Conference of Pattern Recognition Systems, pp. 1–6 (2017)
14. Ronneberger, O., Fischer, P., Brox, T.: U-Net: convolutional networks for biomedical image segmentation. In: Medical Image Computing and Computer-Assisted Intervention, vol. 9351, pp. 234–241 (2015). https://doi.org/10.1007/978-3-319-24574-4

3D Right Ventricle Reconstruction from 2D U-Net Segmentation of Sparse Short-Axis and 4-Chamber Cardiac Cine MRI Views

Lennart Tautz[1,2](✉)(iD), Lars Walczak[1,2](iD), Chiara Manini[1],
Anja Hennemuth[1,2](iD), and Markus Hüllebrand[1,2](iD)

[1] Charité – Universitätsmedizin Berlin, Berlin, Germany
[2] Fraunhofer MEVIS, Bremen, Germany
lennart.tautz@mevis.fraunhofer.de

Abstract. We reconstruct a 3D model of the right ventricle from short- and long-axis image data and evaluate the benefits compared to quantification based on the 2D image stack. Deep learning is used to extract short-axis contours. An initial surface representation based on the contours is refined using long-axis images. Using a deformable model, the surface around the basal plane is adapted to image data. The resulting models capture the shape of the right ventricle better than segmentation from short-axis images alone and allow for a more precise volumetry.

Keywords: Right ventricle · Segmentation · Deep learning · Deformable model

1 Background

Right ventricular (RV) function is a strong predictor of mortality in common diseases such as heart failure, pulmonary hypertension, or congenital heart disease [1]. MR imaging provides a non-invasive means for the quantification of RV dimensions. The RV is connected to the right atrium (RA) and the pulmonal artery (PA) via the tricuspid valve and the pulmonary valve (Fig. 1). In contrast to the left ventricle, its complex shape and high variability between subjects proves RV image analysis to be challenging. The boundaries to surrounding structures, such as the left ventricular outflow tract (LVOT), are often not well defined.

With deep learning approaches having proven useful for the segmentation of other cardiac structures, such as the left ventricle (LV) and left atrium (LA) [6,14], the MICCAI 2021 M&Ms-2 challenge, a follow-up of the first M&M challenge [6], provides a large multi-disease, multi-view, and multi-center data set to evaluate the state of the art of machine learning models for RV segmentation.

Most recent right ventricle segmentation approaches utilize machine and deep learning techniques and are trained on suitable reference image data [3,5,7,10,15], while a minority uses model-based segmentation for the task to integrate knowledge on topology and appearance [2,8].

E. Puyol Antón et al. (Eds.): STACOM 2021, LNCS 13131, pp. 352–359, 2022.
https://doi.org/10.1007/978-3-030-93722-5_38

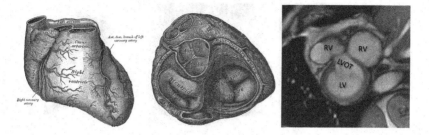

Fig. 1. Anatomy of the right heart. Left: Overview. Note the wedge-like shape between right atrium, pulmonary artery and apex. Middle: View from atria. Note the orientation difference between tricuspid and pulmonary valve. Images from public domain. Right: Basal slice of case 002 (ED phase). Note the weak contrast between RV and LVOT.

For quantification, typically a stack of short-axis (SAX) images is segmented to calculate end-diastolic and end-systolic volumes. The apex is often not covered well in this stack, and the identification of the right ventricular outflow tract (RVOT) and tricuspid valve plane is challenging, in particular when the short-axis planes are oriented along the left ventricle long axis [12]. The correct identification of the ventricular-atrial boundary has a large effect on the measured volumes [9]. The challenge data set contains long-axis (LAX) images to improve the detection and inclusion of these boundaries. Because the slices of the stack, as well as the long-axis image, are acquired in different breathholds, the cardiac structures can show misalignment. The volumetry from sparse 2D segmentations inherently contains a bias. A 3D segmentation of the RV would produce more accurate volumes, but would ideally require high-resolution imaging without motion artifacts.

2   Methods

The challenge image data consists of cine MRI short-axis stacks and corresponding long-axis views. The long axis, extending from the apex through the mitral valve, was used to acquire four-chamber (4CH) views, depicting both ventricles and atria at once. The perpendicular short axis shows cross-sections of the ventricles from apex to the valve plane (Fig. 2).

Our goal is to derive a 3D model from the dual-view image data to evaluate whether quantification benefits from such a representation when compared to quantification of the 2D image stack. First, contours are extracted from short-axis images using a deep learning model. Next, a coarse surface model is created from these contours, and extended towards apex and basal plane using information from the long-axis image. Finally, the surface is locally adapted to the image data around the basal plane.

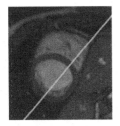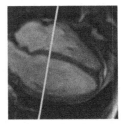

Fig. 2. Challenge image data orientations. Left: Short-axis view. Middle: Long-axis (4CH) view. Yellow shows the respective other orientation. Right: Relative orientation of both planes. (Color figure online)

2.1 Automatic 2D Contour Extraction

The automatic segmentation uses two 2D U-net models trained on the SAX and LAX images, respectively. As the testing set contains unseen pathologies, we included in-house data in the training to obtain a more generalizing model. We perform additional refinement training on the M&M data to adapt and fine-tune the model to the challenge data.

For both segmentation tasks, we employ a 2D U-net architecture. U-net architectures consistently outperformed other methods in recent cardiac segmentation challenges [4,6]. In our configuration, we use 4 layers, resampling to $1.5\,\text{mm}^2$, batch normalization, drop-out, a batch size of 18, and a patch size of 172×172 voxels (Fig. 3). Initial training used both M&M and in-house data, with a learning rate of 0.005. The refinement training used only the M&M data, with a learning rate of 0.0005 and approximately 30000 iterations.

The M&M challenge training data comprises 160 patients, with manual segmentations for LV (endocardium and epicardium) and RV on two time points, end-diastole (ED) and end-systole (ES). SAX images have a slice thickness of 10 mm, and all labels were used for training. A separate set of 40 cases was available as validation data. Our in-house training data contains 99 cases (including four children) with manually refined segmentations on 20–30 time points each. The data covers different LV diseases, and includes data from the ACDC challenge[1] [4]. As the ACDC data does not include RV ground truth, the segmentations were also created by experts at our institution.

2.2 3D Model Generation

A 3D surface model is generated from the 2D segmentations to improve quantification and volumetry. An initial model is triangulated directly from the short-axis segmentation contour points. Apex and basal plane are estimated from the long-axis segmentation result. The apical point is identified in the LAX segmentation as the lowest point with respect to the short axis planes (Fig. 4).

---

[1] https://acdc.creatis.insa-lyon.fr/.

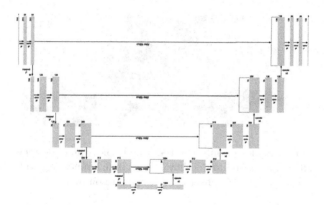

Fig. 3. Architecture of proposed 2D U-Net.

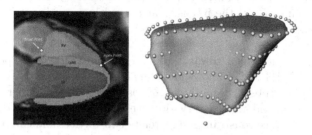

Fig. 4. Left: Segmentation result with apical and basal orientation points. Right: Initial surface model $M_{initial}$ with sparse contour points used for triangulation.

Triangulation between the apex and the contours produces the initial model, $M_{initial}$.

To obtain an approximate basal plane, the atrium-facing part of the LAX segmentation contour is determined. First, the basal and apical junction regions between the two ventricles are identified from distance transforms on the label images. Distances transforms are calculated for the left heart, right heart, and background labels. The sum of these transforms is thresholded with $\tau_{junction}$ (5 mm), yielding areas that are close to each of the three major regions. The shortest connection along the contour between these regions marks the interventricular septum. Then, starting from the basal junction region, the remaining contour points are added iteratively to the basal boundary until the tortuosity of the boundary exceeds a threshold $\tau_{boundary}$ (1.1) (Fig. 5). The vector between the end points of this boundary together with the image normal defines the basal plane, p. The largest distance between boundary points and plane forms the estimated ventricle offset, k, of the ventricle.

$M_{initial}$ is restricted using the plane p, discarding parts above the plane. As the right ventricle can extend further, we adapt the restricted model, $M_{restricted}$, to the image data and information from the LAX images. Points on the center line of the most basal segmentation contour are moved along the short-axis

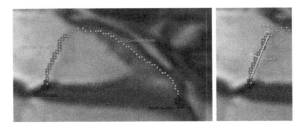

Fig. 5. Basal boundary detection. Left: Basal and apical junction regions and septum points (red), all candidate points (yellow), and final boundary points (blue). Right: Detected plane p, and ventricle offset k. (Color figure online)

image normal by the estimated offset k, but at least by the slice thickness, to form target points T_n. If the intensity at a resulting position is below the 10% quantile of the SAX histogram under the segmentation mask, the target point is discarded.

$M_{restricted}$ is deformed towards the target points T_n using position-based dynamics (PBD) [11]. PBD allows modeling soft bodies using sets of constraints. Constraints relate subsets of mesh vertex positions with each other. Here, we use distance, bending, and area conservation constraints to model the right ventricular material. Distance constraints, e.g., relate two vertices and can be interpreted as an elastic spring. Bending constraints can be used to smooth the surface. Area constraints preserve the overall surface area. Each type of constraint uses a parameter to model compliance. This model has been previously applied to simulate mitral valve behavior [13]. For deformation, a force in the direction to the target points T_n is applied to the nearby vertices (cf. yellow vertices in Fig. 6). The simulation loop uses Newton's second law of motion to integrate forces and uses a Gauss-Seidel type solver to iteratively satisfy the constraints. Here, we run the simulation for a fixed number of ι iterations (400).

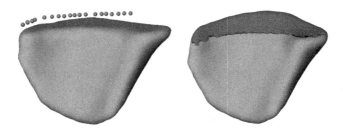

Fig. 6. Left: Model restricted by basal plane estimation ($M_{restricted}$), and target points. Right: Final model adapted to image around basal plane. Deformed region in orange for emphasis. (Color figure online)

Table 1. Segmentation results: Dice coefficients and Hausdorff distances. All values rounded to two significant digits.

View	Phase	Dice			View	Phase	Hausdorff [mm]		
		Mean ± SD	Min	Max			Mean ± SD	Min	Max
SAX	ED	0.88 ± 0.09	0.00	0.97	SAX	ED	17.02 ± 10.11	2.76	57.26
	ES	0.84 ± 0.12	0.00	0.96		ES	17.80 ± 9.99	3.34	57.90
	Both	0.86 ± 0.11	0.00	0.97		Both	17.41 ± 10.04	2.76	57.90
LAX	ED	0.84 ± 0.20	0.00	0.97	LAX	ED	16.43 ± 34.50	2.45	300.00
	ES	0.81 ± 0.21	0.00	0.96		ES	18.40 ± 45.31	2.50	300.00
	Both	0.82 ± 0.20	0.00	0.97		Both	17.42 ± 40.22	2.45	300.00
Total	Both	0.84 ± 0.17	0.00	0.97	Total	Both	17.41 ± 29.29	2.45	300.00

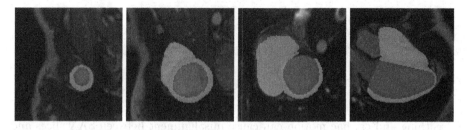

Fig. 7. Selected segmentation results for case 006 (ED phase). Left to right: Apical, mid-ventricular, and basal short-axis slices, long-axis image. Left ventricle in red, left myocardium yellow, right ventricle red. (Color figure online)

3 Results

3.1 Automatic 2D Segmentation

The quality of the 2D segmentation was quantified using the Dice coefficient and the Hausdorff distance (see Table 1). Examples for the results are shown in Fig. 7.

3.2 Comparison of 2D and 3D Volumetry

We compared the volumes of the 2D SAX segmentations and 3D models for the 40 challenge validation cases, and for one high-resolution whole-heart image (Fig. 8). For the SAX segmentations, the volume was calculated as the sum of all slice-wise voxel volumes. For the 3D models, the volume was calculated as the sum of the signed volumes of all tetrahedrons associated with the mesh faces.

For the validation cases, the volumes of the SAX segmentations were in the range from 33218.58 to 315389.05 mm^3, the volumes of the 3D models in the range from 18531.37 to 261955.61 mm^3. The relative differences between SAX and model volumes were in the range of 0.16 to 0.80 (mean 0.30 ± 0.11). For

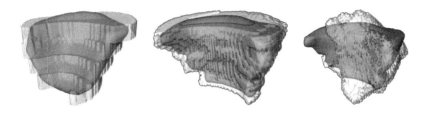

Fig. 8. Masks as 3D volume rendering (yellow) and 3D models (blue), apex at bottom. Left: Case 007 (ES phase). Note the effect of slice thickness, smoothness of the model, and differences in basal and apical regions. Middle: Comparison with automatic segmentation mask for whole-heart case with 1.5 mm slice thickness. Note the higher similarity of covered volumes. Right: Comparison with reference segmentation mask for whole-heart case. Note the differences at apex and around the basal plane. (Color figure online)

the whole-heart, reference volume was 83965.5 mm^3, automatic segmentation volume was 94846.9 mm^3, and 3D model volume was 77855.8 mm^3.

4   Discussion

Breathing and cardiac motion can cause misalignment between SAX slices and between SAX and LAX images, which manifests in the quality of the derived 3D model. The deviation between the orientation of the short-axis stack and the main axis of the RV makes detection of the basal plane challenging, as the necessary image information is blurred due to the large slice thickness. An adapted short-axis orientation with reduced slice thickness, or additional long-axis views may be helpful to obtain a better fitting model.

The volume comparison between SAX segmentations and 3D model shows considerable differences, with consistently smaller volumes from the model. Potential causes for this include the simplified SAX volume calculation which does not reflect the true ventricle shape and results in overestimation, a shrinking effect during surface creation, and the incorrect identification of the basal plane which can impact the volume calculation substantially, as noted by Jorstig et al. [9]. Comparison with the whole-heart case indicates that segmentation in short axis underestimates the RV segmentation.

5   Conclusion

We presented a method to create a 3D model of the right ventricle from sparse dual-view image data. The resulting model captures the shape of the RV better and allows for an improved volumetry. A better alignment of SAX and LAX images will benefit the quality of the reconstruction.

References

1. Amsallem, M., et al.: Forgotten no more: a focused update on the right ventricle in cardiovascular disease. JACC: Heart Failure **6**(11), 891–903 (2018)
2. Arrieta, C., et al.: Simultaneous left and right ventricle segmentation using topology preserving level sets. Biomed. Signal Process. Control **33**, 88–95 (2017). https://doi.org/10.1016/j.bspc.2016.11.002
3. Avendi, M.R., et al.: Automatic segmentation of the right ventricle from cardiac MRI using a learning-based approach: automatic segmentation using a learning-based approach. Magn. Reson. Med. **78**(6), 2439–2448 (2017). https://doi.org/10.1002/mrm.26631
4. Bernard, O., et al.: Deep learning techniques for automatic MRI cardiac multi-structures segmentation and diagnosis: is the problem solved? IEEE Trans. Med. Imaging **37**(11), 2514–2525 (2018). https://doi.org/10.1109/TMI.2018.2837502
5. Borodin, G., Senyukova, O.: Right ventricle segmentation in cardiac MR images using U-Net with partly dilated convolution. In: Kůrková, V., Manolopoulos, Y., Hammer, B., Iliadis, L., Maglogiannis, I. (eds.) ICANN 2018. LNCS, vol. 11140, pp. 179–185. Springer, Cham (2018). https://doi.org/10.1007/978-3-030-01421-6_18
6. Campello, V.M., et al.: Multi-centre, multi-vendor and multi-disease cardiac segmentation: the M&Ms challenge. IEEE Trans. Med. Imaging **40**(12), 3543–3554 (2021). https://doi.org/10.1109/TMI.2021.3090082
7. Chen, J., et al.: Correlated regression feature learning for automated right ventricle segmentation. IEEE J. Transl. Eng. Health Med. **6**, 1–10 (2018). https://doi.org/10.1109/JTEHM.2018.2804947
8. Ghelich Oghli, M., et al.: A hybrid graph-based approach for right ventricle segmentation in cardiac MRI by long axis information transition. Physica Med. **54**, 103–116 (2018). https://doi.org/10.1016/j.ejmp.2018.09.011
9. Jorstig, S.H., et al.: Calculation of right ventricular stroke volume in short-axis MR images using the equation of the tricuspid plane. Clin. Physiol. Funct. Imaging **32**(1), 5–11 (2012). https://doi.org/10.1111/j.1475-097X.2011.01047.x
10. Li, J., et al.: Dilated-inception net: multi-scale feature aggregation for cardiac right ventricle segmentation. IEEE Trans. Biomed. Eng. **66**(12), 3499–3508 (2019). https://doi.org/10.1109/TBME.2019.2906667
11. Müller, M., et al.: Position based dynamics. J. Vis. Commun. Image Represent. **18**(2), 109–118 (2007). https://doi.org/10.1016/j.jvcir.2007.01.005
12. Strugnell, W.E., et al.: Modified RV short axis series-a new method for cardiac MRI measurement of right ventricular volumes. J. Cardiovasc. Magn. Reson. **7**(5), 769–774 (2005). https://doi.org/10.1080/10976640500295433
13. Walczak, L., et al.: Using position-based dynamics for simulating the mitral valve in a decision support system. In: Kozlíková, B., et al. (eds.) Eurographics Workshop on Visual Computing for Biology and Medicine. The Eurographics Association (2019). https://doi.org/10.2312/vcbm.20191242
14. Xiong, Z., et al.: A global benchmark of algorithms for segmenting the left atrium from late gadolinium-enhanced cardiac magnetic resonance imaging. Med. Image Anal. **67**, 101832 (2021). https://doi.org/10.1016/j.media.2020.101832
15. Zhang, L., et al.: Fully automatic segmentation of the right ventricle via multi-task deep neural networks. In: 2018 IEEE International Conference on Acoustics, Speech and Signal Processing (ICASSP), pp. 6677–6681. IEEE (2018). https://doi.org/10.1109/ICASSP.2018.8461556

Late Fusion U-Net with GAN-Based Augmentation for Generalizable Cardiac MRI Segmentation

Yasmina Al Khalil$^{(\boxtimes)}$, Sina Amirrajab, Josien Pluim, and Marcel Breeuwer

Eindhoven University of Technology, Eindhoven, The Netherlands
y.al.khalil@tue.nl

Abstract. Accurate segmentation of the right ventricle (RV) in cardiac magnetic resonance (CMR) images is crucial for ventricular structure and function assessment. However, due to its variable anatomy and ill-defined borders, RV segmentation remains an open problem. While recent advances in deep learning show great promise in tackling these challenges, such methods are typically developed on homogeneous data-sets, not reflecting realistic clinical variation in image acquisition and pathology. In this work, we develop a model, aimed at segmenting all three cardiac structures in a multi-center, multi-disease and multi-view setting, using data provided by the M&Ms-2 challenge. We propose a pipeline addressing various aspects of segmenting heterogeneous data, consisting of heart region detection, augmentation through image synthesis and multi-fusion segmentation. Our extensive experiments demonstrate the importance of different elements of the pipeline, achieving competitive results for RV segmentation in both short-axis and long-axis MR images.

Keywords: Cardiac MR segmentation · Medical image synthesis · Late-fusion network · Domain generalization

1 Introduction

Right ventricular (RV) structure and function assessment is vital for the management of most cardiac disorders, such as pulmonary hypertension, coronary heart disease, cardiomyopathies and dysplasia [7,22,27]. Magnetic resonance imaging (MRI) has become a standard tool for RV function evaluation, as its noninvasive nature and good contrast allow for an accurate computation of cardiac functional parameters [3]. A prerequisite to this assessment is obtaining a correct segmentation of the RV cavity in MR images. However, due to the complex heart motion and RV anatomy, RV segmentation is challenging, typically resulting in lower performance and larger susceptibility to variation [10,12]. Particular aspects impacting the RV segmentation performance include: 1) unclear cavity borders due to blood flow and partial volume effects, 2) presence of trabeculations in the cavity, exhibiting the same gray levels as the surrounding myocardium, 3) the

---

Y. Al Khalil and S. Amirrajab—Contributed equally.

© Springer Nature Switzerland AG 2022
E. Puyol Antón et al. (Eds.): STACOM 2021, LNCS 13131, pp. 360–373, 2022.
https://doi.org/10.1007/978-3-030-93722-5_39

complex shape of the RV, which significantly varies throughout the volume, 4) the significant influence of pathology on morphological variations in the RV, and 5) the use of short axis (SA) image orientation, which is efficient for the analysis of both heart ventricles, but is not fully optimized for the RV [17,22,30]. As a consequence, there is a lack of research on automating RV segmentation, resulting in RV delineation still performed manually in clinical routines, which comes with a number of well-known challenges [8,18].

A number of algorithms have been developed in recent years to tackle the complexity of RV segmentation, ranging from model-based [5,11,20,31] to image-driven approaches [4,23,28,30]. While model-based algorithms can be quite powerful, image-driven algorithms, such as deep learning-based methods, are more prevalent due to better robustness against pathology and acquisition variations. However, data scarcity presents significant difficulties to segmentation, often encountered due to technical, ethical and financial constraints [14,24,26]. This often leads to algorithms trained and evaluated only using samples from single centers, where similar scanner models and imaging protocols are employed, posing the question of their applicability to more heterogeneous data [29]. Attempts to tackle segmentation generalization have been very recent, with the latest one being a M&Ms-1[1] challenge, organized to evaluate the performance of various segmentation strategies for SA CMR images [6], acquired from multi-site and multi-vendor data. However, the focus of this challenge, as well as the majority of the research, is on the joint segmentation of cardiac chambers, with relatively fewer attempts focused on the RV, especially RV pathologies. In this work, we address the problematic aspects of RV segmentation in both LA and SA images by applying a number steps aimed at tackling issues appearing in multi-center, multi-disease and multi-view cardiac MRI data, such as:

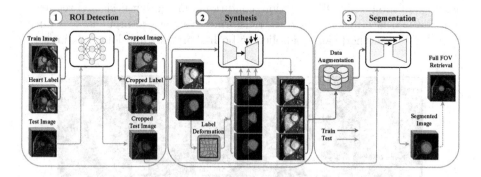

Fig. 1. Proposed pipeline including the ROI detection module (left), image synthesis module (middle), and image segmentation module (right). (Color figure online)

---

[1] https://www.ub.edu/mnms/.

1. **Variations in the field of view (FOV)** caused by changes in resolution and voxel size, which directly affects both the size of the heart in the image, as well as the visibility of the background tissue.
2. **Variations in contrast and intensity** present an additional challenge to DL algorithms as they primarily rely on good contrast between tissues, consistent texture patterns and clear boundaries in regions of interest.
3. **Limited data** hampers the performance of DL methods, while curating a diversified and balanced data-set with annotations is challenging.
4. **Basal and apical slices** are particularly challenging to segment due to the complex structure and size.
5. **Presence of pathologies** altering heart tissue intensity levels and shape can drastically affect the segmentation performance, especially of the RV.

To address the above challenges, we propose synthesis and segmentation pipelines designed to minimize the disruptions caused by variations present in multi-vendor, multi-site and multi-disease data. To reduce the effect of variations in the FOV and heart size, we introduce a heart region detection module (Sect. 2.2). We utilize conditional synthesis to generate a large number of highly realistic images with variations in contrast, heart appearance and pathology (Sect. 2.3), with an emphasis on basal slices, as well as intensity transformations to emphasize the tissue shape (Sect. 2.4). Finally, to regularize the segmentation network performance and produce a robust model, we introduce a late fusion approach to training (Sect. 2.5).

2   Materials and Methods

The segmentation pipeline proposed in this work is depicted in Fig. 1, consisting of region of interest (ROI) detection, conditional image synthesis and segmentation through late fusion, utilizing transformed versions of input images during training. The method can be applied to both short-axis and long-axis images.

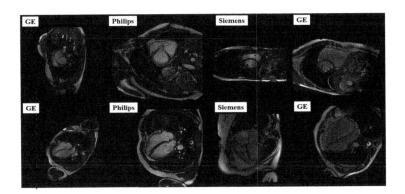

Fig. 2. Variations in field-of-view, image contrast and appearance, anatomy, and pathology for SA and LA images in the training set.

2.1 Data

The M&Ms-2 challenge data is comprised of 360 subjects with various RV and LV pathologies as well as a control group, distributed as shown in Table 1. The data was acquired using four different 1.5T scanners from three different manufacturer vendors (Siemens, GE, and Philips), undergoing variations in contrast and anatomy (see Fig. 2). The training subset includes 160 cases with expert annotations for RV and LV blood pools and LV myocardium (MYO). The validation set contains 40 cases, including two pathologies not present in the training set. The final algorithm is evaluated on an unknown test set containing 160 cases, with the appearance of all pathologies outlined in Table 1.

Table 1. The M&Ms-2 challenge data

Pathology	Training	Validation	Testing
Dilated right ventricle	0	5	25
Tricuspidal regurgitation	0	5	25
Tetralogy of fallot	20	5	10
Interatrial communication	20	5	10
Congenital arrhythmogenesis	20	5	10
Dilated left ventricle	30	5	25
Hypetrophic cardiomiopathy	30	5	25
Normal	40	5	30

2.2 Heart Detection

The first stage of the proposed pipeline consists of automated heart detection, whereby a bounding box is detected that encompasses the complete heart in SA images and the region of LV cavity, myocardium and RV cavity in LA images. The detected bounding boxes are then used to crop the input images of the heart for both segmentation and synthesis training, as well as at segmentation inference time. We generate the training labels from the ground-truth masks by computing the smallest bounding box that fits the entire heart in the FOV and expanding it by five voxels in each direction. A CNN is trained to output the adjustment parameters so that the bounding box, initialized at the center of the image, better fits the heart. A total of 1000 2D mid-cavity slices extracted from all training images are used for training and inference of the SA heart detection model and all available LA images for the LA heart detection model. Appendix A.1 contains more details about the training procedure.

2.3 Boosting Data Augmentation with Image Synthesis

The training data for LA and SA segmentation is further boosted using the image synthesis module depicted in Fig. 3a. This module includes a ResNet-based [13] encoder coupled with a mask-conditional generator that uses SPADE normal-ization layers [21] throughout the network architecture. The ResNet encoder is designed to extract style information of the input image and provide it to the SPADE generator that preserves the content of the input label map.

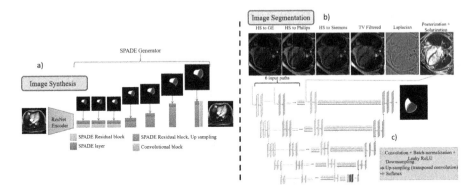

Fig. 3. a) Image synthesis model including residual-encoder and SPADE generator, b) 6 intensity transformations of each image at training and test time, and c) Late fusion multi-encoder U-Net used for the segmentation of LA and SA MR images. An example for LA images is shown, but the same approach is used for SA images.

Previous works [1,2] have shown the effectiveness of using SPADE based generators in translating input segmentation masks to realistic CMR images with plausible anatomy. In contrast to their work, our approach alleviates the need for providing multi-tissue segmentation mask for high-quality synthesis by adding the encoder network. Moreover, label manipulation is proposed to intro-duce anatomical variations in heart geometries of the synthesized images. Upon training the image synthesis model, indicated with blue arrows in Fig. 1, random elastic deformation and morphological dilation are applied on the segmentation masks to create new labels for synthesis. Subsequently, the deformed label paired with its corresponding synthesized image are used for augmentation. More infor-mation about image synthesis can be found in Appendix B.

2.4   Contrast Transformations

To address variations in contrast and intensity homogeneity, we generate six contrast-transformed versions per training image (Fig. 3b), using histogram standardization, edge preserving filtering, solarization and posterization, and a Laplacian filter. More details about their application and pre-processing can be found in Appendix A.2. The obtained transformations per image are fed into the network in a multi-modal fashion and applied at inference time on test images.

2.5 Segmentation and Post-processing

Inspired by multi-modal CNNs and late fusion approaches, we modify a typical U-Net to include multiple encoder layers processing each transformed image fed at the input (Fig. 3), whose outputs are fused in the bottleneck, with dense connections between the pairs of layers within the same path. This allows the network to learn complementary information between different transformations (see Sect. 2.2) of each image as well as a better representation of their inter-relationships, compared to early fusion (or multi-channel approaches). We compare this method to a single-channel network, fed with different variations of each image obtained through applying the same six different transformations at training time. Our hypothesis is that processing the same image represented by different features leads to the network being more robust to typical variations observed in multi-site/vendor and pathology-influenced images. As described above, in addition to the original training set, we augment the training using synthesized images, pre-processed to obtain six of their variants at train time. Both models for SA and LA segmentation are designed in a 2D fashion according to nnUNet [15] guidelines, with extensions to achieve a multi-encoder setup and trained by utilizing a 5-fold cross-validation approach, where the ensembled average output is taken as the final prediction. Models are trained with a total of 320 real images (ED and ES) and a total of 2000 synthetic images. We further apply connected component analysis to remove any false positive predictions. More details about the training can be found in Appendix C.

3 Experiments and Results

3.1 Experimental Setup

We compare the proposed approach to a number of different models containing a different set of pipeline components proposed in this work to evaluate where the most improvement is obtained. Thus, we train the following models: (i) **U-net + HS**, a regular single encoder U-Net with Histogram Standardization (**HS**) applied to all training images, without any Bounding Box (**BB**) detection; (ii) **U-Net + HS + BB**, a model similar to (i) but with heart region detection module added for both LA and SA segmentation; (iii) **U-Net + HS + BB + IT + Synth**, similar model to (ii) with added Synthetic data (**Synth**) at training time, obtained through the procedure described in Sect. 2.3, as well as a set of six Intensity Transformations (**IT**) per each image; (iv) **LF-U-Net + HS + BB + IT**, a late fusion (**LF**) approach combined with a bounding box detector for pre-processing; and (v) **LF-U-Net + HS + BB + IT + Synth**, the proposed model, which is a late fusion approach combined with a bounding box detector and the addition of synthetic data.

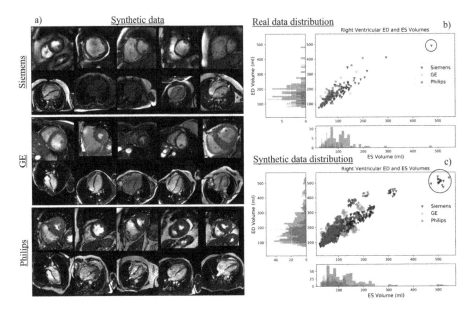

Fig. 4. a) Synthesized LA and SA examples per vendor and b) real versus c) synthetic data distribution corresponding to their RV volumes. Elastic deformation and dilation are applied on the labels to synthesize subjects with manipulated anatomy. Synthetic data distribution shows RV ED and ES volumes for 1000 synthesized cases per vendor. (Color figure online)

3.2   Image Synthesis and Segmentation Results

Figure 4a visualizes some of the synthesized examples per vendor, with elastic deformation and dilation applied on the ground truth labels during synthesis for both LA and SA views. Real (b) and synthetic (c) data distributions in Fig. 4 depict ED and ES volumes for real cases in the train set and synthesized cases in the augmented training set, respectively. For addressing imbalanced number of cases per vendor in the real train set, we generate 1000 synthesized cases per vendor to balance the data. Furthermore, within each vendor set, we synthesize more underrepresented cases scattered at the tail of the distribution by applying more deformations on their ground truth labels. The effects of label manipulation on the ED and ES volumes of RV can be seen from the deviations of the heart volumes for synthetic cases in Fig. 4b and c. For one real case in the Siemens train data (blue) indicated with a black circle in b), we have 15 corresponding synthetic cases with altered heart anatomies in c), indicated with a black circle.

Table 2. The performance of segmentation networks on the validation set

Segmentation model	Dice(avg)	Dice(SA)	Dice(LA)	HD(avg)	HD(SA)	HD(LA)
U-Net + HS	0.886	0.884	0.888	18.859	25.966	11.752
U-Net + HS + BB	0.899	0.898	0.899	11.079	12.141	10.018
U-Net + HS + BB + IT + Synth.	0.902	0.901	0.903	9.092	10.981	7.202
LF-U-Net + HS + BB + IT	0.909	0.909	0.908	8.614	10.602	6.625
LF-U-Net + HS + BB + IT + Synth.	**0.918**	**0.919**	**0.916**	**7.684**	**9.600**	**5.761**

HS: Histogram Standardization, BB: Bounding Box, IT: Intensity Transformation, Synth.: Synthetic data, LF-U-Net: Late-Fusion U-Net

Figure 5 depicts segmentation results obtained by different models on some of the examples in the validation set. Table 2 displays the performance scores obtained from the challenge submission platform on the 80 images in the validation set, in terms of Dice and Hausdorff distance (HD). We observe that the addition of the BB detection module significantly aids in decreasing the HD, particularly for SA images, as it largely eliminates false positives appearing due to large and varying FOV. Moreover, the addition of transformations and synthetic data further help the model in handling examples with unclear boundaries, low tissue contrast and distinct presence of pathology. However, the major improvement is seen with the employment of the late-fusion model combined with synthetic data, which allows the model to extract more variable image features through different encoding paths processing different representations of each image. The addition of synthetic images introduces more examples of hard cases and helps with network regularization and generalization. On the unseen test set, the model yields the scores of 0.912 and 0.895 in terms of Dice and 11.74 and 7.85 in terms of HD, for SA and LA images respectively.

4   Discussion

In this work, we propose a framework consisting of three deep neural networks in the context of Multi-Disease, Multi-View & Multi-Center RV Segmentation in Cardiac MRI (M&Ms-2) challenge. Our framework is comprised of ROI detection, image synthesis, and late-fusion segmentation, carefully designed to address various aspects of model generalization. The ROI detection module is developed to predict a tight bounding box around the organ of interest and crop the images so that the heart is at the center of the field of view. This naturally reduces the effects of complex organs visible in the background on the segmentation performance and takes care of the variation in the FOV and image resolution. The image synthesis model is a mask-conditional GAN that learns the mapping from a segmentation label to a corresponding realistic-looking image. To introduce variation in the geometry of the heart for each generated image, morphological operations such as elastic deformation and dilation are applied on the labels to synthesize subjects with altered heart anatomy. Physiologically-based deformations using mechanistic heart models could be applied on the heart labels to create anatomically plausible segmentation masks for generating more pathological cases.

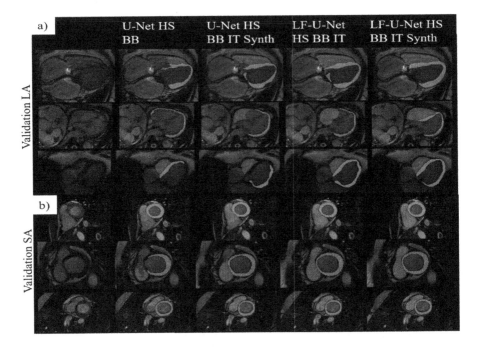

Fig. 5. Qualitative comparison of segmentation results for a) LA images and b) SA images in the validation set obtained using different models. The proposed model with late-fusion, synthetic data augmentation and all intensity transformations shown in last column achieves the best performance. Refer to Sect. 3.1 for model descriptions.

While our synthesis approach introduces more outlier examples influenced by the presence of pathology to the network, we additionally handle variations in contrast by introducing a set of image transformations that have been previously shown to aid DL-based model training in various tasks. By doing this, we attempt to diversify the type of images the network sees during training, as well as put more emphasis on tissue shape, encouraging its adaptation to multi-source images. Combined with the late fusion approach, the network can learn a large set of features that help improve its robustness to unseen data.

5   Conclusion and Future Work

In conclusion, we show that having a diverse training data-set plays the main role in obtaining a robust and generalizable segmentation model. However, since this is typically not possible, especially with pathology, we present an approach that utilizes recent advances in image synthesis and classical pre-processing methods, showing their impact on the segmentation performance in challenging scenarios, without needing to physically acquire more data. Future work consists of improving anatomy deformation to handle more pathological and outlier cases, as well as an in depth study on contrast transformations to handle variation in intensity distribution. Finally, we plan to explore other late fusion approaches, such as dense connections and inception modules in [9].

Acknowledgments. This research is a part of the openGTN project, supported by the European Union in the Marie Curie Innovative Training Networks (ITN) fellowship program under project No. 764465.

Appendix A. Pre-processing Stage

A.1. Heart Detection module

As presented in Sect. 2.2, the first stage of our pipeline is a heart region detection module, consisting of a regression-based neural network that locates and extracts the heart in both SA and LA images, similar to the approach used in [25]. Before generating the training labels, we resample all SA images to a median spatial resolution of $1.25 \times 1.25 \times 10 \, \text{mm}^3$ and all LA images to a spatial resolution of 1.25×1.25 before cropping. We use a simple CNN designed for a regression task, where the output consists of 6 continuous values. The inputs to the network are 2D (256×256) mid-cavity SA slices extracted from the training data-set and all LA slices, respectively, normalized to have the intensity values in the range of [0,1]. The outputs consist of parameters that define the bounding box, namely x and y directions of the center of the initialized ROI and its lower left corner, as well as the scaling factors for the width and height of the initial ROI.

The CNN consists of five convolutional layers, followed by two fully-connected layers with a linear activation. Each convolutional layer uses 3×3 kernels, followed by a 2×2 max-pooling layer. Batch normalization and leaky ReLU activations are used in each layer, except for the output. Dropout with the probability of 0.5 is used in the fully connected layers. The network is trained for 2000 epochs with a batch size of 32 and early stopping (assessed from the validation accuracy), by minimizing the mean squared error between the computed transformation and the actual transformation (estimated from the ground-truth) using the Adam optimizer. We start with an initial learning rate of 0.001 but decrease it by a factor of 0.5 every 250 epochs. All image dimensions and scaling/displacement parameters are normalized in a way to generate translations that are in the range from −1 to 1.

After prediction, all the parameters are de-normalized to reflect the original image scale. On-the-fly data augmentation is applied to the training images, consisting of random translation, rotation, scaling, vertical and horizontal flips, contrast augmentation and addition of noise. At inference time, we again use mid-cavity slices from the SA test images to obtain the adjustment parameters of the ROI (not needed for LA). The predicted bounding boxes on mid-cavity slices of SA images are then propagated through the whole 3D volume, from which these slices were extracted. This procedure is not applied for LA images, where direct detection is possible (both ED and ES LA images consist of a single slice only). The obtained cropped SA and LA images using the predicted bounding box are post-processed to be of the size 128×128 voxels and

176×176 voxels, respectively. These images are then used for training the cardiac cavity segmentation and synthesis networks.

A.2. Appearance Transformations for Targeting Variation in Contrast and Intensity

One of the main challenges of deploying a segmentation algorithm on heterogeneous data is its performance in the presence of extensive contrast and intensity variations. By exploring the provided training and validation sets, we observe that not only the data acquired from different vendors varies in contrast, but that the presence of pathology largely influences proper tissue visibility and often occludes tissue boundaries. Applying image appearance transformations can help with improving both the contrast and tissue visibility, as well as put more emphasis on tissue shape, rather than appearance. To achieve this, we select a set of six transformations per image, where each is fed into a separate encoding path during the training of the late fusion model, namely:

1. **Histogram standardization**: We standardize the intensities of images to those representative of each scanner vendor, by utilizing the algorithm in [19], which detects the landmarks on image histograms in the training set and averages them to form a standard landmark set per vendor. When a new image is acquired, the detected landmarks of its histogram are then matched to the previously computed standard positions by linear interpolation of intensities between the landmarks. A similar approach is applied at inference time using landmarks calculated from the training data. Thus, for each image, we generate its three counterparts, standardized to the landmarks extracted from GE, Siemens and Philips-acquired images.

2. **Edge preserving filtering**: To emphasize the shape of the heart cavities and discard high frequency features, we apply total variation filtering (TVF) on the original input image. TVF is typically used for denoising and produces images with flat domain separated by enhanced edges [16].

3. **Solarization and posterization**: Solarization can be defined as "partial" inversion of light and dark intensity values, with the total solarization being the negative of the image. Posterization retains the general appearance of the image, but gradual transitions are replaced by abrupt changes in shading from one region to another. This emphasizes edges, flattens the image, and is typically used for contour tracing.

4. **Laplacian filter**: The Laplacian of an image highlights regions of rapid intensity change and is therefore often used for edge detection.

Appendix B. Image Synthesis Models

Two identical image synthesis models are trained using LA and SA cardiac MR images. To augment and balance the data using these trained synthesis models, the following strategies are devised;

i) For each vendor-specific subset, the outlier cases are identified based on the end-diastolic or end-systolic volume for the RV calculated using the ground truth label of the SA images. These outlier cases, separated from the rest of the population, are used for image synthesis by applying random label deformations. For balancing the ratio, we apply different number of deformations in a way that we eventually create 1000 synthesized cases for each vendor including 50% outliers and 50% the rest of cases.

ii) For each subject, the ratio of the number of mid-ventricular and apical slices to the number of basal slices is not balanced in the SA stacks; there are typically 2–3 basal slices compared to 6–8 mid-ventricular and apical slices. The basal slices may not be frequently seen by the segmentation network during training compared to other slices. This could account for network failure on these challenging slices. To increase the occurrence of these examples, we utilize the labels of three most basal slices of all cases and randomly deform them 10 times for image synthesis.

Appendix C. Cardiac Cavity Segmentation Architecture and Training Procedure

The architecture of the late fusion U-Net segmentation model aims at learning separate convolutional encoder paths per each transformed image, whose features are fused at their higher layers (or the bottleneck). Here, we assume that higher-level representations from different transformations of each image are more complementary to each other, while containing distinctive features that aid the segmentation process. Each encoding path consists of five convolutional blocks, with four max-pooling layers. Each convolutional block consists of 3×3 kernel convolutional layers, batch normalization and leaky ReLU activation. We apply batch normalization to improve regularization and help the network be less susceptible to noise and intensity variation. Moreover, we apply dropout regularization, with a rate of 0.5, after each concatenating operation to further avoid over-fitting.

To increase robustness and cover a wide range of variations in terms of heart pose and size, we additionally augment the training set by applying data augmentation. Namely, we apply random vertical and horizontal flips ($p = 0.5$), random rotation by integer multiples of $\frac{\pi}{2}$ ($p = 0.5$), random scaling with a scale factor s $\in [0.8, 1.2]$ ($p = 0.2$), random translations ($p = 0.3$) and mirroring ($p = 0.5$). All augmentations are applied on the fly during training. At inference time, besides normalization and in-plane re-sampling, we apply a set of six transformations to generate six images at the input to the model. After pre-processing, each encoding path is fed with batches of 144 128×128 images for training for the SA segmentation model and batches of 64 256×256 images for the LA model. We use a validation set to track the training progress and identify overfitting, where the same augmentation approach is applied to the validation set and the mean Dice score is calculated per each epoch. To train the network, we use a weighted sum of the categorical cross-entropy and Dice loss. We use Adam for optimization, with an initial learning rate 10^{-4} and a weight decay of $3 \cdot e^{-5}$. During training, the learning rate is reduced by a factor of 5 if the validation loss does not improve by at least $5 \cdot 10^{-3}$ for 50 epochs. We apply early stopping on the validation set to avoid overfitting and select the model with the highest accuracy. We train each model (LA and SA) using a five-fold cross-validation on the training cases and use them as an ensemble to predict on the validation or testing set. The training of all models runs for 1000 epochs.

References

1. Abbasi-Sureshjani, S., Amirrajab, S., Lorenz, C., Weese, J., Pluim, J., Breeuwer, M.: 4D semantic cardiac magnetic resonance image synthesis on XCAT anatomical model. In: Medical Imaging with Deep Learning, pp. 6–18. PMLR (2020)
2. Amirrajab, S., et al.: XCAT-GAN for synthesizing 3D consistent labeled cardiac MR images on anatomically variable XCAT phantoms. In: International Conference on Medical Image Computing and Computer-Assisted Intervention, pp. 128–137 (2020)
3. Attili, A.K., Schuster, A., Nagel, E., Reiber, J.H., van der Geest, R.J.: Quantification in cardiac MRI: advances in image acquisition and processing. Int. J. Cardiovasc. Imaging **26**(1), 27–40 (2010)
4. Avendi, M.R., Kheradvar, A., Jafarkhani, H.: Automatic segmentation of the right ventricle from cardiac MRI using a learning-based approach. Magn. Reson. Med. **78**(6), 2439–2448 (2017)
5. Bai, W., et al.: A probabilistic patch-based label fusion model for multi-atlas segmentation with registration refinement: application to cardiac mr images. IEEE Trans. Med. Imaging **32**(7), 1302–1315 (2013)
6. Campello, V.M., et al.: Multi-centre, multi-vendor and multi-disease cardiac segmentation: the M&MS challenge. IEEE Trans. Med. Imaging **40**(12), 3543–3554 (2021)
7. Caudron, J., Fares, J., Vivier, P.H., Lefebvre, V., Petitjean, C., Dacher, J.N.: Diagnostic accuracy and variability of three semi-quantitative methods for assessing right ventricular systolic function from cardiac mri in patients with acquired heart disease. Eur. Radiol. **21**(10), 2111–2120 (2011)
8. Chen, C., et al.: Deep learning for cardiac image segmentation: a review. Front. Cardiovasc. Med. **7**, 25 (2020)
9. Dolz, J., Desrosiers, C., Ayed, I.B.: IVD-net: intervertebral disc localization and segmentation in MRI with a multi-modal UNet. In: International Workshop and Challenge on Computational Methods and Clinical Applications for Spine Imaging, pp. 130–143 (2018)
10. Grosgeorge, D., Petitjean, C., Caudron, J., Fares, J., Dacher, J.N.: Automatic cardiac ventricle segmentation in MR images: a validation study. Int. J. Comput. Assist. Radiol. Surg. **6**(5), 573–581 (2011)
11. Grosgeorge, D., Petitjean, C., Dacher, J.N., Ruan, S.: Graph cut segmentation with a statistical shape model in cardiac MRI. Comput. Vis. Image Underst. **117**(9), 1027–1035 (2013)
12. Haddad, F., Hunt, S.A., Rosenthal, D.N., Murphy, D.J.: Right ventricular function in cardiovascular disease, part i: anatomy, physiology, aging, and functional assessment of the right ventricle. Circulation **117**(11), 1436–1448 (2008)

13. He, K., Zhang, X., Ren, S., Sun, J.: Deep residual learning for image recognition. In: IEEE Conference on Computer Vision and Pattern Recognition, pp. 770–778 (2016)
14. Hosny, A., Parmar, C., Quackenbush, J., Schwartz, L.H., Aerts, H.J.: Artificial intelligence in radiology. Nat. Rev. Cancer 18(8), 500–510 (2018)
15. Isensee, F., Jaeger, P.F., Kohl, S.A., Petersen, J., Maier-Hein, K.H.: nnU-Net: a self-configuring method for deep learning-based biomedical image segmentation. Nat. Methods 18(2), 203–211 (2021)
16. Li, B., Que, D.: Medical images denoising based on total variation algorithm. Procedia Environ. Sci. 8, 227–234 (2011)
17. Marchesseau, S., Ho, J.X., Totman, J.J.: Influence of the short-axis cine acquisition protocol on the cardiac function evaluation: a reproducibility study. Eur. J. Radiol. Open 3, 60–66 (2016)
18. Martin-Isla, C., et al.: Image-based cardiac diagnosis with machine learning: a review. Front. Cardiovasc. Med. 7, 1 (2020)
19. Nyúl, L.G., Udupa, J.K., Zhang, X.: New variants of a method of MRI scale standardization. IEEE Trans. Med. Imaging 19(2), 143–150 (2000)
20. Ou, Y., Doshi, J., Erus, G., Davatzikos, C.: Multi-atlas segmentation of the cardiac MR right ventricle. In: Proceedings of 3D Cardiovascular Imaging: A MICCAI Segmentation Challenge (2012)
21. Park, T., Liu, M.Y., Wang, T.C., Zhu, J.Y.: Semantic image synthesis with spatially-adaptive normalization. In: Proceedings of the IEEE/CVF Conference on Computer Vision and Pattern Recognition, pp. 2337–2346 (2019)
22. Petitjean, C., Zuluaga, M.A., et al.: Right ventricle segmentation from cardiac MRI: a collation study. Med. Image Anal. 19(1), 187–202 (2015)
23. Ringenberg, J., Deo, M., Devabhaktuni, V., Berenfeld, O., Boyers, P., Gold, J.: Fast, accurate, and fully automatic segmentation of the right ventricle in short-axis cardiac MRI. Comput. Med. Imaging Graph. 38(3), 190–201 (2014)
24. Rumsfeld, J.S., Joynt, K.E., Maddox, T.M.: Big data analytics to improve cardiovascular care: promise and challenges. Nat. Rev. Cardiol. 13(6), 350 (2016)
25. Scannell, C.M., et al.: Deep-learning-based preprocessing for quantitative myocardial perfusion MRI. J. Magn. Reson. Imaging 51(6), 1689–1696 (2020)
26. Shameer, K., Johnson, K.W., Glicksberg, B.S., Dudley, J.T., Sengupta, P.P.: ML in cardiovascular medicine: are we there yet? Heart 104(14), 1156–1164 (2018)
27. Simon, M.A.: Assessment and treatment of right ventricular failure. Nat. Rev. Cardiol. 10(4), 204–218 (2013)
28. Wang, C.W., Peng, C.W., Chen, H.C.: A simple and fully automatic right ventricle segmentation method for 4-dimensional cardiac MR images. In: Proceedings of MICCAI RV Segmentation Challenge (2012)
29. Yan, W., Huang, L., Xia, L., et al.: MRI manufacturer shift and adaptation: increasing the generalizability of deep learning segmentation for MR images acquired with different scanners. Radiol. Artif. Intell. 2(4), e190195 (2020)
30. Yilmaz, P., Wallecan, K., Kristanto, W., Aben, J.P., Moelker, A.: Evaluation of a semi-automatic right ventricle segmentation method on short-axis MR images. J. Digit. Imaging 31(5), 670–679 (2018)
31. Zuluaga, M.A., Cardoso, M.J., Modat, M., Ourselin, S.: Multi-atlas propagation whole heart segmentation from MRI and CTA using a local normalised correlation coefficient criterion. In: Ourselin, S., Rueckert, D., Smith, N. (eds.) FIMH 2013. LNCS, vol. 7945, pp. 174–181. Springer, Heidelberg (2013). https://doi.org/10.1007/978-3-642-38899-6_21

Using Out-of-Distribution Detection for Model Refinement in Cardiac Image Segmentation

Francesco Galati[✉] and Maria A. Zuluaga[iD]

Data Science Department, EURECOM, Sophia Antipolis, Biot, France
{galati,zuluaga}@eurecom.fr

Abstract. We introduce a new learning framework that builds upon the recent progress achieved by methods for quality control (QC) of image segmentation to address the poor generalisation of deep learning models in Out-of-Distribution (OoD) data. Under the assumption that the label space is consistent across data coming from different distributions, we use the information provided by a QC module as a proxy of the segmentation model's performance in unseen data. If the model's performance is poor, the QC information is used as feedback to refine the training of the segmentation model, thus adapting to the OoD data. Our method was evaluated in the context of the Multi-Disease, Multi-View & Multi-Center Right Ventricular Segmentation in Cardiac MRI Challenge reporting average Dice Score and Hausdorff distance of 0.905 and 10.472, respectively.

Keywords: Out-of-Distribution Detection · Quality Control · Semi-Supervised Model Refinement · Cardiac Segmentation

1 Introduction

Deep learning (DL) techniques have demonstrated the capability to reproduce the analysis of an expert in cardiac image segmentation from cardiac magnetic resonance (CMR) imaging [3]. Despite their success, they still suffer from two major drawbacks that hinder their translation into clinical practice.

First, unlike experts, DL methods may generate anatomically impossible segmentation results [3], which are an important risk in clinical use. Automated quality control (QC) tools have been proposed [14,15] to assess the quality of a segmentation in the absence of ground truth and flag erroneous segmentations, so they can be discarded from further clinical analysis. The information about the detected erroneous segmentations, which is a surrogate measure of the segmentation model's performance, is generally not incorporated as feedback to the model.

Supported by the French government, through the 3IA Côte d'Azur Investments in the Future project managed by the National Research Agency (ANR) (ANR-19-P3IA-0002).

E. Puyol Antón et al. (Eds.): STACOM 2021, LNCS 13131, pp. 374–382, 2022.
https://doi.org/10.1007/978-3-030-93722-5_40

Second, DL methods still fail to generalise to out-of-distribution (OoD) samples, i.e. data from other domains than the one of the training set [2]. Unfortunately, this may often be the case at inference time, where new samples may come from a different scanner, acquisition protocol, or population demographics. Labeling data from the unseen OoD domain to re-train the original model is straightforward and yet expensive, labour-intensive, and not scalable to clinical scenarios. In fact, it is hardly viable to obtain an annotated training set that can faithfully represent anatomical variability, different demographics, pathologies, protocols, and scanners. Recently, multiple works have explored alternatives [5,7,10,12,13] to improve the generalisation capability of CMR segmentation models, typically, under the assumption that it is possible to know if the new data is OoD. In practice, this is not necessarily the case.

We hypothesize that segmentation quality measurements are a proxy of a model's generalisation capabilities. Therefore, this information can be used to improve a model's performance in OoD data. In this work, we propose to use quality measurements to refine a segmentation model's generalisation capabilities, when there is no knowledge about the distribution of the testing data. Under the assumption that the label space is consistent across different distributions, we train a QC assessment module to learn the variability of the ground truth training data. At inference, the QC module provides estimates of a segmentation model's performance on unseen data. The OoD detection information obtained from the QC module is used to refine the segmentation model, allowing it to improve its generalisation capabilities. The proposed method is evaluated in the context of the MICCAI Multi-Disease, Multi-View & Multi-Center Right Ventricular Segmentation in Cardiac MRI (M&Ms-2) Challenge.

Related Work. Previous works on CMR segmentation have addressed poor generalisation by reducing the model's complexity through regularisation [10] or by reducing the number of network parameters [12]. Although these techniques are very effective to tackle overfitting when the training sets are small, there is no guarantee they can mitigate poor generalisation to OoD data.

Data augmentation has been explored to enlarge the training set by simulating various possible data distributions across different domains, applying geometrical operations to the source training data. This technique, however, has shown variable performance across different scenarios [5]. Domain adaptation techniques propose enlarging the training set by combining labeled in-distribution data with OoD data [7,13] using adversarial training, which is prone to instabilities due to problems, such as mode collapse and non-convergence [1]. Moreover, these methods assume that it is possible to discriminate between in- and out-of-distribution data, which in practice is not necessarily possible.

Our framework differs from previous works in the fact that, first, it is model agnostic. Any given network can be used for segmentation. Second, being formulated as a semi-supervised problem, our QC module avoids the adversarial setup of domain adaptation techniques, thus leading to improved robustness and stability. Finally, differently from all previous approaches, it makes no assumptions about the nature of the distribution of the unseen data.

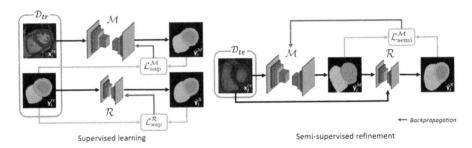

Fig. 1. During the supervised phase, the segmenter $\mathcal{M}$ and the QC module $\mathcal{R}$ are trained independently using $\mathcal{D}_{\text{tr}}$. At inference, if the QC module detects OoD data, a semi-supervised refinement step takes place. The segmenter $\mathcal{M}$ is used to segment unlabelled images from $\mathcal{D}_{\text{te}}$ that are then fed to $\mathcal{R}$. The difference between the reconstruction $\widehat{\mathbf{y}}^R$ and the model's segmentation $\widehat{\mathbf{y}}^M$ is backpropagated to update $\mathcal{M}$.

2   Method

Figure 1 presents an overview of the proposed framework. It consists of a segmentation network ($\mathcal{M}$) and a QC module ($\mathcal{R}$). The information provided by the QC module plays two roles. First, it detects erroneous segmentation masks. Second, the QC information from erroneous segmentation masks is used to refine the segmentation network in a semi-supervised setup to achieve increased performance on OoD data. In the following, we provide a detailed description of these two components (Sect. 2.1), the learning phases (Sect. 2.2), and the mechanism used for OoD data detection and quality control (Sect. 2.3).

2.1   Framework Components

Our framework consists of two elements: a segmentation network and a QC module. The segmentation network or segmenter $\mathcal{M}$ predicts segmentation masks using CMR images as input. The QC module measures the quality of the predicted segmentation masks in the absence of ground truth.

The Segmenter $\mathcal{M}$. The segmenter $\mathcal{M}$ learns a function $\mathbf{f}_M : \mathbf{X} \to \mathbf{Y}$, which is used to predict a segmentation mask $\widehat{\mathbf{y}}^M = f_M(\mathbf{x})$. In this sense, it is a standard segmentation network trained in a supervised setting with a training set $\mathcal{D}_{\text{tr}}$. It should be noted that the functioning of the framework is not conceived to depend on a specific segmenter architecture, for which several options in the literature are tailored to cardiac image segmentation [6].

The QC Module $\mathcal{R}$. In our framework, we use the QC for image segmentations proposed in [9]. It consists in a convolutional autoencoder, which is trained using in-distribution samples to reconstruct the input segmentation masks through a function $f_R : Y \to Y$, $\widehat{\mathbf{y}}^R = f_R(\mathbf{y}) \approx \mathbf{y}$. In our framework, the in-distribution samples are the ground truth masks from $\mathcal{D}_{\text{tr}}$, i.e. samples without segmentation errors.

Under the assumption that the space Y is consistent across domains, $\mathcal{R}$ is used to obtain $\widehat{\mathbf{y}}^R = f_R(\widehat{\mathbf{y}}^M)$, where $\widehat{\mathbf{y}}^M$ is a predicted segmentation mask and $\widehat{\mathbf{y}}^R$ its reconstruction. Following [9], we use the degree of similarity between $\widehat{\mathbf{y}}^M$ and $\widehat{\mathbf{y}}^R$ as a surrogate measure for QC of the segmentation, which is also a measure of the performance of $\mathcal{M}$ on unseen data. Where the degree of similarity is low (i.e. $\widehat{\mathbf{y}}^R \not\approx \widehat{\mathbf{y}}^M$), the segmentation is considered poor and it is flagged as a potential OoD sample. On such samples, the QC information is backpropagated to refine $\mathcal{M}$. The procedure is detailed in the following section.

2.2 Learning Phases

Supervised Learning. During the supervised phase, $\mathcal{M}$ and $\mathcal{R}$ are trained individually on the available training data $\mathcal{D}_{\mathrm{tr}}$. $\mathcal{M}$ is trained to minimise a loss function measuring the dissimilarity between the ground truth masks $\{\mathbf{y}^{\mathbf{tr}}\}$ and the model's prediction $\widehat{\mathbf{y}}^M$, i.e.

$$\mathcal{L}_{\mathrm{SUP}}^M = \mathcal{L}_{\mathrm{GD}}(\widehat{\mathbf{y}}^M, \mathbf{y}^{\mathbf{tr}}) + \mathcal{L}_{\mathrm{CE}}(\widehat{\mathbf{y}}^M, \mathbf{y}^{\mathbf{tr}}), \tag{1}$$

with $\mathcal{L}_{\mathrm{GD}}(\cdot)$ the generalised Dice loss [16] and $\mathcal{L}_{\mathrm{CE}}(\cdot)$ the cross-entropy loss. Differently from standard supervised training, we keep the best m_{best} segmenters, not the single best, according to their performance in a validation set $\mathcal{D}_{\mathrm{val}}$.

The QC module uses the loss $\mathcal{L}^R = \mathcal{L}_{\mathrm{MSE}}(\widehat{\mathbf{y}}^R, \mathbf{y}^{\mathbf{tr}}) + \mathcal{L}_{\mathrm{GD}}(\widehat{\mathbf{y}}^R, \mathbf{y}^{\mathbf{tr}})$, where $\mathcal{L}_{\mathrm{MSE}(\cdot)}$ is the mean squared error loss.

Semi-supervised Refinement of the segmenter $\mathcal{M}$ takes place whenever $\mathcal{R}$ detects poor segmentation quality, i.e. when an unseen segmented sample $\widehat{\mathbf{y}}^M$ is flagged as OoD. The refinement phase (Fig. 1) estimates and backpropagates a semi-supervised loss $\mathcal{L}_{\mathrm{SEMI}}^M$, measuring the similarity between the predicted and the reconstructed masks, i.e.

$$\mathcal{L}_{\mathrm{SEMI}}^M = \alpha\, \mathcal{L}_{\mathrm{WGD}}(\widehat{\mathbf{y}}^M, \widehat{\mathbf{y}}^R) + \mathcal{L}_{\mathrm{SUP}}^M, \tag{2}$$

where $\mathcal{L}_{\mathrm{WGD}}$ is the weighted generalised Dice loss, giving more importance to the RV, and $\mathcal{L}_{\mathrm{SUP}}^M$ the supervised loss from Eq. 1, which differently from $\mathcal{L}_{\mathrm{WGD}}$ relies on annotated training data.

Finally, the scaling factor α controls the reliability of the pseudo ground truth $\widehat{\mathbf{y}}^R$. When the semi-supervised refinement process starts, $\widehat{\mathbf{y}}^R$ is highly reliable. As it advances, $\mathcal{R}$ becomes a less reliable source for QC and $\widehat{\mathbf{y}}^R$ should be less trusted. To account for this, we set $\alpha = 1/k$, where k is a learning epoch.

2.3 OoD Detection and QC-Based Candidate Selection

At inference time, the selected m_{best} segmenters predict candidate segmentation masks $\widehat{\mathbf{y}}^M$ from unseen data $\mathcal{D}_{\mathrm{te}}$, whereas $\mathcal{R}$ reconstructs $\widehat{\mathbf{y}}^R$ to provide a quality measure for each $\widehat{\mathbf{y}}^M$. Following [9], we use the Dice Coefficient (high is best) and the Hausdorff Distance (low is best) as segmentation quality pseudo-measures. We denote them pDC and pHD, respectively.

For increased robustness, we determine the OoD detection thresholds based on the distributions of pDC and pHD. For each set of estimated pseudo-measures, we obtain the first and third quartile, Q_1 and Q_3 and define the following lower and upper thresholds:

$$th_{\text{low}} = Q_1^{\text{pDC}} - 1.5(Q_3^{\text{pDC}} - Q_1^{\text{pDC}}), \tag{3}$$

$$th_{\text{up}} = Q_3^{\text{pHD}} + 1.5(Q_3^{\text{pHD}} - Q_1^{\text{pHD}}). \tag{4}$$

For a test image $\mathbf{x}^{\text{te}} \in \mathcal{D}_{\text{te}}$ with m_{best} predicted segmentations, if pDC $< th_{\text{low}}$ or pHD $> th_{\text{up}} \ \forall \ \{\widehat{\mathbf{y}}_i^M\}_{i=1}^{m_{\text{best}}}$, the sample is considered OoD and model refinement takes place. The procedure is performed until no more OoD samples are detected in $\mathcal{D}_{\text{te}}$ or until a number of semi-supervised refinement epochs K_{semi} is reached.

After semi-supervised refinement, the best $\widehat{\mathbf{y}}^M$ is chosen through a QC-based selection procedure balancing pDC and pHD. The procedure is as follows:

1. The m_{best} candidate models are ranked according to the pDC of their prediction $\widehat{\mathbf{y}}_i^M$.
2. The top ranked candidate model is assessed. If pHD $< th_{\text{up}}$, its prediction is selected.
3. Otherwise, we discard the model and consider the next best ranked model.
4. Repeat points 2–3 until a segmentation is chosen.
5. If none of the m_{best} models meets the requirements, $\widehat{\mathbf{y}}^M$ is set to the average between the best prediction according to the pDC and to the pHD.

We highlight that QC-based candidate selection is completely independent from semi-supervised refinement. While we use the thresholds th_{low} and th_{up} to both determine when $\mathcal{M}$ should be refined and select a model during inference, the latter procedure could be directly applied to already static trained models where no refinement is desired (equivalent to setting $K_{\text{semi}} = 0$).

3   Experiments and Results

3.1   Experimental Setup

Data & Setup. The proposed method was evaluated in the context of the M&Ms-2 challenge. The goal of the challenge was to segment the right ventricle (RV) from CMR images in two different views: short axis (SA) and long axis (LA). The challenge cohort was composed of 360 patients with different RV and left ventricle (LV) pathologies as well as healthy subjects, who were scanned in four clinical centres in two different countries using four different magnetic resonance scanner vendors. The training set contained 200 annotated images, of which 160 were used for training and 40 for validation on the challenge website. The remaining data were used for testing.

The accuracy of the segmentation masks was measured using the Dice Similarity Coefficient (DC) and the Hausdorff Distance (HD). Further details about the data and the challenge can be found in the challenge website[1].

---

[1] https://www.ub.edu/mnms-2/.

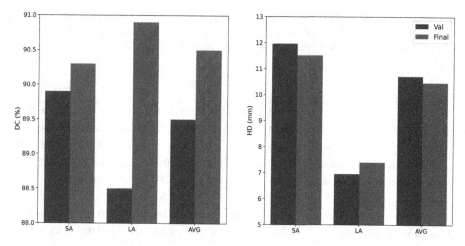

Fig. 2. Short axis (SA), long axis (LA) and average (AVG) dice score (DC) and Hausdorff Distance (HD) obtained on the validation (Val) and final submissions.

Implementation. We chose to use rather simple segmenter models $\mathcal{M}$. In particular, we tested our framework plugging 2 different 2D U-Net architectures, i.e. the 2D models from the 2D-3D U-Net ensembles winning respectively the M&Ms Challenge (BN1) [8] and the ACDC Challenge (BN2) [11]. For the QC module, we used the implementation from [9]. We obtained separate segmenter/QC module paired sets for each view. We set $m_{\text{best}} = 3$. The networks were implemented in PyTorch and trained on Google Colab Pro, alternating Tesla P100 and Tesla V100 GPUs. All our code is publicly available[2].

3.2   Results

Figure 2 presents the performance achieved by the model of each view and the overall average performance (AVG) in the validation and final submissions of the challenge. Despite using a simple segmenter model $\mathcal{M}$ (BN1), our results show competitive performances. In particular, performance in the SA is comparable to that reported for the RV in the ACDC Challenge [3] and M&Ms Challenge [4] using more complex 3D segmenter models.

Table 1 displays the performance of our final model stratified according to the conditions present in both the training and test sets. As expected, the performances look quite homogeneous, without any evident drop on the two conditions, tricuspidal regurgitation and dilated right ventricle, that were not present in the training set. Only the LA-view for dilated right ventricle shows a drop in performance indicating poor generalisation.

We performed an ablation study to gain an understanding of the properties of our framework. We analysed performance in three scenarios: 1) using the best

---

[2] https://github.com/robustml-eurecom/MnMs2.

Table 1. Short axis (SA) and long axis (LA) dice score (DC) and Hausdorff Distance (HD) per pathology. Tricuspidal Regurgitation and Dilated Right Ventricle were excluded from the labelled training set by the challenge organisers.

	SA		LA	
	DC	HD	DC	HD
Normal	0.902 ± 0.064	10.89 ± 5.16	0.918 ± 0.062	7.05 ± 6.34
Tetralogy of fallot	0.899 ± 0.065	12.80 ± 6.22	0.916 ± 0.038	8.33 ± 4.67
Interatrial communication	0.869 ± 0.137	12.09 ± 4.98	0.912 ± 0.083	8.03 ± 7.48
Congenital arrhytmogenesis	0.911 ± 0.069	12.70 ± 10.1	0.929 ± 0.029	5.60 ± 1.72
Hypertrophic cardiomyopathy	0.908 ± 0.083	10.67 ± 5.91	0.916 ± 0.058	6.30 ± 3.30
Dilated left ventricle	0.897 ± 0.077	11.99 ± 6.53	0.906 ± 0.053	8.13 ± 9.30
Tricuspidal regurgitation	0.909 ± 0.051	12.10 ± 8.08	0.910 ± 0.041	6.06 ± 2.98
Dilated right ventricle	0.913 ± 0.045	10.83 ± 4.82	0.881 ± 0.062	9.68 ± 6.58

segmenter, according to the validation set, after supervised training ($\mathcal{M}$); 2) using QC-based candidate selection to choose the best model from the m_{best} candidates at inference time without performing refinement ($\{\mathcal{M}\}$+QC); and 3) using the complete framework (full). Figure 3 summarizes the results using the two different base networks (BN1, BN2).

The results reveal the effects of QC on the segmentation performance. Overall, the use of QC information leads to improvements in the segmentation accuracy across different segmenter models, especially for the HD, which by definition is a measure more prone to anomalies. Figure 4 illustrates this through an example of failure in the segmentation process, solved after the proposed semi-supervised refinement. However, in one case (LA segmentation with segmenter

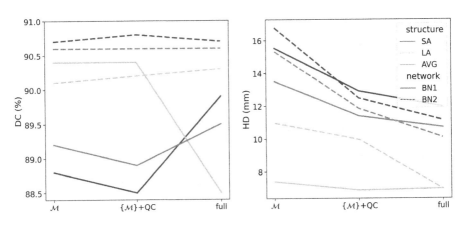

Fig. 3. Ablation study. DC (↑) and HD (↓) results for two segmenters (BN1, BN2) under three configurations: standard supervised learning ($\mathcal{M}$), QC-based model selection ($\{\mathcal{M}\}$+QC) and the full framework (full).

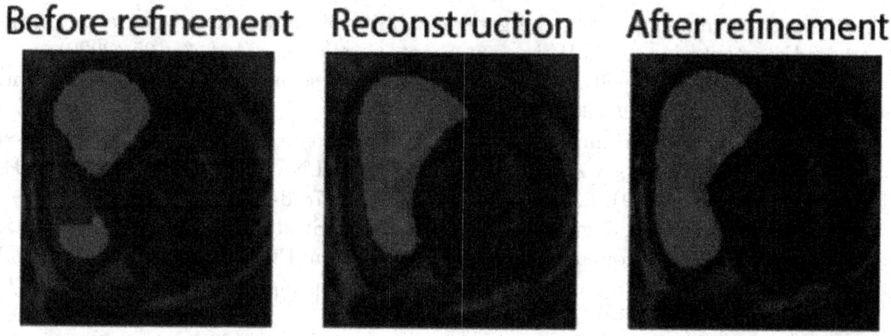

Fig. 4. A visual example of right ventricle segmentation before (left) and after semi-supervised refinement (right). In the middle, the output of the QC module when plugged with the initial segmentation on the left.

model BN1), the use of the full framework reports a drop in performance. Since using QC to select a model does not degrade the performance, we explain this behavior as a failure of the semi-supervised training stage.

4   Conclusions

In this work, we investigate how to couple QC information to refine a segmentation model to increase its performance on unseen OoD data. Differently from previous works, our framework assumes no previous knowledge about the distribution of the unseen data, yet it is able to determine when the model should be refined (OoD images) or not (in-distribution images). We evaluated our framework within the M&Ms-2 Challenge reporting performances which are comparable to those of similar previous challenges [3,4], while using a simple 2D segmenter model. This encourages us to pursue this line of work by investigating more sophisticated mechanisms to establish the OoD detection threshold. This, in fact, remains an open problem in the anomaly detection literature.

References

1. Arjovsky, M., Bottou, L.: Towards principled methods for training generative adversarial networks. In: 5th International Conference on Learning Representations, ICLR 2017, Toulon, France. OpenReview.net (2017)
2. Bai, W., et al.: Automated cardiovascular magnetic resonance image analysis with fully convolutional networks. J. Cardiovasc. Magn. Reson. **20**(1), 65 (2018)
3. Bernard, O., et al.: Deep learning techniques for automatic MRI cardiac multi-structures segmentation and diagnosis: is the problem solved? IEEE Trans. Med. Imaging **37**(11), 2514–2525 (2018)
4. Campello, V.M., et al.: Multi-centre, multi-vendor and multi-disease cardiac segmentation: the M&Ms challenge. IEEE Trans. Med. Imaging **40**(12), 3543–3554 (2021)

5. Chen, C., et al.: Improving the generalizability of convolutional neural network-based segmentation on CMR images. Front. Cardiovasc. Med. **7**, 105 (2020)
6. Chen, C., et al.: Deep learning for cardiac image segmentation: a review. Front. Cardiovasc. Med. **7**, 25 (2020)
7. Chen, J., et al.: Discriminative consistent domain generation for semi-supervised learning. In: Shen, D., et al. (eds.) MICCAI 2019. LNCS, vol. 11765, pp. 595–604. Springer, Cham (2019). https://doi.org/10.1007/978-3-030-32245-8_66
8. Full, P.M., Isensee, F., Jäger, P.F., Maier-Hein, K.: Studying robustness of semantic segmentation under domain shift in cardiac MRI. In: Puyol Anton, E., et al. (eds.) STACOM 2020. LNCS, vol. 12592, pp. 238–249. Springer, Cham (2021). https://doi.org/10.1007/978-3-030-68107-4_24
9. Galati, F., Zuluaga, M.A.: Efficient model monitoring for quality control in cardiac image segmentation. In: Ennis, D.B., Perotti, L.E., Wang, V.Y. (eds.) FIMH 2021. LNCS, vol. 12738, pp. 101–111. Springer, Cham (2021). https://doi.org/10.1007/978-3-030-78710-3_11
10. Guo, F., et al.: Improving cardiac MRI convolutional neural network segmentation on small training datasets and dataset shift: a continuous kernel cut approach. Med. Image Anal. **61**, 101636 (2020)
11. Isensee, F., Jaeger, P.F., Full, P.M., Wolf, I., Engelhardt, S., Maier-Hein, K.H.: Automatic cardiac disease assessment on cine-MRI via time-series segmentation and domain specific features. In: Pop, M., et al. (eds.) STACOM 2017. LNCS, vol. 10663, pp. 120–129. Springer, Cham (2018). https://doi.org/10.1007/978-3-319-75541-0_13
12. Khened, M., Kollerathu, V.A., Krishnamurthi, G.: Fully convolutional multi-scale residual densenets for cardiac segmentation and automated cardiac diagnosis using ensemble of classifiers. Med. Image Anal. **51**, 21–45 (2019)
13. Ouyang, C., Kamnitsas, K., Biffi, C., Duan, J., Rueckert, D.: Data efficient unsupervised domain adaptation for cross-modality image segmentation. In: Shen, D., et al. (eds.) MICCAI 2019. LNCS, vol. 11765, pp. 669–677. Springer, Cham (2019). https://doi.org/10.1007/978-3-030-32245-8_74
14. Puyol-Antón, E., et al.: Automated quantification of myocardial tissue characteristics from native T1 mapping using neural networks with uncertainty-based quality-control. J. Cardiovasc. Magn. Reson. **22**(1), 1–15 (2020)
15. Robinson, R., et al.: Automated quality control in image segmentation: application to the UK biobank cardiovascular magnetic resonance imaging study. J. Cardiovasc. Magn. Reson. **21**(1), 1–14 (2019)
16. Sudre, C.H., Li, W., Vercauteren, T., Ourselin, S., Jorge Cardoso, M.: Generalised dice overlap as a deep learning loss function for highly unbalanced segmentations. In: Cardoso, M.J., et al. (eds.) DLMIA/ML-CDS -2017. LNCS, vol. 10553, pp. 240–248. Springer, Cham (2017). https://doi.org/10.1007/978-3-319-67558-9_28

Author Index

Printed in the United States
by Baker & Taylor Publisher Services